NEW GEOGRAPHIES OF RACE AND RACISM

New Geographies of Race and Racism

Edited by

CLAIRE DWYER
University College London, UK

and

CAROLINE BRESSEY
University College, London, UK

Routledge
Taylor & Francis Group

LONDON AND NEW YORK

First published 2008 by Ashgate Publishing

Published 2016 by Routledge
2 Park Square, Milton Park, Abingdon, Oxfordshire OX14 4RN
711 Third Avenue, New York, NY 10017, USA

First issued in paperback 2016

Routledge is an imprint of the Taylor & Francis Group, an informa business

British Library Cataloguing in Publication Data
New geographies of race and racism
    1. Racism - Great Britain 2. Minorities - Great Britain -
    Social conditions 3. Great Britain - Race relations
    4. Great Britain - Race relations - Political aspects
    5. Great Britain - Ethnic relations
    I. Dwyer, Claire II. Bressey, Caroline
    305.8'00941

Library of Congress Cataloging-in-Publication Data
New geographies of race and racism / [edited] by Claire Dwyer and Caroline Bressey.
        p. cm.
    Includes bibliographical references and index.
    ISBN 978-0-7546-7085-8 (hardback) 1. Minorities--Great Britain. 2.
Pluralism (Social sciences)--Great Britain. 3. Great Britain--Race relations. 4. Great
Britain--Ethnic relations. 5. Great Britain--Emigration and immigration--Social aspects.
6. Human geography--Great Britain. I. Dwyer, Claire. II. Bressey, Caroline.

    DA125.A1N49 2008
    305.800941--dc22

                                                                2008020165

ISBN 13: 978-1-138-24699-7 (pbk)
ISBN 13: 978-0-7546-7085-8 (hbk)

Contents

List of Figures

List of Tables

Acknowledgements

The cover picture is reproduced from Ingrid Pollard's photo essay 'Belonging in Britain'. We are delighted that she has agreed to allow us to use this photograph on the cover of the book.

Figures 1–4 in Chapter 8 are copyright of The Scottish Executive and are reproduced with permission.

Figure 2 in Chapter 20 is copyright of the Burnley Borough Council and is reproduced with permission.

Contributors

Camila Bassi is a Lecturer in Geography, Sheffield Hallam University.

Alastair Bonnett is Professor of Geography at Newcastle University.

Caroline Bressey is a Lecturer in Geography and director of the Equinao Centre at University College London.

John Clayton is a Research Associate in the School of Health, Natural and Social Sciences, University of Sunderland.

Una Crowley is a Lecturer in Geography at National University of Ireland, Maynooth.

Claire Dwyer is a Senior Lecturer in Geography at University College London.

Peter Geoghegan is a Research Associate in the School of the Built Environment at the University of Ulster and a PhD student at the University of Edinburgh.

Mary Gilmartin is a Lecturer in Geography at National University of Ireland, Maynooth.

Sarah Glynn is a Lecturer in Geography at the University of Edinburgh.

Peter Hopkins is a Lecturer in Geography at Newcastle University.

David Howard is a Lecturer in Geography at the University of Edinburgh.

Peter Jackson is Professor of Human Geography, Sheffield University.

Ron Johnston is Professor of Geography, University of Bristol.

Michael Keith is Professor of Sociology, Goldsmiths College, University of London.

Rob Kitchin is Professor of Human Geography, National University of Ireland, Maynooth.

Jason Lim is a Research Fellow in the School of Environment and Technology, University of Brighton.

Linda McDowell is Professor of Human Geography at the University of Oxford.

Caroline Nagel is Assistant Professor of Geography at the University of South Carolina, USA.

Anoop Nayak is a Reader in Geography, Newcastle University.

Jan Penrose is a Senior Lecturer in Geography, University of Edinburgh.

Deborah Phillips is Reader in Geography, University of Leeds.

Ingrid Pollard is a photographer and artist based in London.

Michael Poulsen is Associate Professor of Geography, Macquarie University, Australia.

Lynn A. Staeheli is Professor of Human Geography, University of Edinburgh.

Dan Swanton is a Postdoctoral Fellow in the Department of Geography, Durham University.

Divya P. Tolia-Kelly is a lecturer in the Departmnet of Geography at Durham University and the Convenor of the *Lived and Material Cultures Research Cluster*.

Chapter 1

Introduction: Island Geographies: New Geographies of Race and Racism

Claire Dwyer and Caroline Bressey

Island Geographies

This volume of essays exploring the geographies of race and racism in the British Isles emerged from conversations we had, back in the summer of 2003, about absences and presences of race and racism in contemporary geography. Questions emerged in part from our own research interests. For Caroline, a key theme of her work has been to make visible the black presence in British history and to challenge historical memories that imagine Britain as 'multicultural' only after the arrival of the Windrush in 1948. For Claire, the emergence of an invigorated stream of 'new' geographies of religion (Holloway and Valins 2002; Kong 2001), raised questions about how religious identifications and religious identity politics were considered alongside the more traditional focus of geographies of race and racism. Like a number of others working in different places (Peake and Schein 2000; Schein 2002) we wanted to open up some space for reflecting on questions of race, racism and ethnicity in geography and indeed to urge the enduring geographical significance of racialised discourses, imaginaries and experiences.

At the Annual Conference of the RGS/IBG in 2005 we organised a session 'New Geographies of Race and Racism' as an opportunity to bring together researchers and open debates. This volume is, in part, an outcome of the two sessions we had at the conference and includes some of the papers which were originally presented there. However, in putting together this volume we have also commissioned a number of additional chapters to complement and extend our focus on understanding geographies of race and racism within the *specific* context of the United Kingdom and Republic of Ireland. While much might be gained from a wider international comparative focus, in this collection we have sought coherence by situating the discussion of race and racism within specific national, historical and geographical contexts. As we suggest below, current discussions of difference, inclusion, exclusion and religious discrimination particularly in the UK, especially post 9/11 and 7/7, are located in discourses of race and racism, even if these terms are avoided in public debates, and this collection offers a timely reflection on the geographies of these discourses.

One inspiration for this collection came from a re-reading of Peter Jackson's pioneering edited collection of essays *Race and Racism: Essays in Social*

Geography published in 1987. That volume was important in signalling a shift from more quantitative studies of social geographies of 'ethnic segregation' to a more politicised, theoretically grounded approach which drew upon Robert Miles' understanding of 'race' as socially constructed discourse producing historically and geographically specific racisms. We are delighted that Peter has agreed to contribute the afterword to this book. A second inspiration, and one which we evoke above in our title for this introduction, is Raphael Samuel's *Island Stories, Unravelling Memory*. The second volume of Samuel's *Theatres of Memory* was published posthumously in 1998. In it he reflects on British histories, particularly echoes of Empire, and explores how these are reanimated in different ways in the political present. In drawing together this volume we wanted to develop an historicised perspective on the geographies of race and racism which recognised both the legacies of contemporary discourses and imaginaries and the ways in which histories are re-worked in different presents.

This collection was also motivated by a recognition of the significance of race and ethnicity in contemporary British politics. The twin themes of terrorism and migration which dominate contemporary British politics and popular public debate are infused with ideas about race and the practices of racism, even though there is a marked effort by successive political leaders to distance themselves from these terms. A backlash against 'multiculturalism' as both discourse and practice and a fear of the mobilisation of right wing political movements produces a careful language of 'integration' and 'cohesion'. However, questions of integration, difference, cohesion and 'Britishness' are at the centre of contemporary politics in Britain, questions which both reanimate and re-work previous racialised discourses (Back et al. 2002).

Of particular salience are *geographical* imaginaries in the political and media framings of the politics of race. For example, explanations of urban unrest in northern Pennine towns in 2001; deaths of young, 'urban' Black men involved in 'gang violence' in South London and Manchester; or resentment towards migrant workers in rural Lincolnshire are framed by particular understandings of place which deploy racialised imaginative geographies. Alongside this mobilisation of the racialised imaginative geographies is a contested political debate about the *geography* of ethnicity. The urban disturbances in the cities of Oldham, Bradford and Burnley in 2001 when young white and Asian men fought each other and the police in street battles provoked new policy responses, just as the urban conflicts of the 1980s had some twenty years earlier, generating a new language of 'community cohesion' (Home Office 2001). Community cohesion discourses mobilise particular geographies of ethnicity. Thus the concept of 'segregation', the cornerstone of the earliest social geographies of ethnicity (Peach et al. 1981), has been reanimated and questions about how segregation might be defined, mapped and measured have brought geographers of ethnicity into the front line of political debate (as the chapters by Michael Poulsen and Ron Johnston, Deborah Phillips and Michael Keith in this volume attest). At the same time, policy debate about how community cohesion might be achieved mobilises a geography of locality and socio-spatial interaction which, while it might draw inspiration from Robert Putnam's social capital theory, also echoes calls for 'convivial urban culture' (Gilroy 2004) or 'everyday multiculture' (Amin 2002), evocations of local geographies of multiculture which are explored further in several chapters in this volume. It is clear that contested geographical imaginaries are central to contemporary British politics of 'race' and ethnicity.

Debating Geographies of Race and Racism

In this section we summarise some of the key issues shaping *new* geographies of race and racism. We explore three themes. First, we consider the complexity of contemporary categories of race and ethnicity reflecting both on geographical approaches and shifting political and policy discourses. Second, we reflect on the different geographies of race and racism highlighting the significance of the microgeographies of everyday life in understanding how ethnicity is lived and how ideas of race are made, mobilised and encountered. Third, and building from these two themes, we discuss shifts in contemporary theorisations of race and ethnicity and the possibilities of a politics of 'post-race'.

Framing Race and Ethnicity

Studies of the geographies of race and racism in Britain have been shaped by an understanding of the social construction of race – that 'race' is a discursive category which, although often depicted as natural or biological is ideological, is made in and through particular times and places (Jackson and Penrose 1993; Jackson 1987). Historical geographies have traced the emergence of specific ideas of race and place (Anderson 1987; Anderson 2007) and linked racialised discourses to processes of colonial expansion and to contested geographies of immigration and nation (Smith 1993). Recent work in historical and post-colonial geography has also done much to emphasise contradictions and instabilities in the histories of race and racism which are commonly evoked, tracing for example the significant presence of a Black British population prior to the post-war new Commonwealth immigration (Bressey 2003) and reflecting on how such historical perspectives might change the ways in which ideas about race and anti-racism are articulated or imagined (Bonnett 2000).

Shifting approaches to studying geographies of race and racism can be set within the context of changing political and discursive contexts. Geographical work in the 1970s and 1980s centred on racialised urban inequalities and social unrest realised in particular through conflicts in the housing market (Smith 1989) and policing (Keith 1993) and focused often on the experiences of an urban black British population. Geographers, like urban sociologists, placed an emphasis on understanding the ways in which racism worked through institutional structures and discourses. Alongside other anti-racist theorists their work was marked by a critique of the concept of ethnicity since it could too easily be deployed as culturalist and essentialised. Yet the late 1980s marked a theoretical shift in the deployment of the term ethnicity, initiated in part by Stuart Hall's call for the recognition of 'new ethnicities' (1988), a focus which emphasised both the creative and strategic deployment of different forms of 'identity politics' and perhaps a more anthropological engagement with the dynamic ways in which ethnic belonging and boundary making is constructed and contested. While understanding racialised discourses and practices remained central to critical analysis, research also engaged with the ways in which individual and group identities and belonging were produced and negotiated. For example geographers explored processes of ethnic identification through key events like Carnival (Jackson 1988; Spooner 1996) or in contested regeneration politics (Jacobs 1993; Keith 2005). Studies highlighted the ways in which the meaning of neither race or ethnicity is fixed but contextual and contingent. To make them an object of study requires an analysis of the power relations through which differences are both made visible and given meaning. Geographers of race and racism negotiate a tension, as Keith (2005, 6) describes in his analysis of multicultural cities, 'between languages of belonging and forces of power that make racial subjects visible'.

One site through which this tension is both made evident, and contested, is through the demographic measures used to categorise ethnicity. Analysis of the measures of ethnicity in the census in England and Wales since its introduction in 1991 reveals a shift from categories in 1991 representing individuals by a form of nationality, such as 'Black-Caribbean', 'Indian' or 'Pakistani', to categories based on hyphenated identities, which enable respondents to define themselves as 'Asian or Asian British', 'Black or Black British'. There was a shift too between 1991 and 2001 to expand the category of 'White' to include 'White Irish' and to include a dedicated 'Mixed' ethnic category. In 2001 the introduction of a religion question in the census also reflects an increased emphasis on religion as a key marker of ethnic identity . It seems likely that by 2011 further expansion will allow other groups such as Romanies and Arabs to identify themselves and will expand the category of 'Asian' to include the Chinese. It could be argued that this plethora of identities offers multiple 'ethnic' identities, while simultaneously narrowing the possibilities of a multicultural identity of Britishness.

The shifting categories deployed in the census can be contexualised against broader demographic processes and shifting definitions of ethnicity. In the 1980s the politics of ethnicity was dominated by 'moral panics' surrounding the marginalisation and assumed disaffection of young Black men (Gilroy 1987). In the 1990s young Asian Muslim men became the new 'folk devil' (Alexander 2000) with Islam racialised in the context of fears of global Islamic terrorism. New migration patterns reflect a shift from labour migration from New Commonwealth countries to asylum seeker-led migration from the Middle East, the Balkans and Africa as well as post-accession migration from Eastern Europe. Policy debates about institutional racism, provoked by the Macpherson Report (following the failures of the police to properly investigate the racist murder of the teenager Stephen Lawrence in April 1993) or discussions of multicultural Britishness foregrounded in the Parekh Report, have been succeeded in the wake of the 2001 urban conflicts and the terrorist attacks of 7/7 by a policy of 'Community Cohesion' variously targeted at ameliorating concerns about Muslim extremism, white working-class racism and tensions produced by new migration.

Geographical work reflects this shifting demographic, political and policy framing of discourses of race and ethnicity. Geographers have sought to unpack the concept of ethnicity and particularly to explore the notion of Whiteness both as a historically and spatially situated discourse (Bonnett 2000; Dwyer and Jones 2000) and as a lived ethnic identity made through intersections of class, place and gender (Nayak 2003; Haylet 2001; see also Byrne 2006). Walter's (2001) study of Irish women in Britain also serves to fracture the homogeneity of the category of whiteness, as does the work of rural geographers seeking to disrupt the image of the countryside as a homogenous white space, free from racial conflict (see Askins, 2006). Attention has also been given to the racialisation and socio-spatial exclusion of other groups including asylum seekers (Grillo 2005; Hubbard 2005) and travellers (Holloway 2003). As Muslim populations have emerged as central to the politics of race and ethnicity in Britain, geographers have sought to understand both the ways in which Muslimness might be expressed as a situated ethnic identity (Dwyer 1999; Hopkins 2007) and the spatialised politics of Muslim identification in the public sphere (Gayle 2003; Naylor and Ryan 2002).

Geographies of Race and Ethnicity

Evocations of locality are central to current articulations of 'Community Cohesion' policy in Britain (even though they might be critiqued for romanticised notions of 'community' (Wetherell 2007)). Writing in response to the policy debates provoked by the urban conflicts in 2001, Ash Amin (2002, 959) emphasises the importance of understanding the 'daily negotiation of ethnic difference' within 'the micropolitics of everyday social contact and encounter'. Understanding how ideas about ethnicity are made in and through local and 'everyday' encounters has been given renewed focus. Geographers have drawn inspiration from anthropologies of urban youth cultures (see particularly Back 1996; Alexander 2000; Gillespie 1995) which illustrate the nuanced, provisional and contingent ways in which ethnic identities are negotiated. Les Back's analysis of youth cultures in a housing estate in South London emphasises that national imaginaries are mediated in and through local understandings and that ethnic identities are made through encounter and in place. Urban ethnographies foreground the possibilities and challenges of 'encounter and contact' central to contemporary 'metropolitan multiculture' (Keith 2005, 2). Their insights lie precisely in some of the contradictions and conflicts which these intercultural or multicultural exchanges produce (a poignant example are the skinheads interviewed by Nayak 1999; see also Fortier 2007). It also offers the possibility of understanding geographies of race and racism through 'extroverted' (Massey 1994) geographical imaginaries which re-entangle local experiences with transnational memories, networks and identifications.

In Anoop Nayak's (2003) ethnography of young white masculinities in North-East England he traces local place and class based identifications as well as the re-working of transnational Black subcultural styles negotiated through nuanced everyday geographies. Such empirically grounded, place-based studies also offer insights into the inadequacy of ethnic categories particularly for young people variously positioned as 'mixed race'. While Back (1996) traces contingent and provisional racialised boundaries, Mahtani (2002, 425) suggests the possibilities for 'complicated performances which ... disrupt oppressive and dichotomous readings of ... racialised identities' (see also Twine 1996).

Geographers, like Nayak, are mindful of the dangers of exploring race and ethnicity only through the celebrated, 'colourful' or spectacular sites such as Birmingham's Balti belt or London's Brick Lane (McGuinness 2000, 225). Watt's study of young people in the 'non-place space' of a Home Counties small town (Watt 1998) and Back and Nayak's (1999) analysis of racism in the suburbs offer alternative approaches to understanding the intertwining of ethnic identities and imaginaries and place – particularly how white identities are forged. There has also been a sustained engagement both with analysing experiences of rural racism and the exclusion of black people from the countryside (Neal 2002; Garland and Chakraborti 2006) and in analysing the ways in which ideas of landscape, ethnicity and nation produce rural Britain as a white space (Askins 2006; Matless 1998; Neal and Agyeman 2006). Geographers have drawn particular inspiration from the provocative engagements of photographer Ingrid Pollard (1989, 1993, see also Kinsman 1995) whose series 'Pastoral Interludes' juxtaposes evocations of dominant discourses of race and nation with images of black ramblers and conservationists. Divya Tolia-Kelly's (2006) work with British South Asian women uses landscape memories, identifications and experiences to reshape ideas about nation and belonging.

Theorising Race

While new geographies of race and racism have engaged with more complex and multiple categorisations of ethnicity and have expanded the geographical sites and spaces through which race and racism might be explored, the tensions around the *making* of the categories of race and ethnicity themselves our objects of study remain. While geographers have written critically about the failure to engage sufficiently with geography's own racism, its critical colonial legacy, its enduring whiteness (Bonnett 1996; Pulido 2002; Shaw 2007) they have also raised questions about doing geographies of race reflecting on positionality (Mohammad 2003), essentialism (Bressey 2003) and ethnographic authority (Nayak 2006). As Keith (2005, 18) reiterates 'notions of 'race' and 'ethnicity' are themselves analytical categories that have been produced within the dominant strands of Western thought'. We can trace the historical specificities through which particular ideas of race emerge within particular times and spaces. For Keith, the 'political debate about race and ethnicity is precisely where and when markers of ethnic difference should be rendered invisible and when they should be actively highlighted' (2005, 18).

For geographers of race and racism these questions of making visible, or representation, are crucial philosophical and epistemological questions. We know that there is no such thing as race, yet processes of racialisation continue to structure unequal power relations and reproduce inequalities. While the dominant paradigm within geography has been social constructionist, in the context of anti-foundational theory the adequacy of this approach has been questioned. As Nayak (2006, 414) writes there is 'an inherent paradigmatic tension in social constructionist approaches to race. This involves the tendency to view race as socially constituted on one hand, yet to continually impart ontological value to it on the other, with the effect that race can take a reified status'. The possibilities of a 'post-race' paradigm are prompted by Paul Gilroy's *Against Race: Imagining Political Culture Beyond the Colour Line* who argues 'action against racial hierarchies can proceed more effectively when it ... [is] purged of any lingering respect for the idea of "race"' (2001, 13). Gilroy's provocative call is reflected in attempts to write geographies of race differently but also produced critique and concern about both the politics of abandoning race *or* the politics of invoking race.

Nayak (2006, 416) suggests that 'post-race writing' adopts 'an anti-foundational perspective which claims that race is a fiction, only every given substance to through the illusion of performance, action and utterance, where repetition makes it appear as-if-real'. This critique offers radical potential for geographers of race to, 'put post-race theory into post-race practice' (Nayak 2006, 424) in the situated and contingent encounters of ethnographies, through a recognition that 'race is something that we 'do' rather than who we are, it is a performance that can only ever give illusion to the reality it purports'. For Nayak the question becomes 'can we re-write race into erasure?' (427). However, it can be argued that these arguments do not take on board fully the reality of living with race and racism on the streets rather than on the page. Writing out race does not necessarily result in the eradication of racism.

Some of the same questions produce a different approach in the work of Arun Saldanha (2006) who argues that instead of foregrounding the epistemological status of race, an approach shaped only through discourse, we should recognise the materialist ontology of race – by engaging with phenotypes apprehended through the body. Saldanha (2006, 10) draws on Deleuze and Guattari's (1987) theory of 'machinic assemblages' to argue that race can be understood as 'unmediated connections'. He suggests: 'far from being an arbitrary classification system imposed *upon* bodies, race is a nonnecessary and irreducible effect of the ways those bodies themselves interact with each other and their physical environment' (2006, 10). While Saldanha's approach to understanding geographies of race might also involve ethnographies there is a refusal to privilege the discursive in their interpretation.

The question of a 'post-race' politics raises difficult questions. For some geographers, as is evident in some of the chapters which follow, anti-foundational theory offers an opportunity to produce alternative readings of the difference that race in everyday encounters, challenging readings which remain within the language of representation. Others remain uneasy that 'post-race' arguments do not take on board fully the reality of living with race and racism on the streets rather than on the page. Can 'writing out race' eradicate racism?

These debates, and the expansion of work on geographies of race and racism to engage more critically with categories of identity and place testify to the depth and dynamism of current work on geographies of race and racism. We hope that the essays in this volume will further these discussions of new geographies of race and racism in a collection in which even the diversity in the calligraphy of languages used to denote race and ethnicity by different authors, is reflective of ongoing debates.

New Geographies of Race and Racism

The book is organised in three sections. *Part 1: Racing Histories and Geographies* foregrounds a focus on the historicity of ideas about race and racism emphasising the echoes and linkages between contemporary discourses and older imaginaries. In the first chapter **Alistair Bonnett** traces a history and geography of Whiteness and 'the West' linking white supremacist thinking of nineteenth century colonial endeavour to the 'new symbolic economy' of whiteness produced by the globalisation and racialisation of neoliberalism. His arguments provide powerful evidence for the remaking of ideologies of race in and through specific histories and geographies. Continuities in histories of race thinking, but also the selective forgetting or disavowing of practices of anti-racism, are explored in the chapter by **Caroline Bressey** which uses a radio broadcast discussion about the re-making of a 1950s film about World War II – *The Dam Busters* – to prompt reflection on how arguments of 'political correctness' surface to suppress histories of anti-racism and foreshorten memories of the presence of black people in Britain. She draws on the political activities of the *League of Coloured Peoples*, active in drawing attention to racism in the media in the 1940s. **Ingrid Pollard**'s chapter is a personal reflection on her own presence in an archive. She shares photographs of her family, along with her father's letters, to reflect and narrate a very personal history of the migrant experience and the makings of multicultural Britain. In the last chapter in this section **Linda McDowell**, like Bonnett, addresses the construction of whiteness, this time through an analysis of the politics of labour migration, ethnicity and nation. She traces continuities between the social construction of labour migrants from Eastern Europe in the post-war period and today exploring how whiteness is scripted in a hierarchy of desirability and European belonging.

In *Part 2: Race, Place and Politics*, the chapters explore both the contested dimensions of identity politics and how ideas about race are framed in and through different national policies of immigration, nation-building or multiculturalism policies. In Chapter 6 **Sarah Glynn** analyses engagements made by the Bengali population with political parties, particularly the Labour Party and Respect, and traces the ways in which ideas about ethnicity and class are mobilised at different junctures. Contested identity politics are also central to the discussion in the chapter by **Caroline Nagel** and **Lynn Staeheli** on British Arab activists which explores their experiences through an analysis of integration, which they argue needs to be interrogated critically as a politics of visibility and invisibility.

The next four chapters develop further arguments about the specificity of place, and of national discourses, in shaping languages and meanings of race. Moving beyond the often English-centred focus of geographies of race and racism in Britain these chapters are based on work located in Scotland, Northern Ireland and also a chapter on the Republic of Ireland. **Jan Penrose** and **David Howard** offer a nuanced discussion of political debates about race in Scotland through their deconstruction of the anti-racism campaign developed by the Scottish Executive 'One Scotland, Many Cultures' tracing the shifting emphasis of nation, race and culture and suggesting a gradual weakening of stronger anti-racist messages to a less demanding call for national unity. **Peter Hopkins** emphasises the specificity of the Scottish national context in understanding racialised experiences drawing on examples from ethnographies of young Muslim men and asylum seekers. **Peter Geoghegan**'s chapter on race relations policy in Northern Ireland emphasises the specificity of the context within which race is produced. His analysis illustrates the ways in which sectarian divisions continue to frame an emergent race relations 'equality' policy. The chapter by **Una Crowley**, **Mary Gilmartin** and **Rob Kitchin** on race and immigration in contemporary Ireland presents an analysis of shifting debates about nationhood, whiteness and immigration. They reveal contested tensions around new cosmopolitan Irish identifications through the construction of racialised immigration policies which ignore Ireland's own immigrant past.

The final three chapters in this section offer contributions in different ways to current contested debates about geographies of segregation, integration and community cohesion in England and Wales. **Michael Poulsen** and **Ron Johnston** remain convinced that geographers need to do more to identify and measure the residential segregation which has been linked to contemporary political debates about community cohesion. They argue that existing methodologies do not adequately measure segregation and suggest that policy makers should remain concerned about levels of ethnic segregation and its socio-economic consequences. **Deborah Phillips** is more sceptical of the evidence for increasing levels of residential segregation for minority ethnic groups. Instead she draws on empirical research with Pakistani Muslim residents in three northern Pennine towns to emphasise the racialisation of place and contextual understanding of residential choices. In the final chapter of this section **Michael Keith** uses debates about racial geographies in the public policy realm and reflections on empirical work on the contested local geographies of East London to question 'the curatorial power of the spatial'. Specifically he draws on Georg Simmel's ambivalence about closeness and distance, strange and familiar, to pose questions about moral or ethical geographies of contemporary multiculture and how 'the politics of the racial imaginary' might be refigured.

The chapters in *Part 3: Race, Space and 'Everyday' Geographies* are linked by an emphasis on more ethnographic engagements and explore the possibilities and limitations of encounter and contact through empirical place-based studies in different sites of multicultural exchange. **Camila Bassi** analyses vignettes from the spaces of British Asian gay clubs to trace 'precarious and contradictory moments of existence' juxtaposing both radical possibilities through re-inventions of Asian cultural forms like bhangra with the limitations imposed by racialised hierarchies of 'respectability'. **Jason Lim** too explores the fleeting encounters of night time leisure spaces, focusing on one moment of interaction between two bodies, drawing on a Deleuzian approach to think about the 'ethics of encountering difference' which foregrounds historicising and remembering practices in the processes of 'racialised sense-making'. **Dan Swanton** is also interested in moments of encounter and contact in an analysis of the 'intense materialisation of race' through his meditations on the different urban multicultures of Keighley. He explores how race sticks to some bodies and things as bodies become differently raced in the conjunction of 'bodies and things in particular settings'.

Anoop Nayak's essay returns to some of the questions about anti-racism initiatives posed in earlier chapters. He draws on discussions with young people from schools in North-East England to reveal the complexities of racialised languages and practices and perceptions of anti-racism. He argues that successful approaches to challenging anti-racism as an anti-white practice requires a cultural pedagogy of place – for example engaging with generational and geographical trajectories of whiteness. **John Clayton** is also concerned with multiculturalism in practice in his analysis of the dynamics of inter-ethnic interaction in the self-consciously multicultural city of Leicester. He traces in particular the marginalities articulated by respondents in white working-class neighbourhoods and challenges the unproblematic emphasis on encounter posited in community cohesion policy in findings which resonate with Michael Keith's analysis of Barking and Dagenham. In the final chapter in this section **Divya Tolia-Kelly** reworks ideas about cosmopolitanism through an analysis of the transnational material cultures of British South Asians and the possibilities these trajectories offer for new forms of diasporic belonging. These possibilities are explored in the context of Burnley's Millennium Arboretum as a site for a new 'cosmopolitan vision of landscape, ecology and culture'.

We leave the final words in this volume to **Peter Jackson** in an essay in which we asked him to reflect on 'New Geographies of Race and Racism' as an invited Afterword.

References

Alexander, C. (2000), *The Asian Gang* (Oxford: Berg).

Anderson, K. (1987), 'The Idea of Chinatown: The power of place and institutional practice in the making of a racial category', *Annals of the Association of American Geographers* 77:4, 580–98.

Anderson, K. (2007), *Race and the Crisis of Humanism* (London: Routledge).

Amin, A. (2002), 'Ethnicity and the Multicultural City: Living with diversity', *Environment and Planning A*, 34, 959–80.

Askins, K. (2006), 'New Countryside? New Country: Visible communities in the English national Parks', in S. Neal and J. Agyeman (eds) *The New Countryside? Ethnicity, Nation and Exclusion in Contemporary Rural Britain* (Bristol: The Policy Press).

Back, L. (1996), *New Ethnicities and Urban Culture* (London: Routledge).

Back, L. and Nayak, A. (1999), 'Signs of the Times? Violence, Graffiti and Racism in the English suburbs', in T. Allen and J. Eade (eds) *Divided Europeans: Understanding Ethnicities in Conflict* (The Hague: Kluwer Law International).

Back, L., Keith, M., Khan, A., Shukra, K. and Solomos, J. (2002), 'New Labour's White Heart: Politics, multiculturalism and the return of assimilation', *The Political Quarterly* 73:4, 445–54.

Bonnett, A. (1996), 'Constructions of "Race", Place and Discipline: Geographies of "racial" identity and racism', *Ethnic and Racial Studies* 19:4, 864–83.

Bonnett, A. (2000), *White Identities: Historical and International Perspectives* (London: Longman).

Bressey, C. (2005), 'Of Africa's Brightest Ornaments: A short biography of Sarah Forbes Bonetta', *Social and Cultural Geography* 6:2, 253–66.

Bressey, C. (2003), 'Looking for Blackness: A researcher's paradox', *Ethics, Place and Environment* 6:3, 215–26.

Byrne, B. (2006), *White Lives: The Interplay of 'Race', Class and Gender in Everyday Lives* (London: Routledge).

Deleuze, G. and Guattari, F. (1987), *A Thousand Plateaus: Capitalism and Schizophrenia* (London: Continuum).

Dwyer, C. (1999), 'Veiled Meanings: Young British Muslim women and the negotiation of differences', *Gender, Place and Culture* 6:1, 5–26.

Dwyer, C.J. and Jones, J.P. III (2000), 'White Socio-spatial Epistemology', *Social and Cultural Geography* 1:2, 210–21.

Fortier, A.M. (2007), 'Too Close for Comfort: Loving thy neighbour and the management of cultural intimacies', *Environment and Planning D: Society and Space* 25, 104–19.

Gale, R. (2003), 'The Multicultural City and the Politics of Religious Architecture: Urban planning, mosques and meaning making in Birmingham, UK', *Built Environment* 30:1, 18–32.

Garland, J. and Chakraborti, N. (2006), '"Race", Space and Place: Examining identity and cultures of exclusion in rural England', *Ethnicities* 6:2, 159–77.

Gillespie, M. (1995), *Television, Ethnicity and Cultural Change* (London: Routledge).

Gilroy, P. (1987), *There Ain't No Black in the Union Jack* (London: Routledge).

Gilroy, P. (2001), *Against Race: Imagining Political Culture beyond the Colour Line* (Cambridge, MA: Harvard University Press).

Gilroy, P. (2004), *After Empire, Melancholia or Convivial Culture* (London: Routledge).

Grillo, R. (2005), '"Saltdene Can't Cope": Protests against asylum-seekers in an English seaside suburb', *Ethnic and Racial Studies*, 28:2, 235–60.

Hall, S. (1988), 'New Ethnicities' in K. Mercer (ed) *Black Film, British Cinema* (London: Institute of Contemporary Arts).

Haylett, C. (2001), 'Illegitimate subjects? Abject Whites, Neoliberal Modernisation, and Middle-class Multiculturalism' *Environment and Planning D: Society and Space* 19, 351–70.

Holloway, J. and Valins, O. (2002), 'Editorial: Placing religion and spirituality in Geography' *Social and Cultural Geography* 3:1, 1–9.

Holloway, S. (2003), 'Outsiders in Rural Society? Constructions of Rurality and Nature-Society Relations in the Racialisation of English Gypsy-Travellers, 1869–1934', *Environment and Planning D: Society and Space* 21, 695–715.

Home Office (2001), *Community Cohesion* (The Cantle Report) (London: The Stationery Office).

Hopkins, P. (2007), 'Global Events, National Politics, Local Lives: Young Muslim men in Scotland' *Environment and Planning A*, 39:5, 1119–33.

Hubbard, P. (2005), 'Accommodating Otherness: Anti-asylum centre protest and the maintenance of white privilege', *Transactions of the Institute of British Geographers* 30:1, 52–65.

Jackson, P. (ed.) (1987), *Race and Racism: Essays in Social Geography* (London: Allen and Unwin).

Jackson, P. (1988), 'Street Life: The politics of carnival', *Environment and Planning D: Society and Space* 6, 213–27.

Jackson, P. and Penrose, J. (eds) (1993), *Constructions of Race, Place and Nation* (London: UCL Press).

Jacobs, J. (1996), *Edge of Empire: Postcolonialism and the City* (London: Routledge).

Keith, M. (2005), *After the Cosmopolitan? Multicultural Cities and the Future of Racism* (London: Routledge).

Keith, M. (1993), *Race, Riots and Policing: Lore and Disorder in a Multi-racist Society* (London: Routledge).

Kinsman, P. (1995), 'Landscape, Race and National Identity: The photography of Ingrid Pollard', *Area* 27, 300–10.

Kong, L. (2001), 'Mapping New Geographies of Religion', *Progress in Human Geography* 25:2, 211–33.

McGuinness, M. (2000), 'Geography Matters? Whiteness and Contemporary Geography', *Area*, 32:2, 225–30.

Massey, D. (1994 [1991]), 'A Global Sense of Place', in D. Massey, *Space, Place and Gender* (Oxford: Polity).

Matless, D. (1998), *Landscape and Englishness* (London: Reaktion).

Mohammad, R. (2003), '"Insiders" and/or "Outsiders": Positionality, theory and practice', in M. Limb and C. Dwyer (eds) *Qualitative Methodologies for Geographers* (London: Arnold), 101–20.

Nayak, A. (1999), '"Pale Warriors": Skinhead culture and the embodiment of white masculinities', in A. Brah, M. Hickman and M. Mac an Ghaill (eds)*Thinking Identities: Ethnicity, Racism and Culture* (London: Macmillan).

Nayak, A. (2003), *Race, Place and Globalization: Youth Cultures in a Changing World* (Oxford: Berg).

Nayak, A. (2006), 'After Race: Ethnography, race and post-race theory', *Ethnic and Racial Studies* 29:3, 411–30.

Naylor, S. and Ryan, J. (2002), 'The Mosque in the Suburbs: Negotiating religion and ethnicity in South London', *Social and Cultural Geography* 3:1, 39–59.

Neal, S. (2002), 'Rural Landscapes, Representations and Racism: Examining multicultural citizenship and policy-making in the English countryside', *Ethnic and Racial Studies* 25:3, 442–61.

Neal, S. and Agyeman, J. (2006), *The New Countryside? Ethnicity, Nation and Exclusion in Contemporary Rural Britain* (Bristol: The Policy Press).

Peach, C., Robinson, V. and Smith, S. (1981), *Ethnic Segregation in Cities* (London: Croom Helm).

Peake, L. and Schein, R. (2000), 'Racing Geography into the New Millennium: Studies of 'race' and North American Geographers', *Social and Cultural Geography* 1:2, 133–42.

Pollard, I. (1989), 'Pastoral Interludes', *Third Text* 7, 41–6.

Pollard, I. (1993), 'Another View', *Feminist Review* 45, 46–50.

Pulido, L. (2002), 'Reflections on a White Discipline', *The Professional Geographer* 54:1, 42–9.

Saldanha, A. (2006), 'Re-ontologising Race: The machinic geography of phenotype', *Environment and Planning D: Society and Space* 24, 9–24.

Schein, R. (2002), 'Race, Racism and Geography', *The Professional Geographer* 54:1, 1–5.

Shaw, W. (2006), 'Decolonising Geographies of Whiteness', *Antipode*: 38, 851–69.

Smith, S.J. (1989), *The Politics of 'Race' and Residence* (Cambridge: Polity Press).

Smith, S.J. (1993), 'Immigration and Nation-building in Canada and the United Kingdom', in P. Jackson and J. Penrose (eds).

Spooner, R. (1996), 'Contested Representations: Black women and the St Paul's Carnival', *Gender, Place and Culture* 3(2), 187–203.

Tolia-Kelly, D. (2006), 'Mobility/Stability: British Asian cultures of 'landscape and Englishness', *Environment and Planning A* 38(2), 341 58.

Samuel, R. (1998), *Theatres of Memory II: Island Stories, Unravelling Britain* (New York: Verso).

Twine, F.W. (1996), 'Brown Skinned White Girls: Class, culture and the construction of white identities in suburban communities', *Gender, Place and Culture* 3, 204–24.

Walter, B. (2001), *Outsiders Inside: Whiteness, Place and Irish Women* (London: Routledge).

Watt, P. (1998), 'Going Out of Town: Youth, "race" and place in the South East of England', *Environment and planning D: Society and Space* 16, 687–703.

Wetherell, M. (2007), 'Community Cohesion and Identity Dynamics: Dilemmas and challenges', in M. Wetherell, M. Lafleche and R. Berkeley (eds) *Identity, Ethnic Diversity and Community Cohesion* (London: Sage).

PART 1
Racing Histories and Geographies

Chapter 2

Whiteness and the West

Alastair Bonnett

Introduction

My subject is the decline of whiteness. It may seem a surprising theme. Whiteness is, after all, still very much with us. Indeed, the close relationship between the globalisation and racialisation of neoliberalism has produced a new symbolic economy with whiteness at or near its centre. Whiteness appears to have the capacity to be re-invented, as well as sustained, as a feature of 'global racisms' well into the twenty-first century.

However, the history of whiteness is one of transitions and changes. This history is also a geography. It concerns the way the world has been imaginatively seized, its parts compared and its centre and periphery established. One of the most intriguing moments of transition that we can detect from this diverse scene concerns the development and impact of a crisis of whiteness at the end of the nineteenth century and the beginning of the twentieth. It was an international and transnational crisis that took a variety of routes in different places. However, in Western Europe and, to a lesser extent North America and Australia, this crisis saw the language of race begin to be challenged and the rise of the geopolitical labels of 'the West' and 'Western'. By the end of World War I, the explicit affirmation of white supremacism had begun to appear both crude and unnecessarily offensive to many within 'the West', most especially to many within the political and intellectual elite of countries that could, with some justification, be called 'liberal democracies', such as Britain. Indeed, with reference to post-World War I Britain, Pannikar (1953, 201) has argued that, 'With the solitary exception of Churchill, there was not one major figure in any of the British parties who confessed to a faith in the white man's mission to rule'.

Today, notions of the 'white man's mission' along with the once common category of 'white civilisation', appear to be sinister anachronisms. However, global hegemony and Eurocentric conceit are far from finished. Indeed, I would argue they are more powerful today than in the past. Nineteenth-century imperial endeavour was framed as a diligent *struggle* to subdue and manage subject peoples. It was imagined that, unless continuously controlled and cajoled, the resentful masses would seek to throw off the yoke of foreign rule. By contrast, today's Western supremacists assume that resistance is merely the residuum of old fears. They argue that Western power is so successful that it no longer requires force and that it can no longer be viewed as alien or intrusive. It is a soft power: people *want* to live Western lives; they *seek* the freedoms of the West. Thus contemporary titles tell a story of unmatched confidence: *The Triumph of the West* (Roberts 1985); *Why the West has Won* (Hanson 2000; see also *The West has Won* (Fukuyama 2001)); *The Ideas that Conquered the World* (Mandelbaum 2002), leaving us with *The End of History* (Fukuyama 1992).

This chapter is an attempt to explore some of the ingredients that went into the geopolitical shift from white to Western in the late nineteenth and early twentieth centuries. Drawing on the work of key, mostly British, figures in this transition (notably Charles Pearson and Benjamin Kidd) I argue that a 'crisis of whiteness' can be identified. I also suggest that within this crisis we can find some of the reasons why the idea of 'the West' gained ground.

These arguments imply the decline but not the death of whiteness. The use of whiteness as a central symbol within the cultural economy of neoliberalism shows us how adaptive and resilient this form of racialisation continues to be (Bonnett 2000). As this implies, this chapter, should be understood as an intervention in the development of the much wider debate on the relationship between whiteness, occidentalism and modernity (Bonnett 2004). Central to this debate is the idea that, although racialisation is still important in the contemporary world, scholars in 'racial studies' need to widen their horizons: 'race' and 'racism' are not sufficient categories with which to understand ethnic exclusion in the twentieth or twenty-first centuries. Indeed, 'race scholarship', if it is not brought into dialogue with other identities and processes, can end up contributing to the reification of both 'race' and 'racism'. As this implies, it is important to emphasise that the shift from white to Western cannot be grasped as merely a shift to euphemism by a diabolically clever and unchanging essential white racism. This position also implies that, whilst 'Westerner' can and does sometimes operate as a substitute term for 'white', it also operates within new landscapes of power and discrimination that have new and often fragile relationships with the increasingly widely repudiated language of race.

The Crises of Whiteness

Today *National Life and Character* (first published 1893) by Charles Pearson lies unread in library stores across the English-speaking world. This neglect belies the book's importance. For *National Life and Character* was a harbinger of a new genre of 'white crisis ' literature. A contemporary review describes Pearson as 'Chief among [the] prophets' of racial pessimism (Giddings 1898, 570). Pearson's principle explanation of why white expansion was at an end and white supremacy in retreat rests on demographics (notably Chinese and African fertility), geographical determinism (the unsuitability of the 'wet tropics' for white settlement) and the deleterious consequences of urbanisation on human 'character'. Moreover and crucially the economic ascendancy of those who Inge, following Pearson, was later to term 'the cheaper races' (Inge 1922, 227), meant that the white 'will be driven from every neutral market and forced to confine himself within his own' (Pearson 1894, 137).

These themes had not been synthesised in such resolutely pessimistic fashion before *National Life and Character*. However, they need to be understood as part of an intellectually omnivorous debate on the causes of white decline. It is precisely because white supremacism was at its zenith in this period that such symptoms of decline were found to be so worrying and diverse. The ludicrous heights of racial arrogance to which the jumble of contradictory conceits that formed 'white supremacism' had risen led to a sense of racial vulnerability. Although more and more of the world was passing into white control, by the last years of the nineteenth century, there had emerged a ready market for those who were feeling fretful about the quality of military recruits, the poisonous influence of city life, the rise of feminism, the spectre of intra-European rivalries, the falling birth rate of the middle classes ... and many more other things beside. These manifold worries were grouped together as a white racial crisis. Whiteness was opened out and made an object of popular worry by these discourses.

The Dean of St Paul's Cathedral, William Inge (1922), argued that by 1901 'the tide had really begun to turn' against the white world. The significance of this year is not explained by Dean Inge, although he does alight upon the 1897 Diamond Jubilee celebrations as the 'culmination of white ascendancy'. For Inge, the 'magnificent pageant' of the Jubilee also sounded a death-knell for white power, for

> the spectators ... could observe the contrast between the splendid physique of the coloured troops and the stunted and unhealthy appearance of the crowds who lined the streets. (214)

White self-doubt appeared to be substantiated by a stunning military defeat. The rout of Italian forces by the Ethiopians at Adowa in 1896 had been greeted with consternation in some quarters (Lyall 1910). However, across much of 'the white world' it was the outcome of the Russo-Japanese war that produced the real shock. In 1904 it was generally expected that the Russo-Japanese war, begun that year, would be speedily settled once the Russian Baltic fleet arrived in the Far East. In fact, the Japanese fleet, under Admiral Togo, destroyed all but three of Russia's ships in the Straits of Tsushima in May 1905. With the defeat of Russia a novel phase in international relations began. The 'victory of little Japan over great Russia' explained Basil Matthews in 1924, 'challenged and ended the white man's expansion'. For Matthews it signified 'the end of an age and the beginning of new era' (28); whilst for Inge (1933, 156–7) it marked 'one of the turning points of history'. In *The Rising Tide of Color Against White World-Supremacy*, the American racial polemicist Lothrop Stoddard (1925, first published 1920) phrased the matter in even more cataclysmic terms. With 'that yellow triumph over one of the great white powers' (21), he wrote, 'the legend of white invincibility was shattered, the veil of prestige that draped white civilisation was torn aside' (154).

The significance of Russia's loss also turned on another matter: the formal alliance of Britain with Japan. In *The Conflict of Colour* (1910) Putnam Weale offered a stinging critique of the British government's 'sensational step of allying herself with Japan' (113). For Putnam Weale the Anglo-Japanese Alliance (1902) amounted to a self-defeating form of racial treason:

> The secrets of supremacy have been revealed; and other countries, led by what England has done, are beginning to accept in their extra European affairs what may be called the same clumsy doctrine of <u>pis-aller'</u> [i.e., something done or accepted for lack of anything better] (117)

An ideal of international white solidarity was a logical outcome of the emergence of whiteness as a social and political force. Yet it remained a doomed and crisis-prone ideal, continuously vulnerable to the manifold difficulties inherent in employing a vaguely defined, highly idealised, yet utterly material, category as a significant geopolitical identity. These difficulties are clearly illustrated by the attempt to employ the notion of 'white community' during and in response to the Great War.

Within the literature of white crisis World War I was routinely termed a 'fratricidal war'. The danger the poet Sir Leo Chizza Money (1925) wrote about in *The Peril of the White* 'is not Yellow Peril, or a Black Peril, but a peril of self-extermination' (148): for 'whites in Europe and elsewhere are set upon race suicide and internecine war' (xx). Money's concern with white solidarity led him to attack both Stoddard and Inge for their attention to intra-white racial differences (what Money calls 'Nordiculous theory'). '[I]t is suicidal', he told them, 'to encourage racial scorns, racial suspicions, racial hatreds amongst the small minority that stands for White civilisation' (149). However, both Stoddard and Money were in agreement on the political implications of the war: that the only way white solidarity could be secured was by creating a European political union. 'Europeans must end their differences', argued Money. It is time, he proposed, to 'federate all the States of Europe' (x).

Yet, such a clear solution to the crises of whiteness was immediately undermined by these authors' ruminations on the traitorous nature of huge swaths of white people, most notably Russians and the working classes. Other authors added women and effeminate men to this list of suspects (Whetham and Whetham 1911; Curle 1926; Rentoul 1906). Even the physical environment could not be relied upon to support white ambitions. During the same period academic geographers had become preoccupied with the 'limits of white settlement' across the 'hot tropics' and other climatically unsuitable parts of the colonial world (Trewarthara 1926; Woodruff 1905). The grand aspirations of white dominion and solidarity, and the consequent scale of white vulnerability, made any specific attempt to see a solution to the crises of whiteness appear inadequate. It was within the arena of class conflict, though, that the literature of white crisis exposed the limits of white community most thoroughly.

The literature of white crisis is a literature of white supremacism. Yet it is also a literature in which the mass of white people are treated with suspicion. Despite Pearson's assertion of whiteness as the key to national identity it is clear from *National Life and Character* that not all whites are equally prized racial subjects. In particular 'the city type' (1894, 165) is painted by Pearson in fearful colours.

This paradox provides the clearest evidence that this is not merely a literature about crisis but in crisis: its central myth is constantly found to be failing, to be unworthy. Whiteness is, unintentionally, exposed as an inadequate category of social solidarity. For if the white nation is split between the 'British sub-man' (Freeman 1921) and Stoddard's 'neo-aristocrats' then the idea of white community necessarily appears, at best, a memory of a bond now passed into history.

The problem is compounded by the fact that the suspect nature of most white people is not a minor chord within any of the texts under discussion. It is usually the key site of argument and evidence. For Inge (1922) civilisation is always the property of a small elite: it is 'the culture of a limited class, which has given its character to the national life, but has not attempted to raise the whole people to the same level' (228). Without this cultured few, whiteness is an empty vessel, deprived of intelligence and direction. The 'brainy and the balanced have always controlled our world', agreed Curle; 'when they cease to do so, our White Race must pass into its decline' (1926, 213). The 'sub-man' (Freeman 1921; 1923), or 'Under-Man' (Stoddard 1922), or 'C3' type (Curle 1926) is white yet the enemy of whiteness; an internal enemy who is both a racial throw-back and harbinger of an anarchic future. In *The Revolt Against Civilisation* Stoddard (1922) offers a detailed depiction of the Under-Man as a discrete group, with his own traditions, interests and agenda. '[T]he basic attitude of the Under-Man', says Stoddard, 'is an instinctive and natural revolt against civilisation' (22). The Under-Man 'multiples; he bides his times' (23), waiting for his opportunity. This time, Stoddard concludes, has now come: the 'philosophy of the Under-Man is to-day called Bolshevism' (151), which is 'at bottom a mere "rationalising" of the emotions of the unacceptable, inferior and degenerate elements' (203).

The difficulty of asserting both white solidarity and class elitism was resolved, in part, by asserting that the 'best stock' of the working class had long since climbed upward. Thus the white elite's racial connection to the white masses could be claimed to be real but atrophied. For Ireland

> over a period of several centuries there has occurred a striking and progressive decline in the cultural contribution from the 'lower' classes in the United Kingdom, and, of course, a corresponding relative increase in the contribution from the 'upper' and 'middle' classes. (Ireland 1921, 139)

Two origin myths of the white bourgeoisie were employed by Pearson to secure this argument. One identifies their geographical and social roots in the hardy and muscular country life of pre-industrial rural England. The other locates them as the progeny of natural winners, i.e., as being the inheritors of a fighting stock that was able to demonstrate superiority before the struggle for existence was compromised by state welfare and interference. The former position is commonly encountered through depictions of the degenerative, racially hybrid, nature of the city, a position expressed concisely by Galton in 1883: '[T]he towns sterilise rural vigour' (14; see also Masterman 1901; White 1901; Haggard 1905; Cantlie 1906). Pearson (1894, 164–5) explained that the towns 'have been draining the life-blood of the country districts', the 'vigorous countryman' becoming absorbed into 'the weaker and more stunted specimens of humanity' who fill the towns. Thus the 'racial gift' that rural migrants bring to the town is soon squandered. It is an analysis that both roots the elite firmly within a white, rural past and condemns urbanisation and industrialisation as enemies of the race.

Once the bulk of whites had been dismissed as, in some way, inadequate, the problem of how to construct a positive programme to save white society became acute. Indicative of the seriousness of the problem is the fact that some of the texts under discussion conclude with utopian flourishes; far-fetched proclamations of racial re-birth (Pearson was far too gloomy for such flights of fancy). Freeman's (1921) and Inge's (1922) plans for 'experimental communities' of superior whites are illustrative. Freeman envisaged such settlements in Britain, whilst Inge warned that they would need to be established in remote colonies (he suggests, Western Canada, Southern Chile or Rhodesia) in order to avoid cross-class contamination. In either location, the settlements would consist of non-degraded whites who could live, work and reproduce in isolation. Such plans clearly suggest that the only way of saving 'the race' as to escape white society. In so doing they condemn whiteness as inadequate to the task of defining a meaningful identity for the elite.

At root, the literature of white crisis showed the limits of white supremacism. It illustrated the difficulty of sustaining commitments to racial solidarity, racial supremacism and social anti-egalitarianism as a coherent and stable belief-system. Such a world view is not merely prone to crisis but manacled to it. As such, the potential and the need to signal that whiteness needed to be 'moved beyond' was created wherever this crisis was experienced.

This shift was also enabled by a developing association between racial identification and social conflict. In *The Crisis of Liberalism* Hobson had warned that,

> Deliberately to set out upon a new career as a civilised nation with a definition of civilisation which takes as the criterion race and colour, not individual character and attainments is nothing less than to sow a crop of dark and dangerous problems for the future. (Hobson 1972, 244; first published 1910)

The theme that Hobson stresses – that racial ideology breeds conflict – provided one of the most potent challenges to the explicit assertion of the white ideal. Summarising his research on British attitudes to colonialism in the 1930s, Frank Füredi (1998) notes that 'a clear correlation was drawn between those who were racially conscious and those who were anti-white' (121). What Füredi is highlighting is an increasing tendency to associate 'racial consciousness' with a consciousness of racial oppression. Thus it became the task of British colonial policy, not merely to rhetorically 'deracialise' colonial encounters but, at least to appear, to oppose the meaningfulness of the very idea of racial hierarchy. This process was considerably encouraged by a desire to challenge the global influence of the Soviet Union, whose anti-racist credentials were taken seriously, even by ardent anti-communists. Thus, later attacks on racism – more specifically, on Nazi racism – were able to draw on an existing desire to 'move on' from race as a centre-piece of public discourse. As this implies, opposition to Nazi racism did not create the official acceptance of anti-racism. But it did help secure it. 'There is', noted one senior British official in the wake of the clear opposition to race discrimination offered in the United Nations Charter (1945), 'something like official unanimity of opposition to this species of primitive prejudice' (Corbett 1945, 27).

Western Supremacism

The notion that 'Western society is a unity' (Toynbee 1923, 4), that the West has its own discrete history, that it is 'an intelligible field of study' (Toynbee 1934, 36); that it is, moreover, a 'perspective', an ethno-cultural repertoire, is a creation of little more than the last hundred years. It is a relatively recent invention that exceeds the term's older meanings. The development of this contemporary West can be explained in terms of the impact of specific events. The Bolshevik revolution, the rise of US hegemony, and the loss of colonial power are the most important of these; each acting to make the idea of the West seem more important, more necessary. An emphasis on these three events is favoured by Christopher GoGwilt (1995) in his valuable genealogy of the idea of the West. However, what such an approach tends to miss is that new identities emerge in the context of existing ones. Neither people nor nations are blank slates upon which 'events' are written. Rather it is through and in the context of existing identities that new ones develop. Such a *relational* approach to the topic of the West produces different points of focus depending upon where in the world one is looking. But one of the clearest paths is through whiteness.

It cannot be claimed that the contemporary notion of the West emerged out of the literature of white crisis, certainly not in any direct or simple fashion. However, this old word for a new idea did represent a partial resolution of this literature's failed attempts to marry social elitism with racial solidarity. The idea of the West had clear advantages. Usually defined as a civilisation, rather than a race, the West could connote a socially exclusive cultural heritage as well as a broad territorial community. This function is apparent both within the literature of white crisis and from the emerging literature about the West that also developed from the 1890s.

The idea of the West that developed in the late nineteenth and early twentieth centuries was a varied concoction. It has the bubbling energy of something new and urgent, the miscellaneous character of an idea not yet refined by years of use. The West at this time could be made to mean many things. Before the Bolshevik revolution, its political character was highly mobile: the acme of Western civilisation was imagined to be, amongst other things communism, laissez-faire capitalism, anarchism, authoritarian statism and many other positions besides. However, within English-speaking countries, Benjamin Kidd has a special place in this tumult. GoGwilt claims Benjamin Kidd to be the first English-language author to employ a recognisably contemporary idea of the West. Kidd's reference in *Social Evolution* (1894) to 'our Western civilisation' is the 'first clear instance' of the use of the term 'Western' as a discrete society with its own history, argues

GoGwilt (1995, 54) and, as such, 'an idiosyncratic formulation'. Moreover, Kidd's *Principles of Western Civilisation* appears to be the first serious attempt to define this new entity and, whatever one may think of its leaden prose, it is no less intellectually ambitious that Spengler's later, far more famous and more pessimistic, contribution.

The British civil servant and, in the words of his biographer, 'spiky individualist' (Crook 1984, 3), Benjamin Kidd, made a name for himself with what is, perhaps, the most famous tract of Social Darwinism, *Social Evolution* (1894). Kidd's aloof disdain for the survival of anything and anybody he considered 'unfit' renders *Social Evolution* one of the most pungently racist books of its time. He carried this high-brow strain of viciousness over into *Principles of Western Civilisation*. Yet shorn of the racial rhetoric of *Social Evolution* and employing the full sonorous lexicon of 'Western culture', 'the Western mind' 'Western technology' and so on, this book appears almost contemporary in its concerns. Samuel Huntingdon's (1997) *The Clash of Civilizations* makes no mention of Kidd, yet he is working through and with Kidd's intellectual building blocks.

It is immediately apparent from *The Principles of Western Civilisation*, that the 'principles' to which Kidd refers are as hostile and damaging as anything concocted by the white supremacists of the era. The West is defined as a form of spirit, or consciousness, that is intellectually far-seeing and militarily enforced. Kidd regards the true promise of the West to lie in its potential to subjugate the present in the service of the future. '[T]he significance of Western civilisation' he argues,

> has been related to a single cause; namely, the potentiality of a principle inherent in it to project the controlling principles of its consciousness beyond the present. (289)

> It is to the principle of Projected Efficiency of the social process that every other principle whatever must ultimately stand in subordinate relationship. (396)

Like Pearson, Kidd was inclined towards socialism. Laissez-faire, for Kidd, was 'a surviving form of barbarism' (455) because it was unable to look beyond present needs. Kidd predicted the

> gradual organisation and direction of the State … towards an era of such free and efficient conflict of all natural forces as has never been in the world before. (469)

Kidd's 'western principles', then, are those that ensure the West's total victory in a world of ceaseless struggle and domination: 'We are par excellence the military peoples, not only of the entire world, but of the evolutionary process itself (458).

Kidd's racial vocabulary is vague. It can be inferred that he sees a racial content to being Western and that Western civilisation for Kidd is, for some unstated reason, white. However, Kidd's West is a decidedly non-corporal, non-material entity. The 'Western mind', he writes is destined, sooner or later, to rise to a conception of the nature of truth itself different from any that has hitherto prevailed in the world. (309)

Kidd's propensity for such cloudy abstractions led to his contribution appearing marginal to what was, at the time, the more mainstream debate on whiteness. Inge (1922) accused him of being an 'irrationalist' and *The Principles of Western Civilisation* bewildered and annoyed many of its reviewers (see Crook 1984). Yet Kidd's aversion to empirical detail and fondness for sweeping theorisations enabled him to render irrelevant the contradictions that were causing such anguish within the white crisis literature. By by-passing direct engagement with race, Kidd was able to ignore issues of racial purity, solidarity and sustainability and, hence, questions of class character and quality. For Kidd, 'Western principles' and 'our western civilisation' were transcendental forces whose inherent superiority lies in their orientation to the future, as well as in their, literally, merciless enforcement. The success of the West in 'the modern world-conflict' was thus certain: '[I]t is the principles of our Western civilisation ... and no others, that we feel are destined to hold the future of the world' (340).

It is the *confidence* of Kidd's vision that is so striking. He looks at 'the West', not as something limited by such things as fertility and social conflict, but as a 'big idea' that is turning history into a mirror of itself. 'The West' appears to Kidd to have escaped the bounds of nature: it is not something prosaic but something that has shot out of the earth-bound orbit of traditional culture to become god-like in its judgement.

The idea of the West helped resolve some of the problematic and unsustainable characteristics of white supremacism. Yet it carried its own burden of tensions. One of the most fundamental of these exposes a similarity of outlook between the proponents of whiteness and the West. For these are both projects with an in-built tendency to crisis. From the early years of the last century (Little 1907; Spengler 1926, 1928), through the mid-century (Warburg 1959; Beus 1953; Burnham 1964) and into the present day (Buchanan 2003; Barzun 2001), we have been told that the West is doomed (see also Herman 1997). Although specific causes for this fate are usually at hand, a more general reason may also be adduced. For like whiteness, the idea of the West has been conflated with modernity and global mastery. These vast ambitions create a state of vulnerability. When Western colonialism was as its height, it was said that the West was in its death-throws. When communism spread in East Asia, and as Asian and African countries achieved independence, it was said, perhaps with more justification, that the West was in retreat. Yet even minor phenomenon, like the rise of youth culture or the decline of classical music, have been interpreted as signalling the end of Western civilisation. As with the white crisis literature, almost everything and anything, big or small, has been fed into the omnivorous pessimism of the West's doom-mongers

The dread of decay that arises from the West's global claim closely echoes the panicky sensitivities of the white crisis literature. However, this similarity should not be pushed too far. The literature of white peril was not mirrored by a contemporaneous white triumphalist literature. But this is exactly what we see in the case of the West. For every book announcing its death, another is published claiming its ascendancy. In its own prolix way, Kidd's *The Principles of Western Civilisation* was the first British example. Later, more hesitant fanfares from the height of the Cold War, such as *Must the West Decline?* (Ormsby-Gore 1966) and *Is the Liberal West in Decline?* (Kohn 1957), were contemporaneous with more vigorous statements on *The Rise of the West* (McNeill 1963). However, even McNeill's portrait of the West's flexibility and receptivity to cultural influences is tame compared with the triumphalism of end of century announcements, such as *The Triumph of the West* (Robert 1985) and *Why the West has Won* (Hanson 2000). The contrast with whiteness is stark: only military effort and direct domination would allow the white supremacists a sense of conquest and finality. For the majority of Western triumphalists, though, all that needs to happen is that world 'opens up', begins to see things 'our way' and acts accordingly.

Conclusions

Having become established as the symbol of extraordinary achievement and superiority, as the talisman of world-wide social authority, whiteness was vulnerable to any sign of challenge or social disturbance. The fact that white supremacism relied on the authority of the natural, of biological fact, compounded its unsustainability. For not only did it make it incapable of registering the changing contours of social allegiance (such as the transformation of Russia from a white nation to a red one), it instituted a kind of racist egalitarianism within an anti-egalitarian class structure. In other words, white supremacism implies that *all* white people have to have the characteristics of whiteness: they must *all* be superior, they must *all* be fit to rule. Yet there was no subject that the white supremacists felt more strongly about than the inadequacy of the masses. Their racism demanded egalitarianism; their social elitism demanded something quite different. Something like the idea of the West perhaps?

There is some truth in the latter contention but it is also too neat, too glib. We cannot assume that, because it was in the context of the crisis of white identity that the idea of the West began to become attractive, that this crisis therefore 'produced' or even 'led to' the idea of the West. This point needs to be insisted upon, whilst at the same time the contemporaneous and novel character of the concept of the West that was emerging is recognised. Something new was being born. The literature of white crisis illuminates some of the reasons why, as well as nearly all the reasons why whiteness was inadequate to the challenges, not merely that lay ahead, but of the moment.

The twentieth century saw the rise of 'the West' as a political and social entity with what was claimed to be its own discrete history and traditions. It was employed to narrate the great political clash of the last century, that between communism and capitalism (and also, to a lesser extent, between Nazism and the capitalist democracies). In recent years it has been used to structure the conflict between the USA and its allies and 'Islamic fundamentalism'. Whiteness rarely featured in these debates. Not unrelatedly, the contemporary genre of 'race scholarship' known as 'white studies' tends to avert its eyes from global affairs and geopolitical transformations (Bonnett 2007). As a consequence 'white studies' can seem as curiously parochial and anachronistic as Churchill's once bullish conviction in the 'white man's mission to rule'. Of course whiteness still matters, both within everyday culture and as a structure within the allocation of social roles and material rewards. Yet any attempt to grasp the scope and nature of whiteness, either today or historically, cannot proceed on the assumption that it is comprehensible in isolation. Despite what their apologists say, nether 'whiteness' nor 'the West' are discrete identities with their own history and geography: they must be engaged and examined in relation to each other and other geopolitical ideologies.

References

Barzun, J. (2001), *From Dawn to Decadence* (London: HarperCollins).

Beus, J. (1953), *The Future of the West* (New York: Harper).

Bonnett, A. (2000), *White Identities* (Harlow: Pearson).

Bonnett, A. (2004), *The Idea of the West: Culture, Politics and History* (Houndmills: Palgrave).

Bonnett, A. (forthcoming), 'White Studies Revisited', *Ethnic and Racial Studies*

Buchanan, P. (2003), *The Death of the West: How Dying Populations and Immigrant Invasions Imperil Our Country and Civilization* (New York: Thomas Dunne Books).

Burnham, J. (1964), *Suicide of the West* (New York: Knopf).

Cantlie, J. (1906), *Physical Efficiency: A Review of the Deleterious Effects of Town Life upon the Population of Britain, with Suggestions for their Arrest* (London: G.P. Putnam's Sons).

Corbett, P. (1945), 'Next Steps after the Charter', *Commentary* 1(1), 27–8.

Crook, D. (1984), *Benjamin Kidd: Portrait of a Social Darwinist* (Cambridge: Cambridge University Press)

Curle, J. (1926), *To-day and To-morrow: The Testing Period of the White Race* (London: Methuen).

Freeman, A. (1921), *Social Decay and Regeneration* (London: Constable and Company).

Freeman, A. (1923), 'The Sub-Man', *Eugenics Review* 15 (2), 383–92.

Fukuyama, F. (1992), *The End of History and the Last Man* (London: Penguin).

Fukuyama, F. (2001), 'The West has Won', *The Guardian* 11 October.

Füredi, F. (1998), *The Silent War: Imperialism and the Changing Perception of Race* (London: Pluto).

Galton, F. (1883), *Inquiries into Human Faculty and its Development* (London: Macmillan).

Giddings, F. (1895), 'Review of *National Life and Character: A Forecast.* By Charles H. Pearson', *Political Science Quarterly* 10:1, 160–62.

GoGwilt, C. (1995), *The Invention of the West: Joseph Conrad and the Double-Mapping of Europe and Empire* (Stanford: Stanford University Press).

Haggard, R. (1905), *The Poor and the Land* (London: Longmans, Green and Co.).

Hanson, V.D. (2001), *Why the West has Won* (London: Faber).

Herman, A. (1997), *The Idea of Decline in Western History* (New York: Free Press).

Hobson, J. (1972), *The Crisis of Liberalism* (Sussex: Harvester Press).

Huntingdon, S. (1997), *The Clash of Civilisations and the Remaking of the World Order* (London: Simon and Schuster).

Inge, W, (1922), *Outspoken Essays (Second Series)* (London: Longmans, Green and Company).

Inge, W. (1933), *England: New and Revised Edition* (London: Ernest Benn).

Ireland, A. (1921) *Democracy and the Human Equation* (New York: E. P. Dutton).

Kidd, B. (1894), *Social Evolution* (London: Macmillan).

Kidd, B. (1902), *Principles of Western Civilisation: Being the First Volume of a System of Evolutionary Philosophy* (London: Macmillan).

Kohn, H. (1957), *Is the Liberal West in Decline?* (London: Pall Mall).

Little, J. (1907), *The Doom of Western Civilisation* (London: W.H. and L. Collingridge).

Lyall, A. (1910), 'Introduction', in V. Chirol, *Indian Unrest* (London: Macmillan).

Mandelbaum, M. (2002) *The Ideas that Conquered the World* (New York: Public Affairs)

Masterman, C. (ed.) (1901), *The Heart of Empire: Discussions of Problems of Modern City Life in England* (London: Unwin).

Matthews, B. (1925), *The Clash of Colour: A Study of the Problem of Race* (London: Edinburgh House Press).

McNeill, W. (1963), *The Rise of the West: A History of the Human Community* (Chicago: University of Chicago Press).

Money, L. (1925), *The Peril of the White* (London: W. Collins).

Ormsby-Gore, D. (1966), *Must the West Decline?* (New York: Columbia University Press).

Pannikar, K. (1953), *Asia and Western Dominance: A Survey of the Vasco da Gama Epoch of Asian History 1498–1945* (London: Allen and Unwin).

Pearson, C. (1894), *National Life and Character: A Forecast* (London: Macmillan).

Putnam Weale, B. (1910), *The Conflict of Colour* (London: Macmillan).

Rentoul, R. (1906), *Race Culture; or Race Suicide? (A Plea for the Unborn)* (London: Walter Scott Publishing).

Roberts, J.M. (1985), *The Triumph of the West* (Boston: Little Brown)

Spengler, O. (1926), *The Decline of the West: Volume One: Form and Actuality* (New York: Alfred A. Knopf).

Spengler, O. (1928), *The Decline of the West: Volume Two: Perspectives of World-History* (New York: Alfred A. Knopf).

Stoddard, L. (1922), *The Revolt Against Civilisation: The Menace of the Under-Man* (London: Chapman and Hall).

Stoddard, L. (1925), *The Rising Tide of Color Against White World-Supremacy* (New York: Charles Scribner's Sons).

Trewarthara, G. (1926), 'Recent Thoughts on the Problem of White Acclimatisation in the Wet Tropics', *Geographical Review* 16, 467–78.

Toynbee, A. (1922), *The Western Question in Greece and Turkey: A Study in the Contact of Civilisations*, 2nd edition (London, Constable and Company).

Toynbee, A. (1934), *A Study of History: Volume One* (Oxford: Oxford University Press).

Warburg, J. (1959), *The West in Crisis* (New York: Doubleday).

Whetham, W. and Whetham, C. (1911), 'Decadence and Civilisation', *The Hibbert Journal* 10:1, 179–200.

White, A. (1901), *Efficiency and Empire* (London: Methuen).

Woodruff, C. (1905), *The Effects of Tropical Light on White Men* (New York: Rebman, Co.).

Chapter 3

It's Only Political Correctness –
Race and Racism in British History

Caroline Bressey

Because Political Correctness Didn't Exist Then

On Tuesday 13 December 2005, BBC Radio 4's flagship political programme, *Today*, ran a feature on the re-making of *The Dam Busters* film. *The Dam Busters* is a British film made in 1954 which chronicles the story of the RAF's 617 Squadron, and Barnes Wallis' development of the 'bouncing bomb' which resulted in Operation Chastise and the attack on the Ruhr dams in Germany during World War II. The campaign was led by Wing Commander Guy Gibson and the *Today* programme feature focused on whether any remake of the film should include the name of Guy Gibson's black Labrador, whom he called Nigger.

The first person to comment on this dilemma was Richard Todd, an actor who played Guy Gibson in the 1954 film. Todd outlined his thoughts in a pre-recorded interview:

> Now the nigger question. With political correctness which is a new concept of a way of life in this country and I think all over the world it didn't exist when we made the original film so Nigger was Nigger, but nowadays you can't say that sort of thing.
> ....

Back in the live studio, Carolyn Quinn (CQ) interviewed Jonathan Falconer (JF), an author who has written about the *Dam Busters* film, and George Baker (GB) who starred in the 1950s film alongside Richard Todd:

> CQ – Let's deal with this issue of Nigger the dog. Some people will think it is a bit offensive and ITV recently bleeped the name out, do you think that that should happen when the film's re-made?

> JF – No, erm, it's a question I think of historical accuracy here ... the film and obviously the events are very much part of the time they were made in and took part in and so I think tinkering with the historical accuracy of the film and of the story is a very dangerous and slippery slope to start down.

> CQ – So you think people will accept it as historical accuracy?

JF – Well they ought to. If they are being objective about it then I think they should accept it as historical accuracy, but I can understand why some people may find it offensive.

CQ – George Baker, was there any opinion about the name at the time when you were making the film?

GB – No, none at all. Political correctness wasn't even invented and, er, an awful lot of black dogs were called Nigger.

CQ – And you think the name should be kept in?

GB – Yes, I think if we are seriously going to re-make *Dam Busters* the name of a dog makes very little difference to whether the film was accurate in the first place, whether it was just a wonderful moment in 1955 when it was shown when we were still in austerity and it was terrific uplift to know that we had done something as it was when the bomb was dropped. ...

There are numerous issues that could be raised around race, identity and history from this piece, for example, the question of whether the death of 1,200 German civilians is still something that gives the British nation a sense of enormous uplift – and if so, why. But what I wish to focus upon here is the role of those who use the argument that 'too much' political correctness in Britain has led to a distortion of British history and ideas of Britishness itself. Those who support this position (what I call anti-political correctness, or apc) have used it to dismiss the presence of direct and indirect racism in the teaching and understanding of British history. The case of Guy Gibson's dog is an excellent example of this.

Personally I agree with Jonathan Falconer that 'N' the dog should keep his name. If such a film is to be (re)made, it is important for the sake of historical accuracy, not only in and for itself, but also as a reflection of the ignorance and racism that existed in British society during the 1940s. There is a serious and interesting discussion to be had around these issues, but none of the three men interviewed on the *Today* programme that December morning appeared to have any understanding or knowledge of black history, an essential part of any debate around the historical and contemporary use of racist language. There are numerous historical geographers and historians who could have been contacted to discuss the topic with them. If the subject was being taken seriously they should have been asked to contribute to the debate. The use of the N word five times in less than three minutes perhaps reflects that the subject was not being taken seriously. That the feature appeared in the 8.20am-ish slot usually reserved for lighter items supports this supposition. That such the subject was treated so superficially only compounded the BBC's poor handling of this item. However, it is the fact that the BBC itself was forced to apologise for the use of the N word on its airwaves during World War II that highlights the danger of the ignorance of history as a weapon of the apcs.

The League of Coloured Peoples

Dr Harold Moody was born in Kingston, Jamaica in 1882. In 1904 he came to London to study medicine, and in February 1913 he set up his own successful practice in Peckham. The racial prejudice he faced as a student and then as a qualified doctor motivated his campaign for black civil rights and in March 1931 he became one of the founders of the League of Coloured Peoples. Moody became the League's President, a position he held until his death in 1947. As was stated in the first volume of its journal *The Keys*, the League of Coloured Peoples' aim was to highlight the problems and successes of black people within the British Empire, and to challenge racial discrimination and fight for equality.

However, they also connected their campaign to the civil rights of black people who were part of other nations and empires, as well as those who faced racial discrimination in other forms. The persecution of Jews in Germany, and the lack of concern with which the news of their plight was received by the rest of the world shocked those involved with the League. Later they commented on the Colour Bar Laws that were practised in Paris during World War II, when in 1940, although theatres were reopened, black and Jewish artists were banned from performing on their stages.[1]

The League of Coloured Peoples was concerned with prejudice at home as well as that which occurred throughout the Empire:

> Little is known in England of the legal disabilities under which the African labours in South Africa. The tragic plight of the aboriginal in Australia is a closed book to many at present. The recent persecution of the Jews in Germany, and the apathy with which the news was received by the rest of the world was appalling. [...] On our very doorsteps in Cardiff, Liverpool, London and elsewhere our brothers and sisters are daily meeting with racial discrimination in their search for work. Many served the Empire in the last war but that is of no avail now. [...] Hotels, restaurants, and lodging houses refuse us with impunity.[2]

As the above extract illustrates, within Britain the organisation was mostly concerned with the racial discrimination people faced on a day to day basis, namely the colour bar that was in operation at work, socially and in the media. The difficulties faced by those who applied for jobs in the medical profession (perhaps because this was an issue close to Moody's heart), were often highlighted by the League. For example, in 1933 a woman who applied to twenty-five hospitals found herself refused by every one on the grounds of 'colour'. The Overseas Nursing Association, which had first heard of the woman's case, said that they had applied to eighteen hospitals in London and the provinces and all had said that they could not, or would not, take black probationers.[3]

The colour bar that operated in Britain was not explicit. Black peoples in Britain met with what Una Marson called a 'subtle prejudice'. Una Marson was a Jamaican born feminist writer, poet, some time editor of *The Keys* and the first black woman from the African diaspora to become a programme maker at the BBC (Jarrett-Macauley 1998). She argued that:

In America they tell you frankly where you are and are not wanted by means of big signs, and they don't try to hide their feelings. But in England, though the people will never say what they feel about us, you come up against incidents which hurt so much that you cannot talk about them.[4]

Others saw the colour bar as a more systematic policy of empire. Nyasilie Magxaka from South Africa was one such commentator:

I thought that on leaving South Africa for England, I was at the same time leaving the infamous colour bar behind and was coming to the paradise of freedom. Since my arrival here I have been quite disillusioned on this point. First of all I was refused admission by various hotels because I was coloured. The coloured races in South Africa think that the Colour Bar is a purely local affair and that England, as the centre of the Empire, is the last place where it would be tolerated. The treatment of the coloured people in London almost forces people to believe that the colour bar is the policy of the British Empire.[5]

London, Liverpool and Cardiff were often cited by members of the League as places where black people faced such regular discrimination, and black seamen in Cardiff were the focus of one the League's long standing campaigns. In 1935 the League reflected on what it believed to be its most effective pieces of work. This was its intervention on behalf of black seamen in Cardiff, displaced and unemployed by the operation of the British Shipping Assistance Act. Two researchers, George Brown and Cecil Lewis, spent two weeks in Cardiff getting firsthand information from the seamen. At the end of their visit a public meeting was held at which it was decided to form a branch of the League. By July 1935 it numbered eighty members, and by 1936 Cardiff was home to the majority of the League's members.[6]

The Keys did not only use essays and reports as a means of communicating the racial prejudice people faced. Alongside letters from readers, poems were published. One written by Una Marson and published in 1933. Her poem, 'Nigger', speaks to the very heart of the twenty-first century debate surrounding Guy Gibson's dog .

Nigger

They called me 'Nigger'
Those little white urchins,
They laughed and shouted
As I passed along the street
They flung it at me:
'Nigger! Nigger! Nigger'

What made me keep my fingers
From choking the words in their throats?
What made my face grow hot,
The blood boil in my veins
And tears spring to my eyes?

What made me go to my room
And sob my heart away
Because white urchins
Called me 'Nigger'?

What makes the dark West Indian
Fight at being called a Nigger?
What is there in that word
That should strike like a dagger
To the heart of Coloured men
And make them wince?

You of the white skinned Race,
You who profess such innocence,
I'll tell you why 'tis sin to tell
Your offspring Coloured folk are queer.
Black men are bogies and inferior far
To any creature with a skin made white

You who feel that you are 'sprung
Of earth's first blood' your eyes
Are blinded now with arrogance
With ruthlessness you seared
My peoples flesh, and now you still
Would crush their very soul,
Add fierce insult to vilest injury.

We shall not be called 'Niggers'
Since this was the favourite curse
Of those who drove the Negroes
To their death in the days of slavery.
'A good for nothing Nigger'
'Only one more Nigger gone'
They would repeat as though
He were a chicken or a rat.
That word then meant contempt.
All that was low and base.
And too refined for lower animals.

In later years when singing Negroes
Caused white men to laugh,
And show some interest in their art
They talked of 'Nigger Minstrels'
And patronised the Negro
And laughing at his songs
They could in nowise see
The thorns that pierced his heart.

'Nigger' was raised then to a Burlesque show

And thus from Curse to Clown progressed
A coloured man was cause for merriment
And though to-day he soars in every field
Some shrunken souls still say
'look at that Nigger there'
As though they saw a green bloodhound
Or a pink puppy.

God Keep my soul from hating such mean souls
God keep my soul from hating
Those who preach the Christ
And say with Churlish smile
'This place is not for "Niggers".'
God save their souls from this great sin
Of hurting human hearts that live
And think and feel in unison
With all humanity.

Quoting Marson's poem in full makes it clear that the use of the N word in 1930s Britain was *always* deeply offensive. Her role at the BBC reflects the fact that men and women who challenged these racisms were not entirely marginalised. The decision for people to co-opt the N word as a name for their pets in this context only compounded the derogatory nature of the term. Did all those people who called their black dogs 'N' do so because they were racist, or because they were so ignorant of British racism that they were unaware of the offence they called? Moreover, and perhaps the key to why people are so uncomfortable with such reflective histories, how does such questioning impact upon the heroic worship of Guy Gibson by the media and British society? The failure to acknowledge, let alone answer, these questions, results not only in an unbalanced understanding of history, but discounts the prejudice black people faced in Britain, and in the direct context of the *Dam Busters*, also undermines the role black people played in World War II in spite of the racial discrimination they often faced.

During the war the League received increasing numbers of complaints about the colour bar that was in operation in the British armed forces. Men and women, especially from the Caribbean, began to write to the League outlining their attempts to join the forces. There were stories circulating of blatant racism, young men and women, whose father's had served in World War I, were now being refused entry to the same branches of the forces. Some began to argue that in the face of such racism men and women in the colonies and in Britain should resist all attempts at conscription until some guarantee was given that every branch of the armed services would be opened to black and Asian people when the war had ended.

The league lobbied the armed forces on behalf of men and women who hoped to serve the mother country, and celebrated a limited victory in 1940 when it was decided that, for the period of the war, all British subjects from the colonies who were in the UK would be treated on an equal footing with subjects from the UK when volunteering for enlistment in the armed forces. During the war the League celebrated its tenth year anniversary. *The Keys* had appeared without interruption from its inception until the outbreak of World War II. However, during the war, *The Keys* was suspended, and friends and supporters were asked to content themselves with copies of the monthly *Letters*, published under 'the hum of hostile plans and the boom of friendly guns' (Bressey, 2007). These *Letters* continued to consider the issues of race and racism black people faced, as well as the experiences of black people who were living through the London Blitz.

One article highlighted individuals from the black community in London who had become wardens, Auxiliary Fire Service men, and members of Stretcher Parties, First Aid units, and mobile canteens. It was a contribution to the defence of London that the journal felt should be given the greatest degree of publicity possible. Those who kept London going included men such as 'Buzz' Barton. He was a well known Jamaican born boxer who became a First Aid worker in London during the war. The participation of the black population in proportion to its numbers was reported to be considerable. Theirs was a contribution that the League hoped would not be forgotten, but is a little remembered fact sixty years on.

Throughout the war the League maintained an interest in the representation of black people in the media from the BBC to film, and in 1940 a disheartened review of 'Gone With the Wind' was published. As its author W. Arthur Lewis pointed out, the representation of black people had traditionally suffered from patronising stereotypes in the cinema. He argued that in the book on which the film was based, black people had been portrayed as domestic tyrants, ruling over the most intimate domestic affairs of the white household. In the film black people provided mere light entertainment, giving an impression of a people whose every act was comic – an image reflected in Marson's poem above.

The League also kept a critical eye on those who presented images of black people to the public through the arts, and the organisation was heavily critical of the BBC when one of their announcers used the N word on air in 1940. The letter they wrote was published in the journal along with the Director Secretariat of the BBC's reply and apology:

My attention has been drawn to the fact that one of your announcers, when interpreting some records on the 11th inst., made use of the offensive term 'nigger'.

There is no need for me to remind you that this is one of the unfortunate relics of the days of slavery, vexatious to present day Africans and West Indians, and an evidence of incivility on the part of its user.

I hope, sir, as a public corporation, you will take some steps to repair the damage done. I shall be glad to be advised as to what steps you take so that I may be able to inform my Committee accordingly.

From the Director, Secretariat of BBC to the President of the League.
16/5/40–

Following my earlier letter, I find that our announcer was at fault. The point raised on your letter is fully appreciated, and is one which the BBC is at pains to keep constantly in mind. It was unfortunately overlooked on this occasion, and a reminder on the subject is being given to announcers.

I hope that your Committee will accept the BBC's apology for this slip, which is sincerely regretted.[7]

This example shows that African and Caribbean peoples continued to take grave offence to the use of the N word in the 1940s – well before the original *Dam Busters* film was made or 'political correctness' became part of our vocabulary. It also shows that the BBC acknowledged and understood the offence it caused. That those interviewed on the *Today* programme in December 2005 were allowed to dismiss these political histories of anti-racism as contemporary 'pc-ness', especially when such an obvious example is to be found within the BBC's own archives, reflects two major issues. Firstly, it illustrates the power of historical ignorance as a tool for those who reinforce racism by arguing that they are in fact challenging 'political correctness', and secondly, the general failure of British society to acknowledge black and Asian experiences as a key part of British historiography. The 'memories' of Todd and Baker also reflect a willingness, desire even, to ignore the complexities of history.

Those involved with the *Today* programme piece are not the only commentators to wilfully dismiss black history as an inherent and critical part of British history. In December 2005, Max Hastings wrote an article in the *Guardian* in response to the QCA's assessment of the teaching of history in British schools (the QCA is the Qualifications and Curriculum Authority who act as the regulatory body for public examinations and publicly funded qualifications). It had urged the need to give more positive attention to the part of minorities in Britain's history. The authority's thinking was, Hastings believed, easy to understand:

> to a teenager of West Indian or Muslim background, medieval exchequer practice or 19th century poor law seem remote. Surely we can offer such children knowledge that strikes a chord with their own heritage.
>
> Yet how is it possible to do much of this in a British school without distorting the western experience, which anyone living here is signed up to? Pupils in modern African or Indian schools obviously focus their historical studies on the experience of subject races under foreign rule. But, as a profound sceptic about multiculturalism, I can't see the case for such an agenda, unless the vast majority of British people are to pretend to be something they are not.
>
> History is the story of the dominance, however unjust, of societies that display superior energy, ability, technology and might. If one's own people were victims of western imperialism, it is entirely understandable that one should wish to study history from their viewpoint. But, whatever the crimes of our forefathers, this is the country of Drake, Clive and Kitchener, not of Tipu Sultan, Shaka Zulu or the Mahdi.

Assuming Hastings was referring to Lord Kitchener the Commander-in-Chief during the Boer war in South Africa, rather than Lord Kitchener the calypso artist and composer of the famous song 'London is the Place for me', there are two major flaws in Hastings argument. Firstly, that somehow the place of Drake (a slave trader best known for his exploits at sea), Clive (an eighteenth-century soldier and administrator in India) and Kitchener (inextricably linked to overseas colonial campaigns) in British history is not fundamentally linked to the men and women who they encountered overseas. To understand British history you need to understand these global historical geographies.

Secondly, Hastings implies that black and Asian students would not find histories of the British at home of interest to them. Such a statement can only be based on the assumption that black and Asian students are somehow separate to their white peers, that they do not share their histories. This follows another assumption that black and Asian history in Britain begins some time after the beginning of the twentieth century. There has of course been a continuous presence of African descended peoples in Britain since the 1600s and Asian communities can trace a historical presence back to a similar period. To argue that 'ordinary' histories, such as those of the poor in Victorian Britain do not include a black or Asian presence is incorrect, as a look inside England's Victorian pauper asylums shows us.

A Black Presence in Pauper Britain

In 1898 two pauper women Caroline Maisley and Mary Matthews were admitted to Colney Hatch Asylum. An Italian style structure, it admitted its first patients in July 1851, and was built as the second pauper lunatic asylum for the County of Middlesex, in order to cope with overflows of the pauper insane (Ackroyd 2000). On 1 November 1898 Caroline Maisley became a patient here. We do not know very much about her. What we do know leads us to believe that she was a very ordinary member of London's Victorian working class. When she arrived at the hospital she was 27 years old and married to a dock labourer. We do not know where the couple lived as her previous address was given as the Stepney workhouse. She was thought to be suffering from 'mania' and this attack, which was officially the first one she had experienced, had lasted a week. As a result she had initially been admitted to the infirmary at Stepney Workhouse on 18 October 1898. We do not know which symptoms Caroline suffered from, how she was treated for them, or how she felt about that treatment. What the records do tell us is that Caroline 'recovered,' and left the hospital on the 1 September 1899.[8]

Mary Matthews was admitted to the same asylum on 14 November 1898. She too was a pauper in receipt of poor relief from Poplar Union. Mary also seems to have been an ordinary member of London's working class until her illness took her thorough the doors of the asylum. She was 32 years of age and lived with her husband James Matthews at 36 Broomfield Street, Bromley-by-Bow. However, Mary's order states that at the time of her illness she was living at the Poplar Union Workhouse, although this was because she had been admitted to the workhouse infirmary, rather than her family falling on particularly hard times.[9] She had tried to commit suicide, and it seems likely that it was at this time that James sought help for his wife from the authorities. Dr John Lamont examined Mary on the 12 November at the Poplar Union Workhouse and he concluded that she was 'a person of unsound mind'. As with Caroline there are no further records of Mary's stay in the asylum. Consequently it is not clear how long her attack lasted once she was in the hospital, though she remained in the asylum for almost a year. She eventually 'recovered' and was discharged from the hospital on 29 October 1899.[10] Thus it would seem that she returned to her home in Bromley.

How do these two ordinary women who, thanks to a note on Mary's admission records we know were sisters, challenge Hastings argument? In the surviving records no reference is made to the colour of their skin, but thanks to an album of photographs taken of Colney Hatch patients during their stay we know that the sisters were black women. Their presence in the archives disrupts Hastings argument in two major ways. Firstly, they (and they are just one example) challenge his assumption that only white people lived in Britain during the nineteenth century. Secondly, and perhaps more disturbingly for Hastings, the fact that no reference was made to the colour of their skin means that there are likely to be many more individuals, perhaps thousands, who will never be identified. It means that an absence of colour in British archives can no longer be equated with whiteness, nor should it be assumed that these histories will only be of interest to white Britons (Bressey 2006).

Conclusion

Geographies of race and racism in Britain today are rooted in our understandings of race and racism in Britain yesterday, last year, last century and before. The arguments of those who use 'over political correctness' to ignore the histories of, for example, anti-racism are not questioned enough. As was reported in the report of the London Mayor's Commission on African and Asian heritage, 'Learning about the past and about each other's heritage challenges racism and extremism because racist and extreme views are often based on selective and mistaken views of the past' (Delivering Shared Heritage 2005, 12). Moments of 'white Britishness' such as World War II are imbued with the politics of race, both in their reality (as reflected in the work of the League of Coloured Peoples), its memorialisation (in the memories of actors like Todd and Baker) and its representation (in films like *The Dam Busters*).

It is an important methodological practice for geographers to place contemporary issues of race and racism in an historical context. These will differ through the lenses of national, regional and local experiences. The contexts will not only be connected to the urban. Reconsidering historical and contemporary understandings of race and whiteness in the rural should prove a fruitful avenue for geographer's interested in the politics of race and nation (for example, see Askins 2006). The examples in this chapter illustrate that we do not have to scratch very far beneath the surface to reveal contested histories of language and other everyday practices of belonging. Researching black and Asian histories is not about accommodating 'an other' but focuses on challenging and deconstructing mythologies, and examining the realities of Britain's place in the world – is that not what geography is all about?[11]

Acknowledgements

I would like to thank Claire Dwyer for her comments on this chapter and her support over the years during which this collection has been conceived and completed.

Notes

1   See *The Keys*, 1:1 July 1933, 1–2; Letter no. 10, July 1940, 69–70; Letter no. 11, 13 August 1940.
2   *The Keys*, 1, July 1933, 2.
3   *The Keys*, 1, July 1933, 2; 2:1 (dates unknown as cover page is missing), 17.
4   *The Keys*, 2:1 (dates unknown as cover page is missing).
5   *The Keys*, 2:1 (dates unknown as cover page is missing).
6   *The Keys*, 3:1, July–September 1935, 3–4; 3, October–December 1935, 16–18.
7   Letter no. 9, June 1940, 39.
8   London Metropolitan Archives (LMA), H12/CH/B/01/13.
9   LMA, PO/BG/162/10; LMA PO/BG/154/04.
10 LMA, H12/CH/B/01/13.
11 'Scratch the Surface', an exhibition curated by Jonah Albert at the National Gallery, London in 2007 to mark the bicentenary of the Abolition of the British Slave Trade Act.

References

Ackroyd, P. (2000), *London: The Biography* (London: Chatto and Windus).
Askins, K. (2006), 'New Countryside? New Country: Visible communities in the English national parks', in S. Neal and J. Agyeman (eds) *The New Countryside? Ethnicity, Nation and Exclusion in Contemporary Rural Britain*, (Bristol: The Policy Press), 149–71.
Bressey, C. (2007), 'The Black Presence in England and Wales after the Abolition Act, 1807–1930', *Parliamentary History* 26 Supplement, 224–37.
Bressey, C. (2006), 'Invisible Presence: The whitening of the Black Community in the historical imagination of British Archives', *Archivaria* 61, 247–61.
Jarrett-Macauley, D. (1998), *The Life of Una Marson, 1905–65* (Manchester: Manchester University Press).
The Mayor's Commission on African and Asian Heritage (2005), *Delivering Shared Heritage* (London: Greater London Authority).

Chapter 4

Belonging in Britain – Father's Hands

Ingrid Pollard

You'll have to see London to believe it. Immense, giddy, confusing, a complication of traffic, transport and geography. (From my father's hand)

I arrived in London as a child in the wintry 1950s, one year after my father left Georgetown, Guyana for London and six months after my mother. I am now considerably older than either of my parents were when they set out on their great adventures to London. My father's letters, the inspiration for this chapter, speak of his aspirations, hopes and intentions. His excitement at being in the metropolitan city; places to go, sights to see, hanging out with West Indian friends, jobs, the practicalities of glove wearing, fog, and new dwellings.

Now, as one of the second generation, the maturing children of the 1950s immigrants from the Caribbean, I have reached an age that sees me at the burial of both my parents in UK cemeteries. As custodian of the family archive I know I was fortunate to be brought up with family albums and tales from home; the home of Georgetown and the home of London.

These photographs therefore have a key to play in an act of reconciliation and the process of change. (Sealy, n.d., n.p.)

We were doing a practice run bus journey to my new 'big school'. I wore a red overcoat, it had a wide folding collar, my favourite. I have practically a visceral memory of my father's hand on my 11-year-old shoulder. I can almost remember the weight, the feel of his large adult hand, on that cold autumn evening as he tried to keep his hand warm under my collar. It was just me and my dad.

Every photograph is a certificate of presence. (Barthes, 1980)

I am glad to see letters that mention me, but surprised when he is afraid he will not know me, (or) that I will not remember him. I am reminded of Alina Marrazi's evocative film 'For one more hour with you' (Un'ora sola ti vorrei' 2002) in which the film-maker creates her filmic memories, building a relationship with her dead mother through the viewing of long hidden and forgotten super 8 movies. Films that include visions of Marazzi's childhood holidays with her mother, memories that were new to her.

As if by magic those images projected in front of my eyes seemed to give life back to a mysterious and unknown person to me. (Marrazi 2002, n.p.)

References

Barthes, Roland (1980), *Camera Lucida: Reflections on Photography* (London: Flamingo).

Sealy, Mark (n.d.) *Nothing is Forever*. Catalogue notes Peter Max Kandhola.

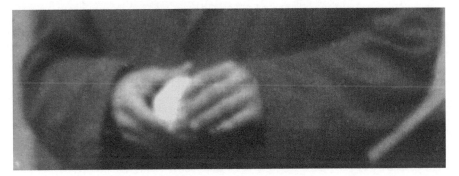

belonging in britain

Many days though the sun comes out bright and cold. Yes cold, it's strange to see the brilliant sunshine without any heat, but that's how it is. It isn't bad though. The alarming tales we hear back home of winter are wildly exaggerated, the only extra clothing I wear is sweater, wrap and coat. Gloves are a nuisance; coats pockets are the best gloves for sticking ones hands in. the worst part of the climate is the strong, cold breeze that lifts my coat tails and searches all over me - brrrrr.

My pal Mitch was here to meet me and I was glad to see him. You'll have to see London to believer it. Immense, giddy, confusing, a complication of traffic, transport and geography.

The places to go the things to do are countless. To give you an idea how happy London can be. Here are some of the places I've been taken to and gone on my own. Cinemas are the least entertainment here big sisters? What beauty they are, lovely decorated with turret balconies and castle like boxes, usher-ettes to show you to your seats, privilege to sit down and see the movie several times over and over again without pay but once. We go in anytime and come out anytime

There are some people who through a bad first impression get a dislike for London that takes along time to cure, sometimes never. Some historical sites and objects are disappointments after reading so much about.

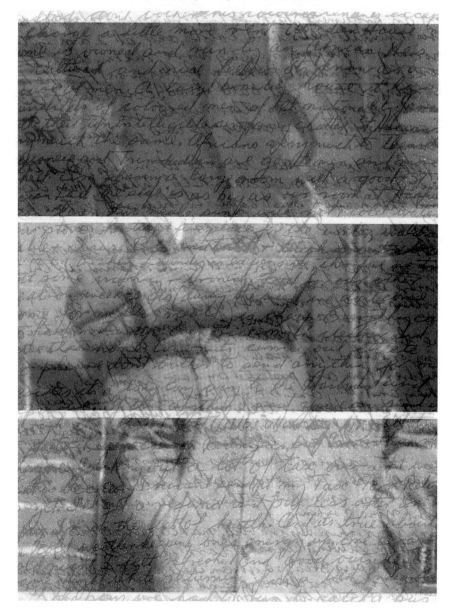

Count your blessings - you have children, mother, brother, family, friends near you. You are in familar surroundings - a good employer, good health ect. Not so with me though, except my few nigger friends I'm stranger in a strange land. lost in asea of white faces in an uncaring and sometimes cruel city.

London is grey and cold. Night falls early. 4 o'clock is lighting time and sometimes much earlier. The morning are sometimes misty. I've been through my first fog about two weeks ago. Could'nt see more than half a corner away. traffic crawls and tied up. Predestrians bump into streams of slow moving traffic and crossing the road was an alarming adventure.

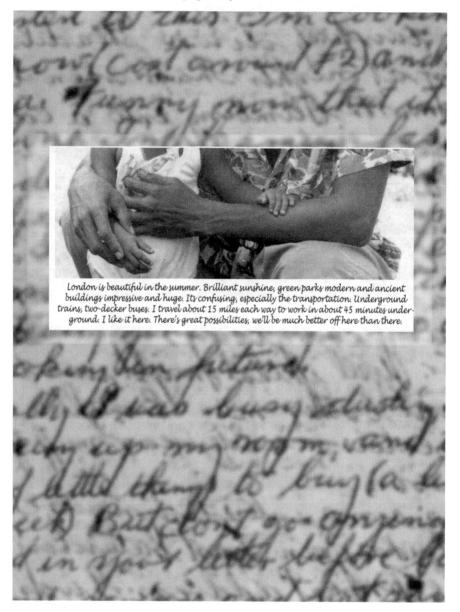

London is beautiful in the summer. Brilliant sunshine, green parks modern and ancient buildings impressive and huge. Its confusing, especially the transportation. Underground trains, two-decker buses. I travel about 15 miles each way to work in about 45 minutes underground. I like it here. There's great possibilities, we'll be much better off here than there.

Get some good snapshots or postcards of Georgetown, the new buildings in the shopping Centre, Queens College, the pictures that show the moderness and beauty of our hometown. To many dunces here (white) think that coloured people come from the jungles. Don't really beleive our country is beautiful or civilised.

The job is interesting. I'm printing bank cheques, very technical and exacting. I plan to hold on and perhaps to do some studies at Printing Trade School in the evenings. These English people are on the whole are nice. There are a few bad 'uns' - who I don't ever hope to meet again.

Chapter 5

On the Significance of Being White: European Migrant Workers in the British Economy in the 1940s and 2000s

Linda McDowell

In this chapter, I examine the ways in which white skins confer differential benefits on economic migrants. I draw on the notion of the 'wages of whiteness' to show how European migrants are preferred by the British state as potential workers rather than economic migrants from other parts of the world. I argue that whiteness is a multiple construct: not all Europeans are equally 'white'. Irish migrants, continental Europeans and white-skinned British-born workers typically are ranked in a hierarchy of desirability by potential employers as well as by the institutions of the state. I draw on two case studies of migrant workers who entered the UK sixty years apart to illustrate these arguments.

Introduction: The Colour Question

The history of immigration to Britain – that story about 'bloody foreigners' (Winder 2004) – has tended to be dominated, at least since 1945, by analyses of the relationships between migration, ethnicity, skin colour and racism. In the last half century or so, the most significant countries sending migrants to the UK were the former colonies: in the Caribbean and South Asia whose inhabitants, at least in the initial post-war decades, were British citizens, tied to the 'home' country through language, schooling and culture. The shock of discovering that to the locals they were all 'black' and so considered by many as less civilised, cultured, or desirable has been well-documented over the succeeding years. However, there is another story about immigration to the UK – that of white migrants. French Huguenots, German, French and Russian Jews, Irish, Italians, Poles, Latvians and, of course, white migrants from Canada and Australasia were also significant among the growing UK population born elsewhere but generally overlooked in post-war analyses of the impact of migration on British society. In part, I suggest, this was because of their white skins, providing these migrants with a cloak of invisibility, unavailable to those visibly-different migrants who landed in the UK from the Caribbean in the 1950s and South Asia in the 1970s. Issues of culture, religion, tradition, gender relations and class position are also part of the story.

Questions about whiteness and migration clearly intersect with other social divisions in the relational construction of identity and in the evaluation of different categories of migrants as suitable and appropriate workers. Labour is differentiated by age, skills, skin colour and gender, selected and directed into particular slots in the labour markets of receiving countries. Most countries, for example, operate immigration policies that restrict incomers to particular categories of jobs, typically dead-end, low-paid, unskilled work in agriculture, construction and the service sector, often for a limited time period. Distinctions are created between categories of migrants on the basis of a combination of their human capital, their willingness to accept low status employment and their status within the receiving country – as citizens or not, as legal or irregular workers. Currently, in the UK, managed migration policies (Flynn 2003) differentiate between a well-educated high-skilled elite among economic migrants regarded as valuable social capital and so awarded a (limited) range of social rights on entry and a low-skilled rump of undifferentiated 'warm bodies' for bottom end jobs.

Further, immigration policies (and patterns) are not gender neutral: a glance at employment statistics, as well as at official governmental schemes to attract particular categories of migrants for shortage sectors in the UK are revealing. Many women migrants, for example, enter the UK to take employment as domestics, caregivers, and nurses, reflecting the gendered attributes of different jobs in Britain (Anderson 2000). These gendered patterns are also connected to social class, ethnicity and stereotypical national attributes, both 'here' and 'there', that become institutionalised in the labour market through segregation and segmentation. As a consequence, immigrants are differently received and socialised depending on their position within racial hierarchies, gender, class background and income/consumption patterns both in their own and in the country of immigration. Thus, both institutional structures and everyday practices position in-migrants as workers (and potential citizens) of differential worth. Clearly constructions of difference – whether based on race, nationality, language or skin colour – are produced and maintained through practices that operate at and across different spatial scales including ideological assumptions, multiple regulatory systems, structures of power and domination and spoken and enacted everyday practices in multiple sites, that operate at both conscious and unconscious levels and open to contestation and renegotiation. This is a complex story and here I want to focus solely on the significance of whiteness at two important periods in the history of British immigration policy – the 1940s and the 2000s. Whiteness has once again become an issue in the UK as the number of new migrants from the expanded European Union continues to grow, greatly exceeding initial official estimates and fuelling a media panic about job competition and community cohesion.

Multiple Constructions of Whiteness

There is now a distinguished body of scholarship in the humanities and in the social sciences about the meaning of whiteness – in, for example, cultural studies (Dyer 1997; Linke 1999), English and literary studies (Babb 1998; Morrison 1992), labour studies (Roediger 1991, 1994), history (Ignatiev 1995; McClintock 1995; Ware 1992), geography (Bonnett 2000) and sociology (Frankenberg 1993, 1997). This work has documented the ways in which whiteness is a social construct, showing how, *inter alia*, white workers distinguish themselves from 'others' on the basis of skin colour and so reap the advantages as members of the white majority society in countries such as the USA, the UK and other European states where in-migrants are often visibly different. In the labour market, while people of colour often suffer an ethnic penalty (Heath 2005), white workers benefit from what Roediger (1991) memorably termed 'the wages of whiteness'.

The 'whiteness' of white workers, however, is not natural, unchanging nor unchallengeable. It is instead a social construct, a variable category and one which is often highly contested. It is a discursive construct, as well as a bodily attribute, deeply implicated in the politics of domination, representing purity rather than earthy embodiment, signifying 'goodness and all that is benign and non-threatening' (Dyer 1998, 45). And despite the expanding literature 'until quite recently, the multiple ways in which whiteness has been politically manipulated, culturally mediated and historically constructed have in large part been ignored' (Linke 1999, 28). I want to argue here that a significant part of this multiplicity lies in the establishment of hierarchies of whiteness. Not all white skins signify equally-valued people.

Although whiteness is the hegemonic characteristic in European societies, some 'white' inhabitants, especially if they are in-migrants, find themselves differentially valued, excluded from or included in the category at different times, in part depending on the demand for labour and the availability of different categories of workers. Furthermore, whiteness is a fluid and mutable concept that both encompasses variation within it and allows elision between categories. The boundaries that are drawn around the terrain of whiteness, despite the assumed rigidity of these criteria of difference, must be viewed as decentred and permeable, thus permitting a challenge to the binary categorisation of white/black.

In an interesting analysis of the history of race and racism in the USA, Ignatiev (1995) documented the ways in which the Irish became part of the 'white race' by triumphing over their perceived 'nativism' and so inferiority, in large part through anti-Black discrimination and violence. These Irish in-migrants to the USA entered a society in which skin colour was already an established, long-standing and significant factor in determining social position. For white migrants to the UK, the situation was somewhat different. The visibly different minority population was relatively small (although accurate numbers are not available). Castles and Miller (2003) mention established communities of Chinese workers and people of African origin, especially in London and in British ports, for example. But in most British cities and in the labour market, the main lines of differentiation were on the basis of social class, gender and possibly religion and language rather than skin colour (which is not to deny racialised acts of intolerance and violence in the early twentieth century). In the immediate post-World War II years, however, when migration became a key part of the government's economic policy and the post-war reconstruction plans, skin colour became a crucial indicator of 'otherness', especially as migration from the Caribbean gathered pace. These post-war years established the conditions both for discriminatory immigration legislation and an ethnically divided working class: a division based on skin colour that has led to the racist attitudes and behaviour that have so disfigured post-war Britain (Paul 1997).

However, while the history of this Caribbean migration to the UK is well-known and well-documented (Carby 1999; Cashmore and Troyna 1990; Fryer 1984; Phillips 1998; Phillips and Phillips 1999; Small 1994), less is known about the white central European in-migrants who also entered the UK in the second half of the 1940s (although see Tannahill 1958; Kay and Miles 1992; Lane 2004). There is another post-war story about migration – one about whiteness – to be told: a story that has renewed significance in the new millennium. In this chapter I want to focus on these white migrants in Britain, assessing the extent to which 'whiteness' itself is not only socially constructed but is a multiple and ambivalent category, in which distinctions *between* white in-migrants place them in an hierarchy of desirability which was both confirmed and challenged in everyday practices in the workplace.

The 1940s: Post-war Labour Shortages and Migrant Recruitment

In 1945, at the close of World War II, the British economy was in severe difficulties (Cairncross 1985; Clarke 1996; Law 1994; Morgan 1990; Whiteside 1999). Weakened by the war years and short of labour, the government turned to in-migration as a potential solution to shortages in basic industries such as iron, steel, coal and agriculture, the construction industry and those parts of the public sector essential to the economy and to the well-being of the population. Transport, the new National Health Service, and, somewhat anachronistically, domestic service in private homes were identified as important parts of the economy short of female labour.

In 1945, three main sources of potential workers were relatively easily available to the British state and to prospective employers (Paul 1997). The first group consisted of refugees and displaced people, demobilised soldiers and prisoners of war, most of whom were homeless and often stateless as a consequence of the upheavals in Europe between 1939 and 1945. During the war itself, for example, over 300,000 German and Italian prisoners of war had been employed in Britain in essential industries, although most of these prisoners were returned to their own countries at the end of the war. In addition sizeable numbers of Polish service-men who had fought with the Allies chose to be demobilised in the United Kingdom. The main source of potential labour identified by the British state, however, was the displaced people, mainly from parts of Europe occupied both by the Third Reich and by the Soviet Union, and living in refugee camps in Germany in 1945.

Many of these refugees were reluctant to return to homelands that had been absorbed into the Soviet Union under the Yalta agreement (Marrus 1985).

This group of potential employees included many young women from the former Baltic States who were seen as ideal workers for female-dominated sectors of the British economy. These women – and the many of the other camp residents – had few, if any, connections with Britain, nor rights of citizenship. Although they were mainly white Europeans, they were aliens by virtue not only of their legal status but also because of their native language, religion and cultural background. They were recruited by the British state as 'European Volunteer Workers' (EVWs) – this term replaced that of 'displaced' as it was regarded as pejorative by British officials. In Cabinet discussions, the term 'European workers' was seen as preferable to that of 'foreign workers', accepting that at that time the British population distrusted foreigners. The term European, it was believed, emphasised the joint heritage of the incomers and the local population, as well as their common whiteness, although as the long debate about membership of the European Community has shown many British subjects still do not see themselves as Europeans. In total 600,000 European 'aliens' (the official designation) were recruited to work in Britain in this post-war period, although only a proportion of them were from the camps in Germany. Between October 1946 and December 1949 about 80,000 people from DP camps were recruited for work in Britain, more than doubling the then foreign-born population of the country.

The second group of prospective workers was from a more traditional source for Britain – Ireland. Although there had been a long tradition of migration across the Irish Sea in the first part of the twentieth century, at the end of the war Irish citizens were particularly encouraged to migrate to Britain. Despite being aliens, they were given the legal rights of citizenship as if they were British subjects under the provisions of the 1948 British Nationality Act. These Irish in-migrants, however, retained their status as Irish citizens and so fell into an ambiguous category in the UK, being neither aliens nor British citizens. In each of the years between 1946 and 1962, approximately 50,000 to 60,000 Irish men and women entered Britain for the first time and began to search for work (Kearney 1990). The third source of prospective workers lay in the Commonwealth dominions. As the territories of what became commonly designated the Old Commonwealth (Australia, New Zealand and Canada) were recruiters rather than sources of white British workers, attention turned to citizens of colour in the 'New' Commonwealth countries. In the early post-war years, residents in the Caribbean, in particular, were identified as new employees. This migration began slowly in the immediate post-war period. Between 1948 and 1952, between 1,000 and 2,000 people entered Britain each year, although the numbers rapidly increased rapidly over a decade to 42,000 in 1957. Although these migrants were British citizens, they were to find themselves regarded by the British population as foreign, as other, as 'coloured' and so as 'less eligible' Britons (Fryer 1984; Gilroy 1987; James and Harris 1993; Phillips and Phillips 1999). Indeed in the official papers discussing Caribbean migration, potential recruits were labelled 'coloured colonial labour' and from 1962 onwards people from the Caribbean increasingly were restrained by the state from exercising their right to independent migration and residence. Interestingly, it seems that Irish migrants escaped the provisions of the 1962 Commonwealth Immigrants Act by virtue of their whiteness (Hickman 1998).

While all three groups were valued as potential contributors to post-war reconstruction, there were clear differences in their recruitment as well as in their status in the UK. The former and the latter group – EVWs and New Commonwealth migrants – were attracted to Britain through recruitment drives, while the Irish migrants moved on an individual, independent basis. The EVWs were aliens and yet, as I have noted, were regarded by the British state as fellow Europeans and as suitable candidates for assimilation. Only the latter group – the Caribbean migrants – came to Britain with full legal rights as British subjects and yet they were to find themselves out of place in post-war Britain. They suffered under a racialised discourse of 'difference', experiencing discrimination in the labour and housing markets (Smith 1996) that continues to this day.

The entry of Irish migrants as well as the continental Europeans was regarded as less contentious than that of Caribbean migrants both by government officials responsible for immigration policy and by the UK public at large because 'they passed an unwritten test of racial acceptability' (Paul 1997, xiii). And yet, while both groups were 'white', Irish in-migrants found themselves discriminated against in the labour market, regarded as less suitable employees than the EVW women who came to Britain in the same years. It may be too large a claim, as the Irish group *The Commitments* (1991) suggested, that 'the Irish are the blacks of Europe', but the common whiteness of Baltic and Irish in-migrants did not prevent them from being ranked in a hierarchy of desirability. Irish women workers had to bear all the connotations that being Irish carries in Britain: stupid, dirty, drunken, belligerent, unreliable, feckless, fecund, or in a less prejorative discourse, as emotional, fey and romantic (Curtis 1984; Hickman 1998; Ignatiev 1995; MacLaughlin 1997; Walter 2001), as well as discriminated on the basis of their religion (Allen 1994) and its association in Britain with 'excessive' fertility. Walter (2001), for example, has documented the long associations with Irish migrants and dirt – from Engels' description in 1845 of the Irish living in 'filth and decay' to contemporary abuse of Irish people. In the late 1940s the British government had a similar concern with the squalid living conditions of Irish migrants in British cities, but assumed that this reflected their earlier conditions in the Irish Republic. A Ministry of Labour memo in 1951 condescendingly noted that the Irish were 'not used to good living conditions' (Paul 1997, 106).

Baltic Cygnets

Displaced persons from camps in Germany were recruited as workers for shortage sectors in industry and the service sector, primarily hospitals, under two schemes: the Baltic Cygnet scheme and Westward Ho! The former is the more interesting in the context of an analysis of whiteness. Despite anxieties about the reception of foreign workers, under the first scheme, the Baltic Cygnet scheme initiated in 1946, more than 1,500 women were recruited. It is interesting to contemplate the image that the recruiters hoped would be associated with the name Baltic Cygnet

– perhaps a vision of young swans, redolent of purity, sailing across the water to the UK or maybe dull stubby cygnets who would blossom into swans under British tutelage. One of the women to whom I talked[1] more dismissively termed the scheme 'that white swan thing', reminding me that cleanliness and sometimes purity were not easy to maintain during exile and residence in overcrowded DP camps. But it is clear that the white skins and an association with cleanliness of Baltic recruits was a key consideration in their selection.

It seems that young Baltic women, often from middle class backgrounds, were the preferred recruits, regarded as superior to women from other nations, typically from less privileged rural backgrounds. An internal minute in the Ministry of Labour (21 November 1945) noted that Polish recruits were mainly of 'the peasant woman type' who might fail to meet the high standards of cleanliness needed in British hospitals, compared to the Baltic women who 'are of good appearance, scrupulously clean in their persons and habits' (Kay and Miles 1992, 48). Once in Britain, the middle-class background of many Baltic EVWs was compared favourably to the peasant backgrounds of Irish women.

After a preliminary visit to a number of camps in Germany, it was suggested in the Draft Report on the Recruitment of Baltic Displaced Persons (1946) that 'an exceedingly good type of woman is available for hospital domestic work in this country'. The report went on to acknowledge the superior appearance, knowledge of English and general educational standards among young Baltic women: 'there is little doubt that the specially selected women who come to this country will be an exceptionally healthy and fit body ... and constitute a good and desirable element in our population'. The focus on the bodies and habits of these women is unavoidable.

Thus right from the start, the purity and cleanliness of Baltic migrants was contrasted both to other women in camps for displaced people and to the 'dirty' Irish. Even so, for both groups, being 'white' in late forties Britain did not mean that they shared equally in the growing levels of prosperity that distinguished Britain in later post-war decades. While Harold Macmillan may have argued in 1957 that the British had 'never had it so good', most Irish migrant workers and many EVWs remained trapped in the bottom-end, dirty and low-paid jobs for which they had initially been recruited a decade earlier (McDowell 2005).

The New European Migrants: Movement from the EU 8 Countries since May 2004

I want now to jump forward and examine some of the similarities between the EVW entrants in the 1940s and the current focus by the British government on the recruitment of economic migrants from the 2004 EU accession states to fill labour market vacancies in an economy characterised by growth rather than by a need for reconstruction. Many of these new EU migrants are from the same countries as the people who found themselves transformed from displaced refugees into economic migrants after the end of World War II. Eight of the ten states admitted to the EU in May 2004 (Czech Republic, Estonia, Hungary,

Latvia, Lithuania, Poland, Slovakia and Slovenia; the other two were Cyprus and Malta) were part of the former Soviet Union/sphere of influence. In January 2007, Bulgaria and Romania were also admitted, although, as I discuss below, under different terms and conditions. Like the European Volunteers workers, the new European migrants to the UK are also predominantly young single people (this was a requirement under the Baltic Cygnet and Westward Ho! schemes and a consequence of freer movement post-2004), although the new economic migrants have been, right from the start, both men and women: currently in a ratio of 60:40.

The major difference between the 1940s and the 2000s is that the new European migrants are neither displaced persons nor, as the EVWs were, tied to specific employment for a number of years as a condition of entry, although they must register and work for a single employer for the first year after entry to qualify for a range of social benefits, providing parallels to the tied labour schemes of the 1940s. On the expansion of the EU in 2004, each of the old fifteen states was permitted to decide its own terms of entry, as well as the employment, residence and other social rights to be extended to new EU citizens for a period of up to seven years after the Accession Treaty. Ireland, Sweden and the UK were the only three of the EU15 member states who granted immediate labour market access to EU8 nationals, followed approximately two years later by the relaxation of restrictions by France and Germany. In the UK, the main impetus behind the liberal entry scheme was severe labour market shortages, especially in the south east of England, mainly in low wage and low skill occupations in sectors such as agriculture, construction, hospitality, transport and the public services as well as in a range of occupations in the health service, including nursing.

Rather than a completely free entry scheme, however, the UK established a Worker Registration Scheme (WRS) for EU8 entrants (and theoretically nationals from these states who were already in the UK) which imposes obligations on both the migrant workers themselves and their employers. After complying with the registration process for a year, workers become eligible for an EEA residence permit and are also entitled to state benefits. By the end of June 2006, the Home Office reported that 447,000 workers had registered (62 per cent of whom were Polish). There is no way of knowing, of course, how many EU8 migrants were working without registering, nor what the total number of EU8 workers in the UK is as the registration requirement applies only to employees and to the first twelve months of employment in the UK. Further, applicants are not required to deregister when they cease employment and/or leave the country. There may be, for example, as many as a million Poles currently resident in the UK (estimate by the Polish news magazine *Polityka* in 2006). Another estimate suggests 350,000 and 500,000 Polish workers have entered post May 2004 (Pidd and Harding 2006). However, there is no doubt that the number of people of central and East European origin now in the UK has increased significantly since May 2004. Although the Poles are the dominant group numerically, Lithuanians and Latvians have the greatest propensity to enter the UK labour market relative to the total population in their home country. These two countries had the lowest GDP per head in 2005 of the A8 states. Like their compatriots two generations earlier, these migrants are attracted/admitted to the UK because of severe labour shortages in the British economy. Many of the pre-2004 migrants from the A8 states first came, for example, on temporary work permits to fill vacancies in agriculture, construction, retail and hospitality and it is clear that many of them were still in the country when their status changed. Research commissioned by the Home Office (Gilpin et al. 2006) found that new entrants work in similar sectors, many of them employed in jobs considerably below their skill levels as were the EVWs sixty years earlier.

What is interesting is that a similar hierarchy of desirability based on whiteness is being re-established in the UK. Many British employers, as well as the state, regard East European workers as preferable to other economic migrants, and indeed to the British working class, because of their apparent or supposed motivation and commitment. Employers argue that these new migrants are generally hard-working and reliable. In a study of workers and employers in the agricultural, construction and hospitality sectors and domestic service, Anderson et al. (2006) found that for employers, the main advantage of accession-state migrants was their capacity for hard work in low paid, demanding and often demeaning jobs with long or anti-social hours that British workers tended to find unacceptable.

Behind this praise for A8 migrants lies an unspoken difference between them and other in-migrant groups relied on by the British state between the early1950s and the new millennium. What unites the old and new European migrants, as well as their commitment to hard work, is their skin colour and (perhaps) their heritage. In a society now riven by racism and marked by discrimination against what were previously termed New Commonwealth migrants and Black and minority ethnic (BME) British-born populations, largely on the basis of skin colour (Solomos 2003), whiteness clearly is a marker of privilege, even at the bottom end of the labour market.

This new large-scale migration from central and eastern Europe thus raises different questions than those addressed by most studies of migration in the intervening years between the entry of the 'old' and 'new' Europeans. Like their predecessors, the new Europeans are less visible because of their skin colour, have European identity in common with their host population and, unlike the post-colonial migrants who came between the two European migrations, they have no previous connections with the UK, entering Britain often speaking relatively little or no English. And just as the EVWs from the Baltic States were preferred to those from elsewhere in Europe, in 2007 a new distinction between white Europeans has been introduced on the basis of nationality. When Bulgaria and Romania were admitted in January of that year, their nationals' access to the labour market was severely restricted. Only self-employed migrants from these states and workers in a limited number of sectors where there are extreme shortages (in the main in agriculture and food processing) are allowed to work. Here we see at work a rhetoric of racialised national stereotypes that distinguishes between 'bad' Bulgarians and Romanians (represented in the popular press as respectively gangsters and Romany people, with the associations of a nomadic lawless way of life (Okely 1996)) and 'good' hard working Others. In the 2000s,

Polish migrants found themselves on the side of the angels, classed with other 2004 A8 countries as desirable, despite the widespread representation in the popular press of the hard-working Polish plumber as a scourge of the local working class. As Colic-Peisker (2005, 622) noted in an interesting paper about the 'whiteness' of Bosnian refugees in Australia: 'Clearly whiteness is not just about skin colour, but also about class, status, language and other features of the individual that can be discerned in social interaction' and imputed/constructed through racialised discourses.

Managed Migration

Further support for the argument that whiteness is of growing significance in understanding the labour market position of contemporary economic migrants lies in a new, or more accurately, extended version of a 'managed migration' policy in Britain. Based on market principles and a version of an 'objective' entry policy, from 2008 potential economic migrants are to be assessed and allocated to one of five categories based on an evaluation of their human capital and an assessment of labour market shortages (Home Office 2006).

The five tier entry-scheme is outlined below.

Tier 1 entrants are those with high skills who are permitted free entry to the UK subject to the possession of a significant number of points calculated on the basis of, for example, educational credentials, age, English language skills and previous career and salary level. These entrants do not have to possess a job offer and are seen as essential contributors to the growth and productivity of the British economy. Scientists and entrepreneurs are examples listed by the Home Office. After two years continuous employment these migrants become eligible for rights of residence.

Under *Tier 2*, skilled workers with a job offer are assessed on a somewhat different points scheme that includes points for sponsor rating (i.e. the employer) and prospective salary. These economic migrants – nurses, teachers and engineers are examples in this category – are also seen as valuable and as potential British workers over a long-time period and so achieve residence rights after five years in the country. These migrants, however, are tied to a sponsoring employer.

Tiers 3, 4 and 5 are different. Migrants in these categories will have no righs to bring dependents nor to remain in the country after their employment/course ends. *Tier 3* consists of limited numbers of workers with low skills to fill vacancies in shortage sectors to be identified by a new Skills Advisory Body. Construction workers for a particular project are the exemplar here. *Tier 4* consists of students and *Tier 5* of a miscellaneous group of temporary workers and youth mobility schemes. The uniting feature of these three categories is that all entrants are admitted only as temporary workers for a limited period.

What is significant for current migration policy and race relations is that tier 3 will be an empty category in the foreseeable future as migrants from the new EU are expected to fill all these vacancies. Thus, there is no doubt that this new managed migration system will have the effect of privileging white migrants. It will radically change access to low-paid employment in the UK for potential migrants from non-EU countries: many of whom are people of colour from the 'South'. Furthermore, the new policy will adversely affect Black and minority ethnic workers already in the UK and/or born here as the vacancies in the low-skilled shortage sectors of the economy that are now to be filled by A8 migrants typically are also sources of employment for BME British and migrant workers. In the future, in all probability, they will find themselves undercut by newcomers prepared to work harder for lower wages.

The new scheme also re-introduces a form of guest, or even indentured, worker scheme similar to the principle that lay behind the EVW schemes in the 1940s. Thus, new questions about the wages of whiteness are raised that could be ignored in the 1940s as the numbers of white in-migrants were relatively small. Perhaps understandably, from the late 1950s onwards, racism based on skin colour seemed to be the most urgent issue facing analysts of migration and ethnicity. Now, however, the multiple construction of whiteness and its consequences demand theoretical and empirical attention.

Conclusions

The immediate post-war years set the template for patterns of ethnic and gender segregation still marked in the British labour market. Women from all classes and backgrounds are less likely to work than men, to be paid significantly less and to work fewer hours, in the main in female-dominated sectors of the economy. This pattern of gender segregation is cross-cut by divisions based on nationality, race and ethnicity and class. White 'native' women, for example, were, at least until the 1970s, predominantly employed on a part-time basis, especially if they were mothers of dependent children, whereas women of colour and other women in-migrants more usually worked on a full-time basis. Irish and Baltic women who came to the UK in the immediate post-war era found themselves doing the sort of dirty, backbreaking, and poorly paid jobs to fill the gaps left by British women leaving the labour market and to support the expanding need for female labour in the growing service sector.

But a distinctive hierarchy developed within these white migrant labourers. As the comparison of Baltic EVWs and Irish women makes clear, 'whiteness' is a multiple category that repays both disaggregation, an analysis of its intersection with other social divisions and a study of the ambivalence that marks social relations between different white groups. Divisions based on ethnicity and religion, as well as on skin colour, prevented migrant women workers uniting to fight for their common interests in the post-war labour market. Further, as the post-war Caribbean migrants found to their cost, and EVWs to their benefit, citizenship did not guarantee preferential (or even equal) treatment in post-war Britain. Common whiteness outbid common citizenship. But it is also clear, that neither was a common white skin an automatic passage to success or equality. The connections between ethnicity, colour, gender and national identity established post 1945 and still evident are more complex than many commentators allow.

In the new millennium, the new Europeans may also find that they are the relative aristocrats among the different groups of migrants now being recruited by the British state and major employers. Like their predecessors, they are young, many are from urban backgrounds and a majority hold post-school educational credentials. And yet, it seems clear that a common European heritage is once again insufficient to prevent the development of a hierarchy of desirability as some new entrants find themselves the victims of a nasty racist rhetoric of difference. While I have sketched a comparison between the 1940s and 2000s here, a detailed study of post-war and contemporary migration from the accession states remains to be undertaken to establish in empirical detail the ways in which degrees of acceptable whiteness are constructed and legitimated on the basis of the intersections between white skins and attributes such as gender, nationality and religion. Although it is clear that many of these new migrants from the accession states are taking low-wage and low status work at the bottom end of the labour market (Stenning et al. 2006), like their predecessors, they have the advantage of a white skin and there is no doubt that a new hierarchy of desirability among migrant workers – both those already established here and new arrivals – is in the process of construction in the UK labour market.

Acknowledgement

This chapter draws on material previously published in my book *Hard Labour* (UCL Press, 2005), in a paper in the *Journal of Baltic Studies* (2007, 38:1, 85–107) and a forthcoming paper (2008) on managed migration policies in the *Journal of Ethnic and Migration Studies*. I am grateful to the editors for their permission to use some of this material as the basis for this chapter. I am also grateful to twenty-five women originally from Latvia who came to the UK between 1946–1949 for telling me their life stories.

Note

1   In 2001/2 I interviewed twenty-five by then-elderly women recruited in DP camps in a study funded by the British Academy (see McDowell 2005).

References

Allen, T.W. (1994), *The Invention of the White Race* (London: Verso).
Anderson, B. (2000), *Doing the Dirty Work* (London: Zed).
Anderson, B., Ruhs, M., Rogaly, B. and Spencer, S. (2006), *Fair Enough? Central and East European Migrants in Low-wage Employment in the UK* (Oxford: COMPAS), http://www.compas.ox.ac.uk/changing status.
Babb, V. (1998), *Whiteness Visible: The Meaning of Whiteness in American Literature and Culture* (New York and London: New York University Press).

Bonnett, A. (2000), *White Identities: Historical and International Perspectives* (London: Prentice Hall).

Cairncross, A. (1985), *The Years of Recovery: British Economic Policy 1945–51* (London: Methuen).

Carby, H. (1999), *Culture in Babylon: Black Britain and African America* (London: Verso).

Cashmore, E. and Troyna, B. (1990), *An Introduction to Race Relations* (Basingstoke: Falmer Press).

Castles, S. and Miller, M. (eds) (2003), *The Age of Migration: International Population Movements in the Modern World*, 3rd edition (London: Palgrave Macmillan).

Clarke, P. (1996), *Hope and Glory: Britain 1900–1990* (London: Allen Lane).

Colic-Peiska, V. (2005), '"At Least You're the Right Colour": Identity and social exclusion of Bosnian refugees in Australia', *Journal of Ethnic and Migration Studies* 31, 615–38.

Curtis, L. (1984), *Nothing but the Same Old Story: The Roots of Anti-Irish Racism* (London: Information on Ireland).

Dyer, R. (1997), *White* (London: Routledge).

Flynn, D. (2003), *Tough as Old Boots? Asylum, Immigration and the Paradox of New Labour Policy* (London: Immigration Rights Project).

Frankenberg, R. (1992), *White Women, Race Matters: The Social Construction of Whiteness* (London: Routledge).

Frankenberg, R. (ed.) (1997), *Displacing Whiteness* (Durham, NC: Duke University Press).

Fryer, P. (1984), *Staying Power: The History of Black People in Britain* (London: Verso).

Gilroy, P. (1987), *There Ain't no black in the Union Jack* (London: Routledge).

Heath, A. (2005), 'Ethnic Penalties in the British Labour Market: The public and private sectors compared'. Paper given at a PSI/TUC conference *Ethnic Employment in the Private Sector*, 22 September, available from the author at Nuffield College, Oxford.

Hickman, M. (1998), 'Reconstructing/Deconstructing "race": British political discourse about the Irish in Britain', *Ethnic and Racial Studies* 21, 288–307.

Home Office (2006), *A Points-based System: Making Migration Work for Britain*, Cm. 6741, Norwich: HMSO.

Ignatiev, N. (1995), *How the Irish became White* (London: Routledge).

James, W. and Harris, C. (1993), *Inside Babylon: The Caribbean Diaspora in Britain* (London: Verso).

Kay, D. and Miles, R. (1992), *Refugees or Migrant Workers? European Volunteer Workers in Britain, 1946–1961* (London: Routledge).

Kearney, R. (ed.) (1990), *Migrations: The Irish at Home and Abroad* (Dublin: Wolfhound Press).

Lane, T. (2004), *Victims of Stalin and Hitler: The Exodus of Poles and Balts to Britain* (London: Palgrave Macmillan).

Law, C. (1994), 'Employment and Industrial Structure', in J. Obelkevich and P. Catterall (eds) *Understanding Post-war British Society* (London: Routledge).

Lewis, J. (1992), *Women in Britain since 1945* (Oxford: Blackwell).

Linke, U. (1999), *Blood and Nation: The European Aesthetics of Race* (Philadelphia: University of Pennsylvania Press).

MacLaughlin, L. (ed.) 1997, *Location and Dislocation in Contemporary Irish Society* (Cork: Cork University Press).

Marrus, M. (1985), *The Unwanted. European Refugees in the Twentieth Century* (Oxford: Oxford University Press).

McClintock, A. (1995), *Imperial Leather: Race, Gender and Sexuality in the Colonial Context* (London: Routledge).

McDowell, L. (2005), *Hard Labour: The Forgotten Voices of Latvian 'Volunteer' Migrant Workers* (London: UCL Press).

McDowell, L. (2007), 'Constructions of Whiteness: Latvian women workers in post-war Britain', *Journal of Baltic Studies* 38: 1, 85–107.

McDowell, L. (2008), 'Old and New European Migrants', *Journal of Ethnic and Migration Studies* (forthcoming).

Morgan, K.O. (1990), *The People's Peace: British History 1945–1990* (Oxford: Oxford University Press).

Morrison, T. (1992), *Playing in the Dark: Whiteness and the Literary Imagination* (Cambridge, MA: Harvard University Press).

Okely, J. (1996), *Own and Other Culture* (London: Routledge).

Paul, K. (1997), *Whitewashing Britain: Race and Citizenship in the Postwar Era* (Ithaca, NY: Cornell University Press).

Phillips, D. (1998), 'Black Minority Ethnic Concentration: Segregation and dispersal in Britain', *Urban Studies* 35, 1681–702.

Phillips, M. and Phillips, T. (1999), *Windrush: The Irresistible Rise of Multi-racial Britain* (London: HarperCollins).

Pidd, H. and Harding, L. (2006), 'Why Would you Leave a Place Like This?', *The Guardian* G2: 6–11.

Roediger, D. (1991), *The Wages of Whitness: Race and the Making of the American Working Class* (London: Verso).

Roediger, D. (1994), *Towards the Abolition of Whiteness: Essays on Race, Politics and Working Class History* (London: Verso).

Small, S. (1994), *Racialised Barriers: The Black Experience in the United States and England in the 1980s* (London: Routledge).

Smith, C.D. (1996), *Strangers at Home: Essays on the Long-term Impact of Living Overseas and Coming 'Home' to a Strange Land* (New York: Aletheia Publications).

Solomos, J. (2003), *Race and Racism in Britain*, 3rd edition (London: Palgrave Macmillan).

Stenning, A., Champion, A., Conway, C., Coombes, M., Dawley, S. and Richardson, R. (2006), 'Migration from the "New" Europe to the UK Regions: Migrant workers beyond the global city'. Paper given at the Annual Conference of the Association of American Geographers, Chicago, 7–11 March, available from the first author at the Department of Geography, University of Newcastle, UK.

Tannahill, J. (1958), *European Volunteer Workers in Britain* (Manchester: Manchester University Press).

Walter, B. (2001), *Outsiders Inside: Whiteness, Place and Irish Women* (London: Routledge).

Ware, V. (1992), *Beyond the Pale: White Women, Racism and History* (London: Verso).

Whiteside, N. (1999), 'Towards a Modern Labour Market? State Policy and the Transformation of Employment', in B. Conekin, F. Mort and C. Waters (eds) *Moments of Modernity: Reconstructing Britain 1945–1964* (London: Rivers Oram Press).

PART 2
Race, Place and Politics

Chapter 6

East End Bengalis and the Labour Party – the End of a Long Relationship?

Sarah Glynn

The London Borough of Tower Hamlets is a favourite site for research on ethnic minorities because, in its Bengali Muslim population, it is home to an exceptionally concentrated ethnic minority group. It is also an area with a long history of immigration and poverty, in the immediate shadow of the wealth of the City. Post-war immigrants from what was then East Pakistan followed a migration chain pioneered by earlier Bengali seamen, and Bengalis now make up over a third of the borough's population and over half in Spitalfields and Whitechapel. This geographical concentration allows us to examine the interacting forces generated through ethnicity, religion, community, place, idealism and self-interest that together impact on electoral politics in a multicultural society. Inevitably such an examination centres on the Labour Party.

The story of the East End Bengalis and the Labour Party is one of a liaison that has turned sour. And like in many more personal relationships, the forces that finally drove the parties apart are the same that first brought them together. Now, as in the past, the main force behind East End Bengali politics is a community-based pragmatism, and the emergence and subsequent evolution of this can be understood by looking at it in the context of wider developments in progressive politics.

The affiliation of ethnic minorities with the Labour Party was once a truth universally acknowledged (Saggar 1998, 6) and reinforced by the overt racism of the Conservative opposition. In Tower Hamlets, this affiliation was also the best way to achieve real power within the political establishment, as (apart from a two-term Liberal hiatus) the Party dominated local politics from the 1920s to the Iraq War. The Labour Party was the natural recipient of most Bengali votes and the natural forum for most mainstream Bengali political activity.[1]

Beginnings: Looking Back to East Bengal

For the earliest immigrants, however, the focus of political activity, as of their lives more generally, was their East Bengali homeland, compounded in the period following World War II with immediate concerns over immigration. Many of these immigrants had little education, and a prominent role was played by a small intellectual elite of students and young professionals who had learnt their politics in the persecuted left and nationalist movements of their homeland. It might have been expected that this would have encouraged the spread of socialist ideas in the immigrant community more generally, but the political approach taken was in line with the popular front politics propagated by the Communist Party, which specified that political ideology be kept in the background.

Popular Front principles were applied whether doing grass-roots community work or campaigning on major issues such as Bangladeshi independence. Tasadduk Ahmed, who played a pivotal role in British Bengali politics and put the Pakistani Welfare Association onto a firm footing, explained:

> My main experience in the UK has been the experience of how to manage or organise united front activities, keeping my own belief to myself and to my close associates.[2]

Left involvement in community work is meant to bring more than immediate benefits; it is conceived as a way of raising the political consciousness of the people who become involved. However, it is very easy for immediate demands to exclude all other considerations, and for political ideas to get little exposure. A similar approach was taken (by Tasadduk and others) in supporting the independence struggle in 1971. Despite the prominence of far left thinking among the political leaders, popular front politics ensured that it was not Marxist ideas and organisations that gained adherents from the political mobilisation.[3]

In fact 1971 provided an important foundation for the Bengali community's close relationship with Labour. This rested both on the involvement of Labour MPs, who kept up pressure on the Conservative government to support Bangladesh,[4] and on the Party's perceived ideological link with the Awami League, who dominated the new Bangladeshi government. Peter Shore, then Labour MP for Stepney, found his own involvement 'led to an ongoing and very close relationship with the community, and indeed with their leaders'.[5]

Developing links with the Labour Party were cemented by the post-war immigration, when future hopes were focused increasingly on a settled life in Britain, and the Bengalis found they could turn to their MP with immigration and other problems. A few British Bengalis returned to take up positions in the newly independent Bangladesh, but generally the pulls were all in the other direction, and many Bengali men who were already working here began to bring over their wives and families. The early 1970s was a period of settlement and consolidation in the Bengali East End, a prelude to the mushrooming of locally based political activism at the end of the decade.

Fighting for the Bengali Community in London

In the late 1970s and early 1980s East End Bengalis became caught up in two major battles – against an aggressive racism being promulgated by the National

Front, and for decent and safe housing. The housing crisis was exacerbated by the arrival of the families, and the new young generation played a major role in the campaigns. Many of them had witnessed radical political action first hand during the war, and unlike the older generation they saw themselves as permanent immigrants. They were ready to fight for their rights, and it was these battles that mobilised the generation that now makes up the greatest part of the Bengali political establishment.[6]

At the centre of the fight were organisations set up by New Left activists from the Race Today Collective with the ultimate aim of creating a movement for radical black self-organisation.[7] They were by no means the only organisations working in the East End, but through their hard-fought grass-roots campaigns they were able to involve more people and generate more radical forms of protest. Helal Abbas, who was secretary of their Bengali Housing Action Group in his teens, and was later to become leader of the borough council, dates his awareness of the possibilities of political action from his involvement in the housing struggles,[8] and many others, even if they were not themselves involved with these organisations, felt their influence in the campaigns of the late 1970s.

In the event, for most of those taking part there was generally little time given to ideological discussion; but separatist organisation itself discouraged the making of links with wider working-class issues, and encouraged a scepticism towards what *Race Today* called the 'white left' – as well as facilitating greater ethnic concentration within certain wards. Despite the radical rhetoric, black organisation nowhere really challenged the British establishment – only attempted to take its fair place within it (Shukra 1998).[9]

A Community within Mainstream Politics

For most of those involved, community-based activism merged easily into community-based pragmatism. Bengalis were suffering through racism and often appalling living conditions and they felt the need to help their fellow immigrants the most efficient way possible. For many this meant joining the Labour Party; and, despite their move to mainstream politics, they generally continued to see themselves as representatives of the Bengali community.

The overriding reason given by my interviewees for joining the party was to continue campaigning for better and fairer treatment of the Bengali community (though of course there were always also those who thought they could achieve this better outside organised politics). Self-organisation, along 'ethnic' lines, had mobilised a new layer of activists, but now they wanted access to the mechanisms of power. As Abbas put it, 'you can only change so much from outside'.[10]

In general, Bengali Labour politics can be described as pragmatic, even when it is dressed in socialist rhetoric. However, the object of that pragmatism varies from the 'community' to the individual – perhaps portrayed as the true representative of the community. Sunahwar Ali described his own matter-of-fact approach:

I'd rather work with people who can deliver thing[s] for the community ... because sometime[s] ideology doesn't work.[11]

When I asked one prominent Labour councillor to explain his choice of party he told me:

...we thought, OK, the fastest way to get in, in this area, will be through the Labour Party because that's the party in control of the local mechanisms at the time ... it wasn't for any political ideals ... because all of them were just as bad ... the Labour Party had brought some of the worst immigration policies and procedures of that period ... The Tories were far too Right for us to even consider ... and to get the voters to vote for it in an area like this; and the Liberals we thought would never get into power ...[12]

A Revolution in the Labour Party

At first, though, many of this newly politicised generation found the Labour Party far from welcoming. Some early Bengali immigrants appear to have had no problems in joining their ward parties,[13] but many Bengalis were faced with a blank refusal. When Sunahwar Ali and two or three others applied to join the party in Spitalfields, at the heart of Bengali settlement, they were told, 'Sorry, we don't have vacancy, we've got too many member, we cannot allow'.[14] The situation was changed by the intervention of Labour left-wingers, but Sunahwar's explanation exhibits the prevailing distrust of their motives:

a group of certain people ... those who believe in left policies on the Labour Party side, they thought, 'This is the time we can utilise them'. They came and said, 'Why don't you become a member we'll help you?'[15]

The inclusion of people from all 'ethnic' backgrounds can also be seen as flowing directly from socialist ideology, and the view of these events from the left is a little different from many Bengali perceptions. At that time, Labour ruled over Tower Hamlets almost unopposed, and it's entrenched, relatively right-wing leadership clung fiercely to power. Jill Cove, one of a small group of Labour left-wingers, explains that it was not only Bengalis who had difficulties breaking into the Party fortress. She applied to join Spitalfields Ward Labour Party, together with her partner George, in the late 1970s; and they were also told there was a waiting list and were given no further information. In 1979, after more than a year, they got hold of application forms and marched into the ward meeting demanding membership. But there were many others in a similar position, and every month the meetings were lobbied by a crowd of would-be members. On one occasion, when there were some twenty people outside, Jill and George found there were only three other members present, and they proposed opening up the party. The new members admitted that day – who could be described as belonging to various 'ethnicities' and social backgrounds – included Abbas and three other future Bengali councillors.[16]

The 1982 council elections added a new dimension to the sense of Bengali exclusion, and this was met by radical action. The Labour Party had chosen to stand one Bengali candidate in St Katherine's ward,[17] but they had nominated no Bengalis in Spitalfields. John Eade has recorded in detail how, in protest, four Bengalis (two of whom 'claimed to have been members of the Labour party' (Eade 1989, 46)) put themselves up for election as independents, and a fifth stood for the newly formed SDP. Two of the independents were chosen and supported by the People's Democratic Alliance (PDA), which was created for the purpose of the election by delegates from different Bengali community groups. The four independents were of varied ages and experience, but their action was a natural outcome of the politics of self-organisation that had been promoted in the 1970s. Spitalfields elects three councillors. When the results were announced, Nurul Huque of the PDA topped the poll, while the other PDA candidate missed third place by twenty-six votes.

Abbas had accepted his party's decisions and campaigned for Labour, but he confirms that the election had the desired impact:

> I think from that [the] party took the message clearly at the following elections: we saw Bangladeshi candidates – people who were capable and able to represent the local community – were given opportunities to stand as Labour Party candidate[s].[18]

This was also the year after the 1981 Brixton riot, when the neglect of their black constituents had been brought abruptly to the notice of politicians of all kinds (Shukra 1994, 54–5).

In fact, Abbas himself was soon to be selected by the Labour Party to stand in the 1985 Spitalfields by-election. Somewhat perversely, the PDA could not resist putting up their own candidate, who squeezed the Labour majority down to just nine votes. The PDA candidate was able to make use of strong Bangladeshi village networks to mobilise his voters.[19] This kind of clan politics is often explained as an Asian import, but it is probably more accurate to say that a close community with strong patriarchal structures allows for the most efficient use of those non-party ties and networks that are exploited by politicians of all backgrounds. The importance of patronage relations was strengthened by communication difficulties that left those who could translate English and Bengali (and understood political procedure) in a powerful position. Existing patterns of patronage that are found in many ethnic minority communities will inevitably be exploited in politics, and possibly reinforced. In ward meetings leading Bengali members would refer to others as 'my members'; and when Labour canvassers went round Spitalfields they did not bother to knock on every door – Bengali party members knew where to find the community leaders who would be able to deliver perhaps twenty votes.[20]

The first Bengali Labour Party members were quick to recruit more. Rajonuddin Jalal, who became a prominent councillor, recalls:

> ... by the mid 1980s, I think, we realised that there is a need to capture political power if you're going to change policies in the town hall. And that's what we did. We started

recruiting Bengalis actively, and there was a time when the Labour Party got fed up, they did not want any more Bengalis.[21]

Jill Cove comments that it was 'like opening the flood gates'. She recalls that it was nothing to get a hundred and fifty membership applications at a ward meeting, often with the forms filled up in the same handwriting. At AGMs it was possible to sign up at the door, and Bengali members would be outside ready to pay for others to join. Mass memberships were accepted in principle, but subjected to random checks that exposed many names that were invalid – people who had not been aware of what was happening or were away in Bangladesh or even no longer alive. This did not exempt Cove from having to refute accusations of racism for questioning these lists.[22] There were also accusations that ward parties were being packed by supporters to secure election of certain Bengalis to party positions or their nomination for council elections, and one of my (Bengali) interviewees described how this worked:

[A] few people, they got some money together and asked all the village people, all the people that they know, to become Labour Party member of … a particular [ward]. And so if you got fifty or two hundred members and if you can pay two hundred times £5 per year [membership] for couple of years, then you can be chosen for whatever you want to be in that particular party.[23]

Such political tactics are not actually illegal (providing the paper members do exist and are willing participants). Nor, though South-Asian kinship links can be used to make them spectacularly successful, are they confined to ethnic minority communities. At the same time, white politicians are not above making use of Asian patronage systems when they work to their own advantage.[24] Eade notes that the local Labour leader liked to 'work through informal links with local activists', 'largely on a personal basis' (Eade 1989, 41) and politicians today all seek out ethnic minority 'community leaders'.[25]

Eade's 1989 study provides the main published account of the Bengalis' early relationship with the Labour Party. It is concerned with the mechanisms of local politics and their interaction with community groups and networks, and the picture he paints is complex, and at times Machiavellian. His account is not set against an analysis of political practice more generally, and so risks being hijacked by those who present a 'racialised political discourse' in which it is predominantly black, and especially South Asian, politics that is associated with corruption.[26]

White politicians are also objects of racialised discourse: notably in 'the perception that the left have used black representation in order to fulfil their own objectives'.[27] This discourse – demonstrated by the comments above on getting party membership – was common to many of my interviewees. It was encouraged by the separatist ideology of black organisation, and a generally pragmatic (as distinct from idealistic) political understanding.[28] When such discourses operate, every action can be ascribed an underlying racist or supremacist motive, whether or not it actually exists, with the risk both of obscuring actual racism and of perpetuating a division into 'them and us'.

Bengali Representation

Despite these feelings, and the legacy of black separatism, the national debate on Labour Party Black Sections does not seem to have played a very significant part in Tower Hamlets.[29] An Ethnic Minorities Group was formed in 1986 to push for more ethnic minority councillors,[30] but the Bengali activists were already operating as an ethnic lobby and did not feel a strong need for a broader 'black' organisational structure.

The selection of a Labour parliamentary candidate to replace Peter Shore on his retirement in 1997 generated inevitable expectations that there would be a Bengali MP. Competition was bitter,[31] and it has been suggested that the unseemly struggle and the lack of Bengali unity were instrumental in preventing a Bengali from being chosen. However, the selection was also seen by many as an example of party racism, and the anger and frustration over Labour's 'failure' to put forward a Bengali candidate was reflected in the 1997 election results. Oona King was safely elected for Labour, but the Conservative's Kabir Chowdhury bucked the trend to celebrate a 6 per cent swing in his favour.[32] A Bengali MP would be seen as a symbol of community achievement, and the lack of this type of ethnic representation is commonly interpreted as a denial of democratic rights.

As councillors, Bengalis have generally regarded themselves as fighting for the betterment of their community, as raising the community profile, and as role models.[33] However, councillors will also emphasise that they are there to represent all their constituents. Catherine Neveu, who researched Bengali electoral representation in 1989, commented on the contradictory attitudes (from Bengalis and others) that both 'expect Bangladeshi councillors to be representatives of the Bangladeshi population and ... accuse them of acting so' (Neveu 1989, 10). Abbas, as first Bengali leader of the council, attempted to satisfy both views:

> By having a Bengali leader now on the council, I think we are sending very clear messages about equality ... but also able to demonstrate that as a Bengali leader you can represent not only the Bangladeshi community ...[34]

This problem is as old as ethnic minorities in politics.[35]

Most Bengali councillors voted with left party members on measures aimed at lessening deprivation, in which, to quote Abbas, the Bengalis and the left shared a 'natural common agenda',[36] but there was no especial affinity with left issues more generally. Thus, there was a lot of Bengali support for the fight against the Poll Tax, which impacted heavily on the poor and especially on overcrowded households with several adult members, but no particular Bengali consensus on campaigning for the miners. Within the general give and take of political bargaining, Bengali party members could act as a group quite independently of more orthodox left/right divisions, leading to strange changing alliances.[37] Links to different Bengali political parties also play a crucial part in individual political careers and add another layer of bonds and influences.

Other Political Platforms

While there will always be accusations of ethnic voting to ensure Bengali councillors and of white voters not voting for Bengalis, the real electoral battles have been between the politics of the different parties, and the real fight for Bengali representation has been *within* the parties. At a local level that has long been won. The 1997 General Election saw Bengalis standing across the political spectrum from the Referendum Party to Socialist Labour, and Bengalis currently occupy a disproportionate thirty-one out of fifty-one council seats.[38]

Before the eruption of Respect (which will be discussed below), those who had had enough of Labour could choose between the Conservatives, who until recently had not had a single councillor elected, and the Liberals, who came to power locally for eight years from 1986, after a vote-splitting breakaway by the former Labour leadership. The racism of Tower Hamlets's Liberal administration became national news on more than one occasion,[39] but this did not prevent some Bengalis choosing to pursue their political path through the Liberal Party. As one of those who became a councillor made clear to me, this was quite consistent with a pragmatic and community-centred approach focussed on taking part in mainstream decision making; and as the Labour Party was already crowded with Bengalis he found less competition for positions of influence.[40]

The Conservative Party has been less effective as a route to local power, but does attempt to appeal to conservative Muslim family values.[41] However, changes in socio-economic circumstances have been reflected in an increase in Conservative support, as occurred with the Jews in previous generations.

The Image becomes Tarnished

A generation on, as the young Bengali activists of the late 1970s and 1980s settled in to become the new political establishment, many Bengali councillors were being criticised in much the same way as they themselves used to criticise the older council regime and their more cautious seniors in the Bangladesh Welfare Association. One young Bengali woman described them to me as a powerful network of middle-aged Bengali men whose petty politics disgusted many of her generation.[42] And local councillors (like national Bangladeshi politicians) had become figures of fun for the politics class at Tower Hamlets College, where some students argued that they were only interested in a good name and staying in power.[43] For older Bengalis, or recent immigrants, who may have little English, Bengali councillors still provide a vital link to local power structures, but for many others, the presence of Bengalis on the council has made no difference, and some expressly told me they would choose to see a white councillor if they had a problem.[44] As Abbas points out, distrust of politicians is a general phenomenon,[45] but when, in 2005, a young Bengali Labour councillor was caught sequestering regeneration money from a youth group,[46] the case was perceived as a symptom of a wider malaise.

Diversity through Gender

Although the British-born generation are getting more actively involved, there have been relatively few Bengali women in politics. A combination of language problems, fear of racial attack, and a 'modesty' inspired and sometimes enforced by both religion and tradition, still prevents many women from merely going out of their homes; and even when they are backed by a supportive family, women with political ambitions may have problems. When I began interviewing in 2000, the one female Bengali councillor was a young barrister. She told me that although she received tremendous cross-cultural support for her political activities, there were always those Bengalis, especially men, who disapproved of such behaviour in a young unmarried Bengali woman; and they were not above trying to sabotage her chances. On the morning of her first selection meeting an anonymous letter went out accusing her of having had a white boyfriend at law school: an (unfounded) charge that was enough to damage her reputation with the older generation and temporarily prevent her selection.[47]

For Pola Uddin, the main barrier appears to have been the assumption by institutions, starting with her grammar school, that Asian women have no ambitions beyond raising a family. Her political career, which has taken her to the House of Lords as a New Labour Peer, provides a striking example of the interaction of community based pragmatism with the wider political climate. She lived through the Bangladesh war as a child and told me 'I don't understand any Bengalis … in my generation who wouldn't be interested in politics'. When she saw the restrictions on what women and Bengalis could do in Britain, joining a movement 'was just as natural as drinking water or breathing air', and later Jill Cove pushed her into joining the Labour Party. As an outspoken and thick-skinned Bengali woman, in a political environment obsessed with identity and with demonstrating diversity, she found herself much in demand:

> … it was nice to have a [woman] who could stand up in a meeting and say exactly what someone else wanted to say and they couldn't; so … I was often being pushed to represent something. And that was a very useful tool in the Labour Party in the early years, and … within the community sector.[48]

While political women provide important role models to Bengali girls, it would be naïve to think that their involvement is either symptom or cause of more fundamental political change. Pola Uddin herself fitted comfortably into the market-based pragmatic politics of pre-9/11 Blairism; and the superficially radical choice, in 2007, of a young inexperienced Bengali woman as Labour candidate for the next parliamentary election is seen as a strategic victory for the established Labour leadership.[49]

The Era of New Labour

The participation of Bengalis – men and women – was welcomed by Labour as evidence of the party's anti-racist credentials, but multiculturalism can also be used to give a progressive veneer to an administration that has abandoned class-based policies. In addition, the prioritising of cultural identity cuts across class divisions, making this a useful tool for those wanting to turn their back on old labour values.

Centralising New Labour now controls council policy; and new partnership forms of local governance are allowing old forms of representative democracy to be bypassed through the direct involvement of unelected interest groups. Interest group politics – developed in numerous regeneration schemes, and now impacting mainstream services – encourages the further development of ethnic or religious-based organisation, with different groups pitted against each other in a divisive competition for limited resources. Both Conservative and Labour governments actively promoted the Muslim Council of Britain as representatives of British Muslims,[50] and the East London Mosque is used as a channel for the provision of local services.[51]

The Political Lure of Islam, and a New Popular Front

By the turn of the millennium, the only radical force for political change was coming from the Islamists (Glynn 2002). Islamist groups, especially those centred on the East London Mosque, were moving into the vacuum created by the decline of the left, using a combination of ideological argument and grass-roots organisationto present themselves as an active alternative to the alienation of neoliberal capitalism. When, in the aftermath of 9/11, more Bengalis, like Muslims everywhere, began to rediscover their religious identity, they found a thriving network of religious organisations already in place.

British Muslims were encouraged to express their opposition to the wars in Afghanistan and Iraq under the banner of the Muslim Association of Britain.[52] The MAB played a distinct role in the Socialist Workers' Party (SWP) dominated Stop the War Coalition, which was set up to coordinate anti-war protest. In January 2004 this anti-war coalition was formally inaugurated into a new political organisation under the banner of Respect, with ex-Labour MP George Galloway as its most prominent leading figure. From the beginning, MAB support was conditional – they supported individual election candidates according to their assessment of Muslim interests, regardless of party – and they made no compromises in their philosophy,[53] but even this circumscribed MAB involvement in the Respect project soon petered out. Respect, however, continued to be regarded as 'the party for Muslims', and they were welcomed by a leadership increasingly branded as opportunist.[54] Unlike the Islamists, the SWP and its allies were all too ready to compromise. 'Because we're a coalition', George Galloway explained, 'we don't bind a Muslim candidate ... to the explicitly socialist parts of our programme'.[55] The prospect of, even limited, electoral success soon attracted a similar layer of community activists and small businessmen as once saw the Labour Party as the route to community (or personal) uplift. Abjol Miah, leader of the Respect councillors, told me

> ... if Labour didn't take us to a war, then, I think, my point of view could have been a bit different. I still have friends in the Labour Party, and they argue that you can bring change from within the party.[56]

And one of the new Bengali Labour councillors commented that he and his Respect colleagues were largely aiming for the same thing.[57] Traditional community networks have been activated to campaign for the new organisation, and there have even been mass-membership applications,[58] and defections to and from Labour.[59]

Respect's first electoral success was in the council by-election caused by the corruption scandal. This was followed by the well-documented victory of George Galloway over the pro-war Oona King to become MP for Bethnal Green and Bow in the 2005 General Election. This was a sensational battle, but the key issue was always acknowledged to be the Iraq War, which Galloway did not hesitate to describe as a war against Muslims.[60] With an electorate that, according to Respect estimates, was 55 per cent Bengali, both main contenders attempted to appeal to Bengali voters through their Muslim identity (Glynn 2006b), and Galloway even claimed that voting for him to oust Oona King gave the best chance of a Bengali MP next time around.

The Bengali focus of the new politics was confirmed in the 2006 council elections. While Respect councillors follow a more Old Labour – and more popular – agenda in areas such as housing, and do pick up anti-New Labour votes, a large part of the electorate did not seem to recognise the new party as representing more than a community-based interest group. The Respect councillors elected were all Bengali and there was a strong correlation between the percentage of votes cast for Respect in each ward, and the percentage of Bengalis.[61]

Meanwhile, a wounded Labour Party limps on, struggling, as elsewhere, to reconcile social democracy and neoliberalism, but also attempting to salvage its credentials as a Bengali and Muslim friendly party, and the natural place for those who want to get something done.

Old Traditions Adapting in New Political Climates

The SWP and Galloway have tried to portray Respect as part of the great East End socialist tradition – and left wing organisations have frequently attempted to generate among the Bengalis an echo of the pre-war left movements in which Jewish immigrants played a key part – but class-based politics have never really taken root in the Bengali East End. Even in the early years of the immigration, when most of the political leadership professed allegiance to various Marxist groups, there was no developed sense of class-consciousness. Many, such as Tasadduk, were themselves in business; popular-front politics incorporated restaurant and workshop owners into left-led agitations; and leftist memories of that time can slip into using the term 'working class' to include small businessmen.[62] Trade unionism was never a priority.

The great majority of first generation Bengali immigrants were proletarianised in Britain, but most still held plots of land back in Sylhet, where status is endangered by working on land belonging to a different lineage (Gardner 1995, 139), and they did not easily identify themselves as workers. Over the years, money sent back to Sylhet has been built up into sometimes substantial landholdings and other investments, so that families living poor working-class lives in London, may now also be significant property owners (Gardner 1995). For a Bengali to call someone working-class is a big insult,[63] and though the term is used by those who wish to imply some sort of link to wider socialist traditions, this is more as a form of rhetoric than to suggest actual class identification. Non-Bengali leftists who tried to identify the Bengalis as part of a wider working class were often regarded with suspicion and accused of hijacking Bengali issues to pursue their own agenda.[64]

Patriarchal links have always played an important role in Tower Hamlets politics – even transmogrifying into very personal support for Peter Shore when he faced a reselection battle in the mid 1980s.[65] One leading Labour figure told me, 'I always ask my people to vote for the Labour Party', and another explained, 'I create councillor, I don't want to be a councillor'.[66] Non-Bengalis have also not been afraid to make use of these links when it suited them – such as in getting a vote out. The activism of the late 1970s and early 1980s did allow a new generation to come through, but today's partnership politics, with its emphasis on 'community leaders', encourages the strengthening of all kinds of informal network.

Bengali involvement in British politics developed in parallel with the growth of identity politics and the waning of the politics of class, and those who became politically active in Britain were not expected to weaken older community ties. Instead, the very real challenges of establishing an immigrant community in one of the most deprived areas of London were met through pragmatic community-based organisation. This continues to be the basis of British Bengali politics today – even if the community is often referred to as Muslim rather than Bengali, and the chosen vehicle may be Respect rather than the Labour Party. Within its own terms, this strategy may be said to have largely succeeded, in that Bengalis have taken their place at all levels of Tower Hamlets society and Bengali children top the borough's school league tables; but this is still an area of multiple deprivation. Unemployment among the Bengalis stood at 24 per cent at the time of the 2001 census,[67] and large portions of the white working class are facing persistent social exclusion. Increasing competition for resources, brought on by the crisis of the welfare state, new immigration from Eastern Europe, and new pressures on the area's resources from middle-class gentrifiers is increasing tensions between ethnic groups (Dench et al. 2006; Glynn 2006b). Fears of growing xenophobia, and especially Islamophobia, are boosting community-based politics, but what is needed instead is a politics that can cut across community divisions and attack structural inequalities.

Acknowledgement

With grateful thanks to all my interviewees.

Notes

1   This picture of automatic support was made clear in interviews.
2   Caroline Adams's interview with Abdul Mannan and Tasadduk Ahmed 1980s.
3   For a fuller discussion see Glynn 2006.
4   See Peter Shore in the *Silver Jubilee Commemorative Volume of Bangladesh Independence*, 1997, 37. One of those most involved was John Stonehouse. Despite his later notoriety and conviction for fraud, the Bengali consensus is that his role in 1971 was genuinely disinterested and honest.
5   Shore, interviewed 15 February 2001.
6   For an account of their campaigns see Glynn 2005.
7   Through their activism and journal the Race Today Collective was at the forefront of Black Radicalism in Britain. Black Radicalism emerged out of the interaction between Communist popular frontism and anti-colonial and Black rights movements. It was a formative strand of the New Left that developed from criticism of excessive structuralism *within* Marxism, to criticism of Marxism itself (Glynn 2005).
8   Interviewed 10 October 2001.
9   There had also been attempts at black organisation among the Bengalis at the time of the racist murder of Tosir Ali in 1970.
10   Interviewed 10 October 2001. A similar point was made in interviews with Jalal 16 August 2000, Shukur 12 September 2000 and Sunahwar 23 January 2001.
11   Interviewed 23 January 2001.
12   Interviewed September 2000.
13   Sheikh Mannan (interviewed 30 March 2002) joined on his arrival in England in the mid 1960s, though that was not in Tower Hamlets; Abdul Malik joined in 1968 and was active on his EC (Adams 1994, 125); and Ashik Ali was welcomed into St Katherine's Ward in 1976 (interview with Ashik Ali, then the only Bengali Labour councillor, in *Labour Herald* 26 November 1982).
14   Interviewed 23 January 2001. Abdus Shukur (interviewed 12 September 2000) recalls a similar reaction.
15   Interviewed 23 January 2001.
16   Jill Cove, interviewed 16 October 2001.
17   Ashik Ali, who described himself as hard-working and on the left (*Labour Herald* 26 November 1982), topped the poll for the ward.
18   Interviewed 10 October 2001.
19   See comment by Robbie McDuff in Eade 1989, 76–7, and *New Life* 26 July 1985
20   Jill Cove, interviewed 16 October 2001. 1982 was the first time it was necessary to canvas, and Whites and Bengalis canvassed together. For the white members this removed language problems, and it also dispelled the natural fears encountered by immigrants who opened their door to find a stranger with a clipboard.
21   Interviewed 16 August 2000. Jalal no longer lives in the East End and defected to the Liberal Democrats in 2007.
22   Cove also rejects Eade's suggestion that the left used membership disputes to delay entry of Bengalis not sympathetic to their views, claiming that this would anyway

have required a greater knowledge of Bengali community politics than they possessed. For her, it was a question not of whether people were left or right, but whether they actually existed.

23 Interviewed April 2001. He decided to do community work outside the political arena.

24 Kalbir Shukra in Saggar 1998, 128–9; Solomos and Back 1995, 72–4, 80, 99, 103–4 and 106–7.

25 Paul Beasley led the Majority Group in Tower Hamlets Labour Party from the early 1970s until 1984.

26 This term is used and explored by Solomos and Back 1995, 95–113

27 Solomos and Back 1995, 157. Although they do not actually use the term this way round, this perception, whether or not it has some basis in fact, can be described as a racialised discourse too.

28 Eade's reference to 'the tenuous nature of the alliance between white radicals within the Labour party and Bangladeshi activists' (Eade 1989, 168) suggests that he has also taken on this discourse, as the term 'alliance' implies something more deliberate than a coalescence of interests.

29 Peter Shore, interviewed 15 February 2001; Helal Abbas, interviewed 10 October 2001.

30 *Hackney Gazette* 21 March 1986: Eade 1989, 78–9. The group was coordinated by Jalal.

31 Jalal, one of the main contenders, was suspended from the party after being accused of sending a fax to the press indicting the local Labour group leader with racism, forged to make it look as though it came from his rival Pola Uddin. He denied the charge. (*Evening Standard* 20 February 1995; *East London Advertiser* 23 February 1995 and 2 March 1995; *Independent* 27 May 1995)

32 <http://www.bbc.co.uk/election97/constituencies/47.htm>.

33 The first points were spelt out clearly by Shukur, interviewed 12 September 2000. The importance of role models is confirmed by a Bengali careers advisor (interviewed April 2001), who told me that the young people he discusses with do not want to do jobs that they have not heard of other Bengalis doing.

34 Interviewed 10 October 2001.

35 See Dadabhai Naoroji's speech when he was prospective parliamentary candidate for the Liberals in Holborn in 1886, quoted in Visram 1986, 248 note 45.

36 Interviewed 10 October 2001.

37 Compare, for example, *East London Advertiser* 27 April 1995 and 2 May 1996.

38 Seventeen out of twenty-seven Labour councillors, three out of six Lib Dem and all eleven Respect (Autumn 2007).

39 Glynn 2006b. Most Bengalis were guaranteed to vote Labour, so appealing to white xenophobia made electoral sense. The Liberals had a large Bengali membership on paper, but much of this was in thrall to the personal ambitions of their community leader (Sabine Drewes 1994).

40 Interviewed January 2001.

41 Shahagir Faruk, Conservative parliamentary candidate in 2001 and 2005, described himself as born Conservative with conservative family values (interviewed 1 August 2001).

42 Interviewed October 2000.

43 Separate interviews with two students, July 2000.

44 This view was put by three separate interviewees.

45 Interviewed 10 October 2001.

46 *East London Advertiser* 4 March 2005 and 15 March 2006, bbc.co.uk 3 April and 31 July 2006.

47 Jusna Begum, interviewed 19 September 2000.

48 Pola Uddin, interviewed 17 January 2005.

49 *East London Advertiser* 9 May 2007.

50 Hansard 19 July 2007 Column 169WH. The government is now trying to promote an alternative organisation.

51 The mosque works with the health authorities, job centre and local schools.

52 Although the MAB had relatively little presence on the ground, their involvement with the anti-war movement at organisational level made it easy for Muslims to protest against the war as Muslims.

53 The MAB was happy to work alongside secular socialists because they saw Islam as in the ascendant and regarded socialism as no threat – interview with Azzam Tamimi, MAB official spokesperson, 17 January 2005.

54 This phrase was used on Respect's 2004 European election leaflets distributed outside the East London Mosque. Some of the harshest criticisms have come from the left in the *Weekly Worker*.

55 Interview with *Pink News* 21 February 2006.

56 Interviewed 25 July 2007.

57 Conversation at Muslim Community Fair, 28 July 2007.

58 *Weekly Worker* 6 July 2006; SWP Appeal to Respect Members 24 October 2007. <http://www.swp.org.uk/respect_appeal.php>.

59 Long-serving Labour councillor, Ghulam Mortuza defected to Respect in June 2005 and back to Labour that November; Respect councillor Waiseul Islam defected to Labour in May 2007, days before former Labour councillor Kumar Murshid (who had just been cleared of corruption charges) defected to Respect; and in July 2007 Shamim Chowdhury resigned, criticising his Respect Party and causing a by-election.

60 He told the TELCO hustings on 20 April 2005 'If you make war against Muslims abroad, you are going to end up making war against Muslims at home'.

61 $R^2 = 63\%$ and $P<0.001$ – i.e. 63 per cent of the variation is explained by the percentage of Bengalis.

62 For example. by Syed Abdul Majid Qureshi, interviewed by Caroline Adams in the 1980s

63 Sunahwar Ali, interviewed 23 January 2001.

64 Interviews with Anamul Hoque 14 September 2000, Ansar Ahmed Ullah 14 November 2000 and Sunahwar Ali 23 January 2001.

65 As Shore put it, 'There was no question of the Bengali[s] as a group, as it were, being turned against myself or in favour of anyone else' (interviewed 15 February 2001). There was also a Bengali candidate, but Jill Cove was seen as the main rival contender.

66 Interviewed 31 January and 7 February 2001.

67 Tower Hamlets Council *Employment and Income bulletin v 5.2*, section 3.1(iii). Although the figure for white unemployment is only 7 per cent, this will be affected by the large increase in (predominantly white) middle-class residents.

References

All interviews were conducted by the author unless otherwise stated. Caroline Adams' oral history tapes are in Tower Hamlets Local History Library.

Adams, C. (1994 [1987]), *Across Seven Seas and Thirteen Rivers* (London: THAP Books).
Dench, G., Gavron, K. and Young, M. (2006), *The New East End: Kinship, Race and Conflict* (London: Profile Books).
Drewes, S. (1994), 'Ethnic Representation and Racist Resentment in Local Politics: The Bangladeshi community and Tower Hamlets Liberal Council 1986–93' (manuscript copy in Centre for Bangladeshi Studies, Roehampton).
Eade, J. (1989), *The Politics of Community: The Bangladeshi Community in East London* (Aldershot: Avebury).
Gardner, K. (1995), *Global Migrants, Local Lives: Travel and Transformation in Rural Bangladesh* (Oxford: Clarendon Press).
Glynn, S. (2002), 'Bengali Muslims: the new East End radicals?', *Ethnic and Racial Studies* 25:6, 969–88.
Glynn, S. (2005), 'East End Immigrants and the Battle for Housing: A comparative study of political mobilization in the Jewish and Bengali communities', *Journal of Historical Geography* 31, 528–45.
Glynn, S. (2006), 'The Spirit of '71 – how the Bangladeshi War of Independence has haunted Tower Hamlets', *Socialist History* 29, 56–75.
Glynn, S. (2006b), 'Playing the Ethnic Card – Politics and Ghettoisation in London's East End'. Paper given to conference on 'Ghettoised Perceptions versus Mainstream Constructions of English Muslims: The future of the Multicultural Built Environment', University of Central England, Birmingham, <http://www.geos.ed.ac.uk/homes/sglynn/Ghettoisation.pdf>.
Neveu, C. (1989), 'The Waves of Surma have Created Storms in the Depths of the Thames – Electoral Representation of an Ethnic Minority: A case study of Bangladeshis in the East End of London'. Paper given to APSA 85th Annual meeting, Atlanta, GA (manuscript copy in Centre for Bangladeshi Studies, Roehampton).
Saggar, S. (1998), *Race and British Electoral Politics* (London: UCL Press).
Shukra, K. (1998), *The Changing Pattern of Black Politics in Britain* (London: Pluto Press).
Silver Jubilee Commemorative Volume of Bangladesh Independence (1997), (London: The Committee to Celebrate the Silver Jubilee of Bangladesh Independence).
Solomos, J. and Back, L. (1995), *Race, Politics and Social Change* (London: Routledge).
Visram, R. (1986), *Ayahs, Lascars and Princes: The Story of Indians in Britain 1700–1947* (London: Pluto Press).

Chapter 7

Integration and the Politics of Visibility and Invisibility in Britain: The Case of British Arab Activists

Caroline Nagel and Lynn A. Staeheli

Introduction

Britain's multicultural system has been built on the notion that there are clear, visible racial and cultural differences within British society – differences that are categorised, monitored, and represented both through the workings of the state and the efforts of community leaders. In recognising and accommodating differences in the public sphere, British multiculturalism has been viewed as an alternative to 'assimilation', a term that denotes the enforced conformity to dominant norms. But since its inception in the 1960s and 1970s, it has been clear that multiculturalism has its limits and that dominant groups are willing to accept only certain types or degrees of difference as compatible with the wider societal aim of 'integration'.

In the 1980s the limits of multiculturalism were tested by young black men, whose particular visibility was often read as a threat to public order (Gilroy 1987). In more recent years, anxiety about the supposed failures of multiculturalism have focused on Muslim communities, whose tendency to cluster in inner city neighbourhoods and, in some instances, to cover their faces, are seen as indicating unwillingness to embrace British identity and values (Phillips 2006). The language of multiculturalism has not been entirely abandoned in Britain, but in the context of growing concerns about Muslim 'self segregation' and extremism, it seems that multiculturalism is being superseded by an emphasis on integration. Integration, like multiculturalism, can be seen as a more palatable alternative to assimilation – one that suggests 'interaction' rather than 'conformity'. But as employed in current political discourse, it clearly stresses the responsibility of minority groups to embrace dominant norms, rather than the need for the majority of society to accommodate and respect minority cultures and identities (Yuval-Davis et al. 2005). As with multiculturalism, the notion of integration utilises a 'visual lexicon' of cultural difference. Integration, in other words, can be read off the way minority groups look (dress, hair styles, clothing, and so on) and the physical landscapes in which they live. If certain visible differences are regarded as innocuous, acceptable, and even worthy of protection, others are seen to signify an unwillingness to become part of British society.

This chapter looks at integration as a politics of visibility – and invisibility – in which notions of belonging and social membership revolve around conflicting interpretations of a group's physical presence in the public sphere. We contrast this conceptualisation with that offered by mainstream academic discourse, which tends to treat integration as a 'neutral' process that can be gleaned through the objective observation of minority group characteristics. We argue that notions of integration, far from neutral, reflect preconceived, and usually unexamined, ideas of what makes groups the same or different from the 'mainstream'. We then examine the politics of visibility from the perspective of minority groups – in this case, British Arabs, who are often classed among Britain's 'invisible minorities'. We demonstrate that for British Arab activists, integration involves efforts to become more visible – to enhance their stature in the public eye and to make themselves seen and heard – and to submerge particular types of difference that have been stigmatised in public discourse. We illustrate these points by drawing on interviews conducted with forty British Arab activists between 2003 and 2006 in London, Liverpool, Birmingham, and Sheffield.

Integration in Question

Questions relating to minority integration have been central to British geography for many years. One important tradition in the discipline has drawn on concepts and techniques, many devised in the United States context, to measure and to assess degrees of minority integration and, conversely, segregation. Within this tradition, 'integration' signifies both a *process* of becoming similar to the population-at-large and the *endpoint* of this process – that is, the status of being the same as the population-at-large according to a particular set of measures. For population geographers, of course, the measure of interest is place of residence, and attention focuses on the degree to which minority groups are spatially similar (integrated) or dissimilar (segregated or clustered in enclaves or ghettoes) (see, for instance, Peach 1996). Spatial distance within this literature is usually read as social distance, and a high degree of spatial segregation is associated with social isolation and a lack of social mobility among minority groups.

The use of census data to reveal patterns of residential clustering and a variety of other social characteristics conveys an understanding of integration as a process that can be objectively observed, measured, verified. But, as Wright and Ellis (2000) note, the criteria geographers use to assess and measure integration reflect particular assumptions about what indicates sameness. In other words, in choosing particular criteria (e.g. residence, language usage, intermarriage, and so on) and particular benchmarks, mostly based on the experiences of the white middle class, scholars make decisions about what constitutes belonging and integration. In this sense, scholars implicate themselves in wider political debates about the role of minorities in society and the ability of minorities to become part of 'us'.

Population geographers, to be sure, are cognizant of the political nature of their work, and recent studies of minority integration in British cities have recognised the importance of their findings (most of which point definitively to the lack of ghettoisation and self-segregation among minority groups) to current debates on 'social cohesion' (see, for instance, Johnston et al. 2002). Nonetheless, as Graham (1999) has argued, studies of integration are data-driven exercises that utilise census material not out of any explicit theoretical aims, but because these data are available. The result is that the meaning of integration tends to remain unquestioned and unexplored, as do the social categories that underlie the data collection process (see Simpson 2004).

We do not wish to suggest that empirical studies of integration have no purpose or value. Patterns of similarity and dissimilarity can tell us a great deal about socio-economic privilege and disadvantage and about relative access to employment, education, and leisure opportunities. These patterns, in turn, can tell us about interactions between social groups and the ability for particular groups to participate fully within particular spheres in society (see, for instance, Massey and Denton,1998). But there is also a need to understand 'integration' not only as an objectively determinable process or end state, but also as a set of ideas about how certain groups are to be included in the public. Assessments of minority integration, whether made by academics or politicians, always involve claims about which behaviours 'matter'. This understanding of integration requires that we look beyond 'objective' indicators of difference and interrogate how certain ways of *seeing* difference emerge within particular historical and geographical contexts.

The Politics of Visibility and Invisibility

The importance of visual cues in marking and assigning particular meanings to social difference has been an important theme of critical scholarship across a range of disciplines and subfields. Scholars like bell hooks (1989), for instance, have spoken of the importance of the visibility of dark skin in the oppression of people of colour and have noted the simultaneous invisibility of individuals marked as different and their hypervisibility as stereotypes (also, Hogeland 1996). The issue of visibility has been central, as well, to the burgeoning literature on disabled people, which has explored the social anxiety and discomfort created by the presence of visible bodily differences in dominant, able-bodied spaces (e.g. Kitchin 1998). All of these literatures point to the notion that particular bodies, marked by skin colour, disability, cultural affiliation, sexuality, age, or gender, are deemed to be 'out of place' and disruptive of the 'natural' or 'normal' order of things (see Cresswell 1996; McDowell 1997; Sibley 1995).

Such insights into visuality have informed recent studies of immigrant integration and assimilation, a field of study otherwise dominated by census-based and quantitative studies. In his critical account of the assimilation of 'white ethnics' into American society, for instance, Jacobson (1998) speaks of the creation of a 'visual lexicon' that dominant groups use to describe, to translate, and to make sense of difference in their midst. In America in the nineteenth and early twentieth centuries, Jacobson shows, ideas about difference revolved primarily around notions of innate physiological distinctions between 'racial' groups manifested in hair colour and texture, nose size, eye shape, and so on. By the mid-twentieth century, these elaborate racial typologies were largely discredited, and Jews, Italians, Irish, Syrians and others, who had once been considered different racial groups with distinctive, visible characteristics, came to be seen and to see themselves as simply 'white'. This ideological, discursive and visual process of whitening has been deeply implicated in changing patterns of residential location, intermarriage and social interaction.

In Britain, as in the United States, visual cues of difference have been central to discussions about assimilation and integration. For instance, Enoch Powell's infamous 1968 'Rivers of Blood' speech, which led to his removal as a shadow cabinet minister but which resonated with large swathes of Britain's middle class, was filled with references to physical differences and the unwillingness of 'negroes' and other New Commonwealth immigrants to integrate.[1] In today's 'racially sensitive' political context, problematic difference has tended to be read from 'cultural' markers, rather than 'racial' difference. Clothing styles adopted by certain groups, in particular, are often seen to reflect particular sets of values and attitudes in conflict with those of the 'core society' (Dwyer 1999). This is true of several groups, including young working class men in 'hoodies', who are typically seen as anti-social troublemakers; it is also true of Muslim women in veils, who are often perceived in Britain and elsewhere as rejecting modern, liberated lifestyles and as disrupting the rational, public sphere with a stridently religious identity (Dwyer 1999; see also Bowen 2007).[2] Still, physiognomic differences remain important in Britain's lexicon of difference. This is certainly true for Black British people, and also for British Muslims, many who trace their origins to Pakistan, Bangladesh, and increasingly North African and the Middle East. Surveillance practices, which have been greatly expanded by the government and the private sector in recent years, require particular understandings of what potential 'troublemakers' or 'terrorists' look like (see Koskela 2000), and certain visual cues – hair colour, skin colour, facial hair and so on – indicate who is suspicious and who is not.

Bodies, to summarise, are imbued with ideas about difference and sameness. While many visible differences in society receive little attention, others are at the centre of debate about national identity, social cohesion, and integration. Integration, in this sense, needs to be understood as a visual practice and politics that involves identifying particular visible differences as meaningful and placing these differences in wider narratives of belonging and social membership.

It would be a mistake, however, to see the visual lexicon of difference as something written and read exclusively by the dominant society. Contemporary scholarship, especially in Queer Studies and Disabilities Studies, offer insights into the quotidian politics of visibility and invisibility as practiced by subordinate groups. Samuels (2003), for instance, speaks of the difficult choices faced by some disabled people – especially those with non-visible disabilities – about whether to 'pass' as able-bodied or to perform 'the dominant culture's stereotypes of disability' (Samuels 2003, 246). In either case, individuals deal with multiple quandaries: in choosing to make visible a disability, one opens oneself up to scrutiny, pity, and claims of fraud; in 'passing' one may face 'a profound sense of misrecognition and dissonance' (Samuels 2003, 239). Walker's (1993) account of lesbian sexual style further highlights the ambivalence and complexity that surrounds visibility and invisibility. There is, Walker argues, an impulse to reclaim signifiers of difference that have been devalued by dominant groups. But while reclaiming identity is affirming, she argues, it can also replicate practices that use visibility to exclude.

Our aim here is not to draw analogies between the subordinate groups described by Samuels and Walker and our own study participants but simply to indicate the agency (however limited) individuals may exercise in being visible or invisible, and the ambivalence and ambiguity that accompanies this agency. The rest of this chapter examines the diverse efforts of British Arab activists to negotiate membership in British society, focusing on the tensions between visibility and invisibility that inhere in these efforts. To begin this discussion, we give a brief overview of the research on which this analysis is based and a description of Arab-origin communities in Britain.

About the Study

The following analysis is based on intensive, semi-structured interviews conducted with 41 British Arab activists between 2003 and 2006 in London, Birmingham, Sheffield, and Liverpool as part of a study on citizenship, identity, and belonging among Arab-origin groups.[3] Our interviewees were mixed in terms of generation, with about a third born in Britain or raised in Britain from a very young age and the rest emigrating from the Arab world as students or adults. There were roughly equal numbers of male and female respondents. As with most migrants from the Arab world, our interviewees had come to Britain for a combination of reasons: to gain graduate or postgraduate qualifications; to take advantage of economic opportunity mainly in skilled sectors, and/or to flee political instability or statelessness. Almost all of our interviewees were at least nominally Muslim, and the majority traced their origins to Palestine, Lebanon, Iraq, Egypt and/or Yemen.

We identified our respondents through their involvement in Arab-oriented activities – that is, social and cultural clubs, political organisations, and philanthropies dedicated to Arab communities in Britain and/or abroad. In this sense, all of our interviewees could be seen, at least in part, as embracing a public, visible Arab identity, often in conjunction with more specific national (e.g. Palestinian, Yemeni). A few embraced a Muslim identity, as well – a point elaborated in greater detail below. As organisational leaders and activists, these individuals were well educated and immersed in community and organisational life. They were more likely than 'ordinary' people to be cognizant of debates about integration and to be engaged with the local and national state, as well as with multi-ethnic forums. Their views, therefore, should not be taken as representative of British Arabs as a whole. But while perhaps not representative, our interviewees' views shed light on the ways in which individuals interpret and weigh the benefits and predicaments posed by visibility and invisibility. These complex balancing act between difference and sameness, visibility and invisibility, is central to the organisational efforts not only of British Arabs but of many minority groups.

Being Visible and Invisible

Hypervisibility, Invisibility and Non-recognition

We began each of the interviews by asking our study respondents to comment on the biggest challenge or problem facing Arab-origin communities in Britain. Many of our respondents answered this question by expressing their dismay at the constant barrage of negative images from the Arab world in the Western media. They argued that people in the West, having no understanding of Arabs other than that offered by the media, have come to associate Arabs with terrorism, religious fanaticism, violence, and the oppression of women. Interviewees regarded this visibility as compromising the ability of Arab-origin communities to participate fully in British society. Nada,[4] for instance, a British-born activist of Yemeni heritage explained,

> Obviously, I think, since September 11, the issues have become – how do I put it? – [British Arabs] are more noticeable now. ... Just being judged as what the British people see in the media .... They're looking at the Arabs in one direction, which would be the stereotype of a terrorist, the people that blew up the Twin Towers, or suicide bombers.

Yet, at the same time that they lamented the ubiquity of negative media images, interviewees also described the *invisibility* of British Arabs as a major problem faced by the community. Some interviewees argued that Arabs had chosen to make themselves invisible out of a sense of fear or lack of self-confidence, or simply out of apathy. Hassan, a young, British-born man, contended that British Arabs desire to connect with the cultural aspects of Arabness, but not with issues of political injustice affecting Arabs both in the West and in the Arab world. As a result, he argued, British Arabs – unlike Black Britons – had not formed a 'real' community that could represent itself to the wider public:

> I'm not convinced when I say 'the Arab community'. It's a ridiculous statement to make, but I don't feel so ridiculous saying 'the Black community' because I feel that there is somebody to represent them and to speak on their behalf, and there is some kind of lifestyle that is very 'Black Briton'.

Mansur, a young man of Palestinian-Egyptian origin who came to Britain as a stateless refugee, made a similar comparison between British Arabs and British Asians, speaking ruefully, like Hassan, about the lack of clear, visible community identity among British Arabs. He stated: 'We haven't got the numbers of, say, South Asians, so we weren't able to create ghettoes the way South Asians did. We look more like Europeans than South Asians do. We normally know English better, especially if you are middle class, so you can integrate easily.' Mansur, however, did offer a somewhat more sympathetic reading of British Arab invisibility than Hassan, attributing it to 'our collective inferiority complex'.

Several other interviewees linked Arab invisibility to the British state's refusal to recognise Arabs as a cohesive, cultural group, and argued that this refusal had contributed to the political and cultural marginalisation of Arab-origin communities in Britain. For instance, Wajih, a first-generation immigrant in his fifties who is active in a variety of Yemeni, Arab, and Muslim organisations, criticised the British government for 'treating Arabs either as part of the Asian community or of the Muslim community in general'. This attitude among British officials, he contended, was indicative of a wider refusal to acknowledge Arab cultural difference and the particular needs of Arab-origin minorities in Britain – including the need to have local services provided in Arabic and to have state-funded Arabic language instruction for young British Arabs. Many study participants shared Wajih's desire for cultural validation and preservation, and they spoke of cultural recognition as a prerequisite for full membership and participation in society (cf Fraser 1994; Modood 2007). Without this, they reasoned, Arabs and other minorities would always feel alienated and would lack adequate representation and access to political resources. To reiterate an earlier point, many study respondents were critical of British Arabs themselves either for living in isolation as foreigners in Britain or for over-assimilating into British society and forgetting their Arabness; but there was, simultaneously, a sense that validation and recognition must come, in the first instance, from dominant groups, particularly in a context of generally negative media coverage of immigrants and the Arab world.

Recognition and Visibility

Many of our study participants, therefore, described their activism at least partly in terms of changing existing perceptions of Arabness and enhancing the visibility of Arabs in the public sphere. In other words, their response to both the hypervisibility of negative stereotypes of Arabs and the invisibility of Arabs as legitimate cultural and political actors in British society was to foster a more positive kind of visibility in dominant spaces and spheres – one predicated on cultural difference but also cultural equality.

In distinguishing between negative and positive kinds of visibility, our interviewees spoke of the need to prove or to convince the British public of the cultural value and legitimacy of Arabs and the validity of Arab political struggles. Mansur, described above, described this in terms of coming out of hiding and confronting stereotypes – something he felt he could do only after he became a legal citizen in Britain:

> After I got the [British] passport, I came to the realisation that hiding wasn't helping the cause or helping me personally or other people. ... I realise that to help our cause, everyone has got to do their bit no matter how small, even if it is just telling people about the Palestinians. ... Unless you confront the stereotype, it is always going to be there.

Similarly, Nada, a young British-born woman, stated, 'Unlike, say, normal English people or white people or however you want to define them, because we are Arab ... we have to give an image that we're not all terrorists. We have to work that little bit harder to prove to people that we're not what the media portray us as'. For Nada, the quest to transform media images has involved participation in an organisation that sponsors lectures and public events that provide an alternative view of events in Palestine and elsewhere in the Arab world. Hala, a woman raised mostly in Britain in a Palestinian-Iraqi family, mentioned her desire to promote Arab films in Britain in order to alter Western perceptions of the Arab world. Hala was also making plans to start a charity shop that would sell Arab-made products to benefit Arab-oriented charities. In describing her plans, she emphasised the importance of locating the shop on a main street and making its Arab character visible to passers-by. For Hala, the point of such strategies is 'to change the stereotyping, to allow Westerners an insight into the way we see things, the way we look at things, and the way we feel, rather than just seeing us stereotyped as terrorists and all the negative stereotypes that are attached to being an Arab'.

For these individuals and many others with whom we spoke, there was a sense of urgency about 'coming out', as it were, as Arabs – that is, to cease to fear their identity, to project it in a positive way, and to allow the British 'mainstream' to see Arabs as contributing to society. This sentiment is summed up well by Leyla, a woman of Palestinian-Syrian heritage, who founded a magazine for British Arabs. She spoke of the magazine as promoting Arabs and 'sending a message' – a highly visual one filled with beautiful, glossy images of Arab-origin models, designers, pop stars and so on – both to young British Arabs and to the society at large:

> It's all about promoting Arabs. Whether it's the girl who does the makeup for the photo shoots or whether it is the model herself – it's always about, 'Look, these are British Arabs' ... We need to feel that we can live here, be integrated, work here, and not have to fear our identity ... A lot of people associate the words Muslim and Arab with terrorist or wife beater or something derogatory. So let them see: 'Oh, didn't you know that this person is an Arab? Oh, yeah, look ...'.

Envisioning Arabness

While it is clear that many British activists are intent on enhancing the visibility of British Arabs and projecting a more positive image of Arabness, it must be noted that our interviewees are not exposing any single version of Arabness to the public gaze. While almost all of our interviewees identify themselves at least partly as Arab, there is a great deal of diversity among them (and the communities they serve) in terms of class status, religious orientation, and political attitudes.

Therefore, the kinds of imagery they project, and the challenges they face in projecting this imagery, vary significantly. For some, like Leyla, the publisher of the British Arab magazine described above, the image of Arabness to be placed in the public eye is a decidedly modern, cosmopolitan one, reflecting the relatively well-off, highly educated status of many British Arabs in London. The magazine celebrates traditional Arab cultural forms – clothing, food, dance, and music. But in celebrating 'tradition' it tends to avoid making visual reference to, for instance, politicised or Islamist elements of contemporary Arab society. This is not to say that magazine staff and readership are not politicised, but rather, that their understanding of Arabness, and the Arabness they wish to publicise, appeals to a specific aesthetic.

Other interviewees reveal very different elements of the 'British-Arab experience'. Activists in Liverpool, for instance, described their efforts to bring disadvantaged, working-class, and mainly Yemeni communities into wider public view through an Arab arts festival. Few in Liverpool's Arabic-speaking community, they related, had ever been to the city's civic spaces, including the performance hall where the arts festival was scheduled to take place. The activists trained a group of somewhat reluctant children to perform traditional Yemeni dances at the festival, and the children received an enthusiastic response from the audience. Our interviewees described this as major turning point for the community in terms of the willingness for community members to expose themselves to the public gaze and to place themselves squarely within Liverpool's civic space.

Looking at a more individualistic level, it is clear that our interviewees have different views on what aspects of their Arabness should be made public and visible, and which aspects should remain private or hidden. Some of our interviewees, for instance, see Islam as an important part of their identities and their sense of Arabness, but argue that religious belief should not be publicised or 'broadcasted' to the wider public. While these interviewees do not shy away from a visible, public Arab identity, they are dubious of visible markers of religiosity. For example, Manal, a political exile, wore a headscarf for several years but decided to remove it because of the problems it raised for her. She stated, 'Immediately people started to look at me as if I were a different person … People started to be very careful talking to me. They immediately classified backward, terrorist, Muslim. By the end, I was fed up and I took it off'.

Others, however, have been undeterred by the negative stereotypes attached to Islamic garb, and a small number of our respondents cover their heads and/or speak positively about women who do so. One of our interviewees, Rafiya, a young British-born woman active in minority and women's organisations in Sheffield, acknowledged the tensions that arise because of her veil, but maintained that the headscarf, while publicly visible and highly politicised, is not meant to be a public statement about difference, but rather, a symbol of personal commitment to her religion. Her case highlights not only the contested nature of the headscarf (see Gökariksel and Mitchell 2005; Secor 2002) but, more generally, the contested nature of visibility. That is, there are a variety of ways in which individuals and groups envision their own identities and that of their communities, and a variety of judgments they make about the most appropriate way to make those identities visible to a wider public – that is, to visually represent the community.

Conclusion

At the beginning of this chapter, we critiqued understandings of integration that focus on empirical observations of settlement patterns, arguing that such observations seldom tell us about the everyday negotiations that take place as individuals and groups establish their membership in society. We have advocated instead a more critical approach to integration that examines the construction of difference and sameness through a variety of visual practices. Integration, in this sense, is bound up with a politics of visibility and invisibility: it involves the hypervisibility of particular differences but also the invisibility of those who are marked by those differences. We have tried to show how British Arabs, in particular, engage in these politics of visibility and invisibility by contesting and submerging what they feel to be negative images while projecting alternative, positive images of Arabness into the public sphere and into dominant spaces.

We have also tried to reveal some of the ambiguities that emerge from the politics of visibility and invisibility as practiced by British Arabs. Some of our respondents, stressing the need for Arabs to join the 'mainstream' of British multiculturalism, have involved themselves in interfaith coalitions, 'Black communities' forums, and ethnic group coalitions. These kinds of involvements show the desire among many activists to see themselves as a subordinate group and their willingness to relate their own problems to those experienced by other minorities. Yet, at the same time, it is clear that the activists with whom we spoke are intent on putting forth a specifically *Arab* identity.

A number of tensions and exclusions emerge as British Arabs encounter and define themselves in relation to other groups. While our study respondents tend to see themselves as different from the dominant 'white', or 'English', population, they have been reluctant to position themselves within minority categories. They have been especially eager to avoid being classed as 'British Muslims' given the tendency in public discourse to associate British Muslims (i.e. those of Pakistani and Bangladeshi origin) with self-segregation and Islamic radicalism. While many of them see the stereotyping of British Muslims as unfair, they seem less inclined to challenge these stereotypes than to carve out a separate space in British society from which they can make their own set of claims on the public sphere.

Even as they are unified in their efforts to foster a uniquely British Arab identity, our study respondents lack consensus about the kind of Arabness that should be projected. Their divergent visions of Arabness reflect the very different class backgrounds and political commitments among them. Arabness *looks* different in the working-class neighbourhoods of Liverpool and Sheffield than it does in Kensington and Knightsbridge. The politics of visibility and invisibility, in this sense, is a highly fractured one that tends to produce and reproduce differences at the same time that it seeks to undermine them. Far from a natural or organic process, or one that can be clearly observed from residential patterns, integration must be seen as an inherently fragmented and contentious enterprise that lacks any clear path or endpoint.

Notes

1   For instance, in his 1968 speech to his Wolverhampton constituency, Powell stated, 'To be integrated into a population means to become for all practical purposes indistinguishable from its other members. Now, at all times, where there are marked physical differences, especially of colour, integration is difficult though, over a period, not impossible. There are among the Commonwealth immigrants who have come to live here in the last fifteen years many thousands whose wish and purpose is to be integrated and whose every thought and endeavour is bent in that direction. But to imagine that such a thing enters the heads of a great and growing majority of immigrants and their descendants is a ludicrous misconception, and a dangerous one' (Powell's complete speech can be found on multiple websites, including www.sterlingtimes.co.uk, which has dedicated itself to celebrating 'Englishness and patriotism').

2   Thus, in October 2006, prominent Labour politician Jack Straw expressed his concern about the unwillingness of some of his female Muslim constituents to remove their face covering when speaking with him. Straw argued that their face coverings made meaningful communication difficult and, more significantly, that veiling practices created barriers to community life and 'social cohesion'.

3   This research was part of a cross-national study of British Arab and Arab American activists funded jointly by the National Science Foundation and the Economic and Social Research Council.

4   Names of respondents have been changed to protect confidentiality.

References

Bowen, J. (2007), *Why the French Don't Like Headscarves: Islam, the State, and Public Space* (Princeton, NJ: Princeton University Press).

Cresswell, T. (1996), *In Place/ Out of Place: Geography, Ideology and Transgression* (Minneapolis: University of Minnesota Press).

Dwyer, C. (1999), 'Veiled Meanings: Young British Muslim women and the negotiation of difference', *Gender, Place and Culture* 6:1, 5–26.

Fraser, N. (1994), *Multiculturalism: Examining the Politics of Recognition* (Princeton, NJ: Princeton).

Gilroy, P. (1987), *Ain't No Black in the Union Jack: The Cultural Politics of Race and Nation* (Chicago: University of Chicago Press).

Gökariksel, B. and Mitchell, K. (2005), 'Veiling, Secularism and the Neoliberal Subject: National narratives and supranational desires in Turkey and France', *Global Networks* 5:2, 147–65.

Graham, E. (1999), 'Breaking Out: The opportunities and challenges of multi-method research in population geography', *Professional Geographer* 51:1, 76–89.

Hogeland, L.M. (1996), 'Invisible Man and Invisible Women: The sex/race analogy of the 1970s', *Women's History Review* 5:1, 31–53.

hooks, b. (1989), *Talking Back: Thinking Feminist, Thinking Black* (Boston: South End).

Jacobson, W.F. (1998), *Whiteness of a Different Color: European Immigrants and the Alchemy of Race* (Cambridge, MA: Harvard University Press).

Johnston, R., Forrest, J. and Poulsen, M. (2002), 'Are there Ethnic Enclaves/Ghettos in English Cities?', *Urban Studies* 39, 591–618.

Kitchin, R. (1998), '"Out of Place", "Knowing One's Place": Space, power and the exclusion of disabled people', *Disability and Society* 13:3, 343–56.

Koskela, H. (2000), '"The Gaze without Eyes": Video surveillance and the changing nature of urban space', *Progress in Human Geography* 24:2, 243–65.

Massey, D.S. and Denton, N.A. (1998), *American Apartheid: Segregation and the Making of the American Underclass* (Cambridge, MA: Harvard University Press).

McDowell, L. (1997), *Capital Culture: Gender at Work in the City* (Malden, MA: Blackwell).

Modood, T. (2007), *Multiculturalism* (London: Polity Press).

Peach, C. (1996), 'Does Britain have Ghettoes?', *Transactions of the Institute of British Geographers* 22, 216–35.

Phillips, D. (2006), 'Parallel Lives? Challenging Discourses of British Muslim Self-segregation', *Environment and Planning D: Society and Space* 24, 25–40.

Samuels, E.J. (2003), 'My Body, My Closet: Invisible disability and the limits of coming-out discourse', *GLQ: A Journal of Gay and Lesbian Studies* 9:1-2, 233–55.

Secor, A. (2002), 'The Veil and Urban Space in Istanbul: Women's dress, mobility and Islamic knowledge', *Gender, Place and Culture*, 9, 5–22.

Sibley, D. (1995), *Geographies of Exclusion: Society and Difference in the West* (London: Routledge).

Simpson, L. (2004), 'Statistics of Racial Segregation: Measures, evidence and policy', *Urban Studies* 41, 661–81.

Walker, L.M. (1993), 'How to Recognize a Lesbian: The cultural politics of looking like what you are', *Signs* 18:4, 866–90.

Wright, R. and Ellis, M. (2000), 'Race, Region and the Territorial Politics of Immigration in the US', *International Journal of Population Geography* 6:3, 197–211.

Yuval-Davis, N., Anthias, F. and Kofman, E. (2005), 'Secure Borders, Safe Haven and the Gendered Politics of Belonging', *Ethnic and Racial Studies* 28:3, 513–35.

Chapter 8

One Scotland, Many Cultures: The Mutual Constitution of Anti-Racism and Place

Jan Penrose and David Howard

Introduction

Racialised identities and place are mutually constituted. This means that just as the Scottish context will influence how 'race' is constructed and experienced within its borders, so too will constructions and experiences of 'race' influence what constitutes Scotland and Scottishness. In other words, the way that Scots – from all backgrounds – and the Scottish government perceive themselves and their country will have a direct influence on the degree and form of racism and on the content and efficacy of policies that are designed to combat it. This chapter explores this relationship in two main ways. First, it provides a brief overview of how attitudes to race and racism in Scotland have changed over time; particularly in conjunction with recent transitions in Scotland's political status and legal obligations. Second, it analyses the Scottish Executive's showcase anti-racist campaign – *One Scotland, Many Cultures* – to demonstrate how such programmes reveal official understandings of what constitutes racism in Scotland; of how it can be overcome; and of what Scotland ought to be like, without racism.[1] The chapter concludes with an assessment of this campaign (and of the views that inform it), with an eye to capitalising on causes for hope and addressing areas of concern.

Evolving Perspectives on Race and Racism in Scotland

For most of its history, Scotland has been imagined by most 'White' Scots as a place that is devoid of racism.[2] In part, this is because it has been perceived (however inaccurately) as a White country and, consequently, as one in which racism had neither purchase nor relevance. The faulty logic at work here stemmed from two assumptions: first, that the presence of visible minorities was a prerequisite to racism; and second, that the degree of racism in any given society was directly proportional to the size of its non-White community. By these measures, and in comparison with England, it was relatively easy to maintain the illusion that racism did not feature in Scottish society. The fact that 'race relations' were widely recognised to be problematic in England also made it relatively easier for Scots to view racism as an exclusively English problem and, as Bob Miles (1990, 278) has noted, this position was reinforced by the claim that the Scots, unlike the English, were an inherently tolerant people.

To implicate the English even further, many Scots believed that their own tolerance of minority groups was a product of experiences of national oppression at the hands of the English (Miles and Dunlop 1986, 1987). Clearly, this self-image of at least relative racial tolerance involved some collective national amnesia given the significance of Scottish contributions to British imperial expansion that was often grounded in racism (Samuel 1998). This self-image also relied on the artificial geographical separation of Scottish practices abroad from the experiences of Scots in their own country and the larger entity of the United Kingdom. Nevertheless, the general tendency for Scots to define themselves in contradistinction to the English meant that there was little need to deploy racialised minorities in this way. In Scotland, the 'significant other' in terms of self-definition was the English, not Blacks or Asians or any other visible minority group. All of these factors meant that the *racialisation* of Scottish society and politics remained very limited until at least the mid-1980s, especially when compared with England and other parts of the world. For many Scots – especially those who define themselves as 'White' – this situation was easily, but wrongly, confused with an absence of *racism*, despite evidence of racist harassment, abuse and lethal violence (Armstrong 1989). As a result, Scotland's longstanding cultural diversity and individual experiences of racism remained largely hidden phenomena until very recently.

To some extent, it was the work of scholars like those cited in the preceding paragraph that contributed to a gradual change in perceptions of racism in Scotland (see also Peter Hopkins' chapter in this edition). In March 1985, three Scottish Institutions of Higher Education launched the Scottish Ethnic Minorities Research Unit (SEMRU) as a collaborative venture to address the lack of Scotland-specific information on the position of ethnic minority groups and their experiences of racism (GCU Archives). In the early 1980s, estimates of Scotland's visible minority population ranged from 38,000 to 52,000 people; figures that represented between 0.8 and 1.0 per cent of the country's population at the time (Armstrong 1989, 32). Yet, despite comprising such a small proportion of the total, research soon made it clear that the experiences of these people put paid to the assumption that their limited numbers precluded problems of racism.

In 1987, a report produced by David Walsh for the Scottish Ethnic Minorities Research Unit revealed that over 80 per cent of Indian and Pakistani people interviewed in Glasgow had experienced racist abuse, while 20 per cent had been victims of physical attacks (Walsh 1987). In addition, racist attacks were shown to have caused significant damage to the property of visible minorities, and abusive language was a common experience for many of these people. In 1990, Ahmed Sheikh died after being attacked, and in 1992, an elderly man, Niaz Ahmed Khan was brain-damaged and later died following a serious assault, but neither of these events reached the front pages of the Scottish newspapers and no one was convicted for the offences (Nutt 2000). By 1998, the Commission for Racial

Equality had compiled figures which showed that even though Scotland was home to just 2.1 per cent of the UK's people from minority backgrounds, it had recorded 7.3 per cent of racially-motivated incidents (BBC News Online 1999; Commission for Racial Equality 1998). Moreover, Central Scotland recorded the highest rate of racist incidents, some fifteen times as high as in London (BBC News Online 1999). Despite these and other revelations of just how pervasive racism was in Scotland, such issues were viewed widely as the legislative responsibility of Westminster throughout the 1990s – a view that permitted complacency to prevail until the end of the millennium (Williams and De Lima 2006, 499).

In the late 1990s, several factors coincided to both demand and encourage a shift in Scotland's attitudes and responses to racism. First of all, the election of New Labour in 1997 brought with it the promise of more liberal attitudes toward immigration and greater attention to issues associated with cultural diversity, racial equality and social inclusion in the UK as a whole. Second, and less sanguinely, media coverage of the growing number of racist attacks in Scotland nudged a reluctant society towards confrontation of its own prejudice. Key incidents included the death of 15-year-old Imran Khan in Glasgow, in February 1998, when he was stabbed by a 17-year old White youth during a fight that was attributed to trouble between Asian and White gangs in the south side of the city (Scottish Executive 1998). In November of the same year, 32-year-old Surjit Singh Chhokar was murdered in Wishaw (Lanarkshire) and here again, the three prime suspects were all White men. Although both cases raised the spectre of racism in Scotland, the latter case was especially inflammatory because its presentation in two trials (in March 1999 and November 2000), ultimately meant that no one was sentenced to imprisonment for this offence (Seenan 2000). This fundamental failure was laid squarely at the feet of the Scottish judiciary by the presiding judge. Moreover, the fact that police had immediately dismissed racism as irrelevant to either attack led to accusations of institutional racism in this corner of law enforcement as well. In December 2000, the then Scottish justice minister, Jim Wallace, was reported to have admitted that there had been institutional racism in both the crown office and Strathclyde police (Seenan 2000).

The timing of these events was critical because they occurred in the aftermath of the murder of Stephen Lawrence (in April 1993), and alongside the emergence of the far-reaching ramifications of the Macpherson Report, which was published in February 1999. This report constituted both an inquiry into the management of the case and an attempt to advance practical recommendations for avoiding similar problems in the future. Thus, while the murder of Stephen Lawrence had a catalytic effect on British perceptions of race relations, the flawed process of its investigation also forced the country to confront the reality of racism that was embedded in British institutions. In Scotland, the verdict of the first Chhokar trial was delivered immediately after the Macpherson Report had been declared a watershed in British race relations. The fact that the accused was convicted of assault (rather than murder) and that the crown did not move for sentencing meant that he walked free from court. For many people in Scotland this sequence of events seemed to demonstrate that very little had changed and for some, it was provocative incentive to ensure that something did.

The changing political context in Scotland constituted a third key factor in the growing commitment to challenging racism and to developing policies which could combat it effectively. Most obviously, Scottish devolution came into effect in July 1999 when the Scottish Parliament was officially opened. Although the Parliament did not acquire the power to develop race relations legislation, it did allow for the promotion of equal opportunities in innovative and particularly Scottish ways. The new parliament was based on four key principles that addressed just governance, accessibility, accountability and equality (Scottish Office 1998). The latter cornerstone, and the establishment of the Equal Opportunities Committee as a Standing Committee of the Parliament, were especially relevant to new efforts to combat racism in the Scottish context. These developments anticipated that a commitment to all types of equality would be built into the structures of Parliament and would remain constant despite changes in government and in particular policies. Under the first post-devolution government, a dedicated Equality Unit was established within the Executive in order to facilitate the integration of equality commitments into policy development, service design and delivery right across the work of the government.

The opportunity to demonstrate Scotland's official commitment to promoting all forms of equality – including the pursuit of anti-racism – was given added incentive when the Race Relations (Amendment) Act (RRAA) was passed by the UK Parliament in November 2000. This legislation was, to some extent at least, a response to the Macpherson Report and this is another indication of the heightened profile that race relations was acquiring throughout the United Kingdom around the turn of the millennium. The Amendment Act extended key elements of the Race Relations Act 1976 to public authorities by outlawing race discrimination in functions not previously covered and by making chief officers of police vicariously liable for acts of race discrimination by police officers (RRAA 2000: explanatory note 8). Both of these changes were reflections of efforts to acknowledge, address and eliminate institutional racism. In addition, the new legislation introduced a proactive element which placed 'a duty on specified public authorities to work towards the elimination of unlawful discrimination and promote equality of opportunity and good relations between persons of different racial groups' (RRAA 2000, explanatory note 8).

In July 1999, within the new political context of a devolved Scotland, and in anticipation of the new legal responsibilities eventually created by the RRAA, the Scottish Executive set up two working groups to advance an anti-racism agenda. Although the move toward consultation was seen as a 'cosmetic exercise' by some (BBC News Online 1999), the Stephen Lawrence Steering Group and the Race Equality Advisory Forum (alongside the Scottish Executives's 2000 Equality Strategy), did establish the framework of Scotland's strategy for achieving race equality.[3] Their identification of actions that needed to be taken by the government, by criminal justice agencies, and by the wider public sector was an essential first step in efforts to eliminate institutional racism and effect related social change. As this suggests, however, the attempt to implement an anti-racist agenda was based on the dubious assumption that a cohesive dynamic exists between corporate and individual responses. In other words, it was assumed that institutional change would automatically change individual attitudes and behaviours simply because the structures surrounding these individuals were changing. The failure to clarify just how parliamentary structures and government directives might transform individuals' attitudes in practice can be seen as a potential weakness of such institutionally-led anti-racism programmes.

As will become clear below, this weakness may have influenced the effectiveness of a major, multi-phased advertising campaign that was to form a key element of Scotland's anti-racism strategy from 2002 until the present time. In order to have a benchmark against which to assess the success of this campaign, the Scottish Executive, commissioned independent research into attitudes toward race-related issues among Scotland's adult population (NFO System Three 2001a and 2001b). Although the majority of respondents believed that they were not racist (76 per cent), some 49 per cent regarded racism as, at worst, a slight problem and tended to view racist abuse or assault as isolated incidents rather than as evidence of a wider underlying problem of racism in Scotland (NFO System Three 2001a, 7–8). Meanwhile, these perceptions were challenged actively by an increase in reports of racist incidents, from 386 in 1999 to 2371 in 2002 (Scottish Executive 2003). It was against this backdrop and with the close cooperation of the Commission for Racial Equality Scotland that, on 24 September 2002, the Scottish Executive launched its 'groundbreaking campaign to tackle racism' (Scottish Executive 2002a).

One Scotland, Many Cultures: Institutional Anti-racism in Scotland

The *One Scotland, Many Cultures* campaign is produced and managed by the Equality Unit at the Scottish Executive (along with key partners), and it combines two quite different motivations or rationales. The first, as outlined above, is a mandatory legal requirement to comply with the Race Relations (Amendment) Act (2000). This Act makes it incumbent on public authorities in the UK to be pro-active in eliminating unlawful discrimination and to promote equality of opportunity and good race relations. As this suggests, its overriding goal is to ensure that no group or individual is disadvantaged through the development, design and/or delivery of policies and services. The main outcome of this politically-driven legal process, in terms of the Scottish Executive, was the compilation of its Race Equality Scheme entitled, *Working Together for Race Equality* (2002b) (republished in 2005 and 2008). This scheme is primarily comprised of inward looking documentation on how each department of government met, or planned to meet, the legal requirements of the Race Relations (Amendment) Act.[4] At the same time, the report also acknowledged the outward looking remit of the *One Scotland, Many Cultures* publicity campaign as the more exciting and innovative dimension.

As this suggests, the second motivation behind the 'One Scotland' Campaign is the desire to advance one of the Scottish Executive's founding principles – namely, the 'delivery of social justice' through increased social inclusion (Scottish Executive 1999). This dimension of the campaign moves away from the rather cold and impersonal legal imperatives to the more flexible realm of policy initiatives that are intended to have direct personal impact. The policy in question involves raising awareness of racist attitudes and behaviour in Scotland as a prerequisite to 'tackling and eliminating racism altogether' (Scottish Executive 2002b; Scottish Parliament 2006). To advance this policy, an intensive, multimedia publicity campaign has been developed as the most public and visible part of the anti-racism programme.

This campaign, *One Scotland, Many Cultures*, was launched by the Scottish Executive in 2002 and since 2005 it has provided the contextual framework within which five so-called 'super brands' have been developed and promoted. These five core programmes promise to make Scotland a 'healthier', 'safer', 'smarter', more environmentally responsible and less discriminatory society, under the banner of five 'super brands' (Irving 2004).[5] As this function of the campaign suggests, it is designed to be highly integrative, both between specific policies and broader visions and between different scales and contexts. Thus, the campaign builds on Britain-wide initiatives to confront racial discrimination, notably 'Show Racism the Red Card' and 'Rock against Racism', and it also complements or supports other projects led by goal specific and/or local organisations like the Centre for Education for Racial Equality in Scotland and the Glasgow Anti Racist Alliance (GARA).

The *One Scotland, Many Cultures* campaign was launched under the broad remit of raising awareness of Scotland's cultural diversity and actively celebrating it. More specifically, however, the core of the campaign seeks to challenge racist attitudes and behaviour – as well as the language that is used to transmit them (Scottish Executive 'One Scotland', http://www.infoscotland.com/infoscotland). This focus on attitudes and behaviour makes it clear that addressing the 'unrecognised' aspects of racism is fundamental to the campaign. At the same time, it demands that the Scottish public recognise that racialised discrimination is as virulent and as problematic in Scotland as sectarianism, the latter being a phenomenon that is much more commonly acknowledged (and, arguably, accepted as 'traditional'). In addition to promoting an improved understanding of what might constitute racism or racialised abuse (i.e. things that include language use, attitudes and behaviour), *One Scotland, Many Cultures* also challenges the assumption that ethnic or national identities are stable and unchanging, as well as the belief that they are bestowed by parental inheritance at birth. Importantly, this allows the campaign to portray Scottish society as both historically rooted and as an evolving social formation.

Although logically sound and pragmatically necessary, this portrayal of Scotland easily appears paradoxical to both of the key groups that the campaign seeks to address. For White racists (witting or not), this conception of Scotland allows them to cling to the illusion of a past that was exclusively White, and this quality is deeply ingrained in both their sense of themselves and their understanding of their country (see BBC 2002). Even those who accept that Scotland is changing, and who accept that multiculturalism is the reality of its present and its future, are often reluctant to relinquish an image of Scotland's past as defined by people, symbols and events that are, at least implicitly, bound-up with Whiteness. Conversely, those who have suffered from racism (overwhelmingly non-Whites) often feel excluded by the construction of an historical image of Scotland that denies its pluralism and, in the process, their claims to equal membership within it. As we demonstrate below, the campaign attempts to tread softly between these conflicting visions of what Scotland was, but this has the effect of producing some disquieting ambiguity about just what a future Scotland of many cultures might look like and who it might serve.

One Scotland, Many Cultures: Campaign Content

The campaign itself has consisted of a series of posters for offices and schools, billboards and buses; television, cinema and radio presentations; and more recently, message-laden water cooler cups placed in 750 offices across Scotland. The material is disseminated throughout the country, and the accompanying campaign website provides updated statistics and campaign-related teaching materials (http://www.onescotland.com/). As indicated above, the first phase of the campaign ran during the autumn of 2002, and a number of new phases have been launched since then which incorporate new material, often in response to the assessment of previous phases of the campaign, including stakeholder feedback. The change of Scottish government in May 2007 (from Labour to SNP), ushered in a period of policy review, but given the SNP's manifesto commitment to developing the 'One Scotland' campaign (SNP 2007, 66), it is likely that it will continue for some time to come (Dinwoodie 2007).

The overriding message of the campaign is that pluralism and equality are concepts that need to be learned and socialised in order to be sustained. As such, it sets out to teach the Scottish public to recognise what might constitute racial discrimination and to accept multiculturalism as a fundamental characteristic of the Scottish nation, not just in the present but also in the past and the future. Although the campaign is not explicit about all of the qualities of its intended audience, the White majority is probably the main target because their privileged position makes it unlikely that these people will have experienced racial discrimination first-hand. As such, they are most likely to be most in need of education about racism. In addition, young people living in Scotland have also been singled out since evaluations of the first phase of the campaign made it clear that they will be key agents for the transformation of Scottish society (Carr and Smart 2003).

Through the analysis of campaign materials, and of interviews with those involved in the production and ongoing review of the anti-racism agenda, we have identified three key elements that have been drawn into the Scottish Executive's fight against racism and its corresponding pursuit of social transformation. In the context of racialised discourse, the campaign seeks to challenge racist language, behaviour and attitudes. All three of these elements highlight the Scottish Executive's belief that subtle forms of racialised discrimination, including the subconscious acceptance of it, can be as damaging as more overt forms of physical or verbal abuse. In the remainder of this chapter we will show how each of the elements figures in the *One Scotland, Many Cultures* campaign and attempt to show how they work together to advance the dual goals of anti-racism and multiculturalism.

First of all, the campaign confronts Scottish people with the need to assess everyday language and to reject terms that are either motivated by racism or which advance it unwittingly through ignorance. The importance of language to advancing the campaign's goals is probably best illustrated by two separate scenes, both set in the same grocery store, which formed the substance of a television and cinema presentation that was first shown in 2002. The first scene illustrates how the use of the term 'Paki' to describe the shop (and by association, its owners) produces disruptive friction and social disharmony which, in turn, leads to racist graffiti – 'Paki Scum' – being scrawled across the shop's protective steel shutters. The second scene replays the situation but this time it shows how the outcome can be changed for the better when courtesy and equality govern the shop's (and, by implication, society's) transactions. As a whole, the film suggests that the rejection of what might be termed low level or 'shallow' racism can produce a dramatic transformation in everyone's everyday interactions in a multicultural Scottish city (Murji 2006). The centrality of language to this transformation is made explicit when the film ends by superimposing the phrase 'One Word Many Consequences' over the racially derogative slur that appeared in the first scene (see Figure 8.1).

Figure 8.1   'One word has many consequences': still from television presentations entitled *Violence* and *Virtue*, for *One Scotland, Many Cultures*

Source: Scottish Executive (2002c).

While the use of the term 'Paki' in the previous example would have been quite obviously racially offensive to most Scottish viewers, several other advertisements address other common expressions that are frequently used without any intention of expressing overt or conscious hostility toward members of racially defined groups. Yet, as these advertisements make clear, the absence of malevolent intention does not eliminate the capacity for terms to offend those who they denigrate. The television productions that are designed to convey this message all begin by presenting a series of very brief clips in which people from countries overseas are interviewed about their perceptions of Scottish people. In each case, very positive comments about Scots are followed by scenes set in Scotland, where White Scots interact with, or are observed by, Scots from racialised minority groups: in each case the White Scots behave in negative ways that belie their positive international reputation. For example, one segment begins with an interview of two young Asian women in an unidentified Asian city in which one of them describes Scots as follows: 'They are very welcoming, hospitable, I like the way that they talk' (vox pops1, Scottish Executive 2002c). The scene then shifts abruptly to a Scottish pub in which a White man, sitting with a group that includes a woman of Asian descent, downs the last of his pint and asks, 'Right then, who's for a Chinky?' (vox pops1, Scottish Executive 2002c). Clearly, this is not the kind of talk that the young Asian women from overseas had in mind when they were describing Scots, and the statement is met with silence from those at the pub; a silence that seems to be borne of shock for some and confusion for others. Eventually, all of these people turn to the Asian woman to ascertain her response and, in some cases, to offer unspoken apologies (see Figure 8.2). The hurt that is manifested on the young woman's face, and through her body language, leaves the viewer in little doubt that they have just witnessed racism in action, regardless of what the intentions of the speaker were. The film concludes with a still shot of one of the posters featured in the campaign, one that admonishes Scots to 'Live up to Your Reputation'.

Although these television advertisements focus on the element of language, the ways in which language produces racism is shown to be closely linked to behaviour: the second key element of the *One Scotland, Many Cultures* campaign. Thus, they demonstrate that although terms like 'Paki' and 'Chinky' may not seem offensive in their own right (to *some*), they acquire this quality because they are used as vehicles for perpetuating negative stereotypes of racial, and often cultural, difference. These stereotypes, and the behaviour that they legitimise, are shown to exclude actively members of these minority groups from full and equal membership in Scottish society. The message here is that changing linguistic habits is a key step in eliminating behaviours which are racist, whether this was their intention or not. As this suggests, the campaign position advanced here is that racism without malice is still racism.

Both language use and behaviour can be reflections of attitudes, and this is the third key element of racism that the *One Scotland, Many Cultures* campaign seeks to address. Attitudes generally sit deeper in human psyches than do the words and actions that are used to convey them, or indeed, in some instances, to conceal them. Thus, efforts to address racist attitudes must find ways of using language

Figure 8.2   Pub scene from a television advert for *One Scotland, Many Cultures*

Source: Vox pops 1, Scottish Executive (2002c).

and behaviour but also of going beyond them. In general, this has taken two key forms in Scotland. The first, as reflected in our discussion so far, is to change attitudes toward individuals who belong to racialised minority groups in Scotland – they can no longer be considered inferior, or other (i.e. non-Scottish), nor can they be treated as such. The second, and closely associated change in attitudes that is being pursued, is the wider attempt to alter how Scottish people view the social and cultural attributes of the Scottish nation. To this end, the campaign aims to conjure national pride in the past, while directing collective aspirations towards an ever-changing sense of a racially diverse, but nationally coherent Scottish people and place. When set within the framework of the campaign as a whole, slogans such as 'Don't let Scotland down' and 'A small country, not a county of small minds', implicitly suggest that individual expressions of racism do a disservice to the country and that a progressive Scotland is one in which cultural diversity thrives.

The need to recognise and accept cultural diversity in Scotland is reinforced in posters like that which was produced for Refugee Week 2006 (Carr 2008). This poster uses a Russian Matroyska, or nesting doll, as a visual metaphor for the composition of Scottish society (Figure 8.3). Here, the largest figurine – stylised after the Russian original – is decorated with the Scottish flag and this is set alongside six other dolls of descending size. These dolls signify the six main countries of origin of refugees in the UK; they are decorated, accordingly, with the flags of Turkey, Pakistan, Iran, Somalia, Iraq and Afghanistan. Presumably, this visual metaphor was designed to reify the general campaign slogan, placed below it, which asserts implicitly that the notions of 'One Scotland' and 'Many

Cultures' sit easily together. Unfortunately, closer consideration of this visual metaphor begs fundamental questions about just how this complementarity will work in practice.

Figure 8.3   Scotland as a Matroyska doll

Source: Scottish Executive (2006).

Although it was created to publicise Refugee Week, the Matroyska metaphor still has the effect of exposing pervasive ambiguity about just what form the Scottish Executive wants cultural diversity to take in Scotland. On one hand, it can be seen to present Scotland as comprised of diverse cultural entities (nations, as defined by the flags of their states) that are all independent from and, if ranking is anything to go by, subservient to, the dominant Scottish culture. This has the effect of reinforcing the perspective of those Whites who see themselves as 'more Scottish' than non-Whites and, in consequence, as having greater rights to determine the future of 'their' country. It is difficult to see how people who have suffered from racism and/or cultural exclusion in Scotland can locate themselves as equal constituents of Scotland within this metaphor. Unfortunately, other interpretations do not fare much better in terms of advancing visions of Scotland as a country that has been transformed into a pluralist utopia through sensitive inclusion and accommodation of cultural difference. The most progressive possible interpretation of the Matroyska metaphor is to see it as inviting viewers to imagine all of the dolls nested inside of one another, retaining their individuality but comprising the larger entity of Scotland at the same time. Yet, even in this reading, it is not clear whether Scotland is absorbing the cultures which comprise it to produce an entirely new, hybrid, form of Scottishness, or if an established form of Scottishness is simply hiding (and conceivably marginalising) all other cultures.

As this example suggests, the appeal to a notional level of positive and flexible nationalism that encompasses a multicultural vision of unity asks more of Scottish people than changes in their language use and behaviour. It is asking them to imagine themselves and their nation in ways that challenge White complacency with imaginings that have long privileged them over other citizens of Scotland. Clearly, then, the element of attitudinal change, is probably the most complex, but also the most crucial, component of the 'One Scotland' campaign. While non-racist linguistic and behavioural discourse might be 'taught', the mental and emotional will that is required to create a non-racist Scotland cannot be so directly instilled or assessed. This is particularly true if the 'new Scotland' is seen, in some way, to come at the 'cost' of those who are under most pressure to change their attitudes to their country and to their understanding and expression of themselves. As a whole, the subtle rationale of the campaign predicts that non-discriminatory language and behaviour will promote more open attitudes to a racially, culturally and religiously diverse Scottish nation, and that this, in turn, will automatically produce a more efficient and equitable society.

Not surprisingly, the connection of people, tradition and place as core constituents of the Scottish nation is well accepted. The campaign material, however, aims to enlist the entire populace in recognising that these elements and linkages are not static and that 'Scottishness' is dynamic. To this end, the campaign website includes a timeline of Scottish history that is used to reveal social and cultural transitions that have served to shape today's nation. The message here is that Scotland has never been exclusively White and that pluralism and transition rather than static homogeneity have long been the country's defining characteristics. To some extent at least, the Scottish Executive's desire to establish a history of diversity can be seen as a strategy for dissolving fear at ongoing immigration and attendant social change. This is important because of the country's humanitarian obligations to refugees and asylum seekers but also, and more cynically, because of the country's reliance on immigration to maintain economic expansion and the demographic growth on which this relies. Thus, the portrayal of Scotland as a society of immigrants underpinned the 2004 launch of the Scottish Executive initiative called *Fresh Talent: Working in Scotland*, a scheme designed to encourage the relocation of 'bright, talented, hardworking people who can make a positive contribution to the Scottish economy' (Scottish Executive 2005, http://www.scotlandistheplace.com). As these connections suggest, the moral imperatives of the *One Scotland, Many Cultures* campaign are paralleled by the less obvious economic imperatives. Racism and the illusion of cultural homogeneity must be challenged because they are wrong but also because they have the potential to impede seriously the economic development upon which Scotland's future rests.

The *One Scotland, Many Cultures* campaign itself, does not accentuate the economic incentives for anti-racism and a multicultural conception of the Scottish nation. Instead, it accentuates the need for all citizens of Scotland to work together in order to ensure the best possible future for everyone and, quite literally, to keep the Scottish flag flying. This strategy is most obvious in a television advert that aired early in the campaign. With considerable subliminal effect, all but the last segment of the film is done in black and white and it begins, against a background track of rousing guitar rifts, by alternating between images of white and then black hands gripping a massive rope. The focus then shifts to the faces of a whole range of individuals – Black, White and Asian; men and women; old and not so old – which are shown in very quick succession, all straining with the effort of heaving the rope. Ambiguity is cleverly deployed because, at first, it seems that the different groups of people – especially White and non-White – are engaged in a tug-of-war against one another. However, when the camera finally pans back, it becomes clear that all of these people have been working together to raise an enormous Scottish saltire (Figure 8.4).

Figure 8.4   A saltire being hoisted in a scene from the television advert, 'Tug of War', for One Scotland, Many Cultures

Source: Scottish Executive (2002c).

As a whole, the symbolism of this advert insists that the only way to raise the Scottish flag – and by implication the country itself – is if the multicultural society 'pulls together' and, if the size of the flag is indicative, it is also clear that the product of these efforts will be much greater than any individual or cultural group. Perhaps more subtly, the fact that the only part of the film to appear in colour is the blue and white of the saltire could suggest that these are the only colours that matter in the Scotland that is being promoted by the *One Scotland,*

Many Cultures campaign. As the strain on people's faces suggests, the building of a multicultural society might be tough work, but with cohesion rather than division, a united Scotland will result, and it will be one in which skin colour does not matter.

Perhaps ironically, however, the saltire can also be seen to exemplify a more static element of Scottishness, one that contrasts quite sharply with the dynamic pluralism which is the intended campaign message. The flag itself dates from the ninth century, is rooted in a Christian iconography and is the oldest national flag to remain in constant use today. For some then, this advertisement could be seen to depict the use of immigrant labour to resurrect an imagined Scotland of the past and to thereby preserve the inequalities that long privileged White people over all others. Although this is an unlikely reading in today's climate, and given the overall gist of the campaign, it does raise the crucial question of just how much the Scottish nation can change at the core. There is some evidence that language and behaviour are already changing, and this should not be dismissed lightly, but if the Scotland that most people imagine continues to be grounded in an illusory past that is White and Christian, there is a real danger that the long-hidden complacency about racial inequality in Scotland will be perpetuated. Clearly, attitudes about what – and who – Scotland is must change if the behaviour that constitutes these entities and the language that is used to describe them are to do more than provide lip-service to an impossible dream of one Scotland that rests easily with its many cultures.

Conclusion

The dawn of the twenty-first century marked an end to Scotland's long-standing complacency about race and racism within its borders and the *One Scotland, Many Cultures* campaign has played a fundamental role in effecting this change. As we have shown, the campaign has sought to mobilise all Scots such that they recognise racism in multiple guises and that they speak, act and think in ways that expunge it from Scottish society. In the process, and by implicitly eliding race with culture, the campaign has also demanded that Scotland come to be viewed as a place of many cultures. Importantly, the campaign's clarity about how to advance an anti-racist agenda has not been matched by either an understanding of how this is linked to multiculturalism, or by a clear vision of what a multicultural Scotland should be.

To some extent, this should not be surprising because anti-racism is much less threatening than multiculturalism to most White Scots. Anti-racism demands that individuals change their views and treatment of others, or that they at least appear to do so. In contrast, multiculturalism demands that individuals change their understanding of themselves and of the country that helps to define them. For those who are accustomed to positions to privilege, and/or those who experience this change as loss, this can be a very profound demand indeed – it involves changing their identity. Moreover, multiculturalism insists that attitudes to race are not influenced solely by the day-to-day treatment of individuals but also by the way in which these people are situated within any given society. In other words, multiculturalism makes it clear that racism cannot be eliminated through language, behaviour and attitudes if the infrastructures of society continue to marginalise or exclude people who are somehow identified through notions of race.

This assessment of the distinctions between anti-racism and multiculturalism makes it easier to understand why the *One Scotland, Many Cultures* campaign may have focused on anti-racism in the first instance. It also makes this focus a source of concern. This is especially true given changes in the context in which the campaign has been advanced. The early campaign material promoted the largely passive recognition that Scotland has 'many cultures' but it never suggested that the dominant conception of Scottishness should be transformed to reflect this. More worryingly still, any overt pressure to clarify the relationship between Scotland and its composite cultures was reduced in 2005 when the campaign title was changed, ostensibly to fit in with the wider re-organisation and associated re-branding of the Scottish Executive. In this restructuring, 'Many Cultures' was removed from the campaign title: from this point on, 'One Scotland' and its vague allusion to unity, has stood alone to do the work of advancing an anti-racist agenda in Scotland. We have not yet had a chance to explore how this change – from *One Scotland, Many Cultures* to *One Scotland* – has been justified within the Scottish Executive and the likelihood of doing so may have been reduced with the recent election of the Scottish National Party (SNP) in May 2007. If the change in campaign name was a semantic acknowledgement of resurgent nationalism, then it seems reasonable to assume that an emphasis on ill-defined unity, at the expense of active multiculturalism, will continue under SNP rule. It is worrying, for example, that the section of the SNP manifesto which relates to 'Promoting Equality in Scotland' includes the following statement. 'Prior to independence the SNP will not promote or support legislation or policies which discriminate on the grounds of race, disability, age, gender, faith or religion, social background or sexual orientation' (SNP 2007, 66). This wording belies a reactive, rather than a proactive stance to issues of equality – the focus is on what the party will *not* do rather than what it *will* do. Worse still, the opening phrase seems to suggest that even this limited commitment will only be retained until the goal of independence has been achieved. By its very nature, and despite any number of protestations to the contrary, nationalism sits uneasily with pluralism. Thus, even though the new government is legally obliged to continue the pursuit of an anti-racist agenda, we fear that the transformative potential of such an agenda will be lost as long as it remains severed from the active promotion of Scotland as a place of many cultures.

Acknowledgements

The authors would like to thank the Scottish Executive for permission to reproduce images from their advertising campaigns in this paper. We are especially grateful to Rhona Carr for her interview contributions and for her careful reading of an earlier draft of this work.

Notes

1   It is important to note that between the drafting of this chapter and its final submission the Scottish National Party initiated a name change that will see 'the Scottish Executive' replaced by 'the Government of Scotland'. The blatant nation-building agenda behind this transition is discussed to some extent in our conclusion.
2   We have capitalised 'White' and 'Black' throughout this chapter in order to maintain consistency with census categories and to denote both formal categories and popular labels/proper nouns for classification.
3   The Stephen Lawrence Steering Group was chaired by the then Deputy First Minister and Justice Minister, Jim Wallace MSP, and the Race Equality Advisory Forum was chaired by the then Minister for Social Justice, Jackie Baillie MSP.
4   Other key public bodies like the police, health boards and local authorities were also required to publish such schemes.
5   The five 'brand images' are illustrated on the campaign webpage of the Scottish Executive, http://www.infoscotland.com/infoscotland.

References

Armstrong, B. (ed.) (1989), *A People Without Prejudice? The Experience of Racism in Scotland* (London: Runnymede Trust).
BBC News Online (1999), 'Racism Action Plan Attacked', <http://news.bbc.co.uk/1/low/uk/398426.stm>, accessed 14 July 2007.
BBC (2002), Debate on Racism, moderated by Kay Adams, 16 October, 10.40 pm.
Carr, Rhona (2008), personal communication.
Carr, R. and Smart, P. (2003), interview conducted 15 May 2003, Scottish Executive Equality Unit.
Commission for Racial Equality (1998), *Annual Report of the Commission for Racial Equality* (London: Commission for Racial Equality).
Dinwoodie, R. (2007), 'Salmond: Scotland leads way in social cohesion', *The Herald*, 1 August.
GCU Archives. Glasgow Caledonian University, Research Collections: University Archives, Scottish Ethnic Minorities Research Unit (SEMRU) Archive, <http://www.gcal.ac.uk/archives/semru/index.html>, accessed 28 June 2007.
Irving, L. (2004), interview conducted 2 November 2004, Scottish Executive Equality Unit.
Miles, R. (1990), 'The Racialization of British Politics', *Political Studies* 38: 277–85.
Miles, R. and Dunlop, A. (1986), 'The Racialisation of Politics in Britain: Why Scotland is different', *Patterns of Prejudice* 20:1: 23–32.
Miles, R. and Dunlop, A. (1987), 'Racism in Britain: The Scottish dimension', in P. Jackson (ed.), *Race and Racism: Essays in Social Geography* (London: Allen and Unwin), 119–41.
Murji, K. (2006), 'Using Racial Stereotypes in Anti-racist Campaigns', *Ethnic and Racial Studies* 29:2, 260–80.
NFO System Three (2001a), *Attitudes Towards Racism in Scotland. July 2001*, prepared for the Scottish Executive and Barkers Advertising, Reference No. 8337, 14 August.

NFO System Three (2001b), *Attitudes Towards Racism in Scotland. November 2001*, prepared for the Scottish Executive and Barkers Advertising, Reference No. SOS X01, 17 December.

Nutt, K. (2000), 'Where Racism's Brutal Face is a Fact of Life for Asians', *Sunday Herald*, 3 December.

RRAA (2000), *Race Relations (Amendment) Act 2000. Further Extension of 1976 Act to Police and Other Public Authorities* (London: Her Majesty's Stationery Office)

Samuel, R. (1998), *Theatres of Memory: Island Stories – Unravelling Britain*, Vol. II, (London: Verso).

Scottish Executive (1998), 'Crown to Hold FAI's into Hospital Deaths', news release, 2 November.

Scottish Executive (2002a), *One Scotland, Many Cultures*, news release, 24 September.

Scottish Executive (2002b), *One Scotland, Many Cultures: Working Together for Race Equality – The Scottish Executive's Race Equality Scheme* (Edinburgh: Scottish Executive).

Scottish Executive (2002c), *One Scotland, Many Cultures* campaign material, personal communication.

Scottish Executive (2003), Statistics. Recorded Crime in Scotland. Racially Aggravated Offences (including harassment and conduct) by Police Force Area, 12999–2002, Table A7, <http://www.scotland.gov.uk/Publications/2003/05/17154/22027>, accessed 29 June 2007.

Scottish Executive (2005), 'Fresh Talent: Working in Scotland', news release, 16 June.

Scottish Executive (2006), Refugee Week 2006, Matroyska poster.

Scottish National Party (2007), *Manifesto: It's Time*, SNP, 12 April.

Scottish Office (1998), Report of the Consultative Steering Group on the Scottish Parliament: 'Shaping Scotland's Parliament', <http://www.scotland.gov.uk/library/documents-w5/rcsg-00.htm>, accessed 4 August 2007.

Scottish Parliament (2006), Official Report 28 June 2006, <http://www.scottish.parliament.uk/business/officialReports/meetingsParliament/or-06/sor0628-02.htm>, accessed 8 November 2007.

Seenan, G. (2000), 'Stabbing that Exposed Scots Racism', *Guardian*, 8 December.

Walsh, D. (1987), *Racial Harassment in Glasgow* (Glasgow: Scottish Ethnic Minorities Research Unit).

Williams, C. and Lima, P. de (2006) 'Devolution, Multicultural Citizenship and Race Equality: From laissez-faire to nationally responsible policies', *Critical Social Policy* 26:3, 498–522.

Chapter 9

Politics, Race and Nation: The Difference that Scotland Makes

Peter Hopkins

Introduction

A young Muslim woman recently commented to me that she was particularly struck by the fact that, when attending conferences, Muslims from Scotland identified as 'Scottish Muslims'. She observed that Muslims in England 'would not normally self-identify in that way'. This observation led me to reflect critically on the ways in which the Scottish context differs from the rest of the UK in terms of experiences of race, racism and ethnic identity. With this in mind, I was drawn to research about the geographies of race and racism and found that Scotland has largely been ignored, or has been subsumed into research about race and ethnicity in the UK (although for rare exceptions see Hopkins 2004; Miles and Dunlop 1987; Smith 1993). However, there are a small group of active scholars working in education and social policy who have been engaging with issues of race and racism specifically in the Scottish context, and this work has drawn attention to issues connected with racism and social policy in Scotland (Arshad 1999, 2003) and the experiences of ethnic minorities in rural areas (de Lima 2001, 2003, 2004, 2005). There has also been important work commissioned by the Scottish Executive[1] (e.g. Netto et al. 2001) which has sought to explore issues of race and racism in Scotland and identify recommendation for future policies. Building upon this work, I intend to use this chapter to argue that although there are continuities between the Scottish and English situations in terms of experiences of race and racism, there are also important discontinuities between these two contexts. It is the unique aspects of the Scottish context that deserve particular attention in that they structure the everyday lives, opportunities and experiences all communities in Scotland, and in particular those of black and minority ethnic communities.

In order to explore these issues further, I draw upon literature about the geographies of race and racism in Scotland as well as the experience of two research projects I have worked on that partly investigated issues of race, ethnicity and racism in the Scottish context.[2] The first project examined the life and times of young Muslim men in Scotland and covered a number of topics including the young men's constructions of their gendered, racialised and ethnicised identities; their experiences following the events of 11 September 2001 as well as their negotiations of their religious and national identities. The second project sought to explore the needs and experiences of unaccompanied asylum-seeking children and young people in Scotland (Hopkins and Hill 2006). Part of this project assessed the young people's everyday experiences which included their accounts of racial harassment, discrimination and social exclusion. An important consideration in both projects was the extent to which the Scottish context influenced (or not) individual's experiences of racism, senses of identity or feelings of belonging.

To explore these issues, I have structured this chapter around two main sections. First, I explore continuities and connections between the Scottish context and that of the rest of the UK in terms of the geographies of race and racism. Second, I then consider the discontinuities and disjunctures between the Scottish context and that of the rest of the UK. Here, I focus upon issues relating to differences in the diversity, distribution and structure of the black and minority ethnic population, factors connected with Scottish politics and governance as well as ways in which Scottish national identities are articulated in order to highlight these discontinuities and explain the unique aspects of the Scottish context.

Continuities and Connections

There are a number of continuities and connections that can be drawn between Scotland and the other nations of the UK that suggest that there are some generalisations that can be made about the experiences of race and ethnicity in the UK context. For example, much work about the geographies of race and racism has emphasised the ways in which both concepts are socially constructed (Dunn 2001; Jackson and Penrose 1993; Smith 1989). As Smith (2005, 97) explains, the ways in which 'race' comes to have meaning in everyday life relies upon:

> ... the old and erroneous, yet enduring, idea that human populations can be subdivided into racial types on the basis of traits which are written on the body – skin colour, physique, facial features – which are taken to signal a common point of origin (a homeland, and/or a particular family history or ancestry), used to index a range of social practices (language, religion, traditions), and taken-for-granted as an axis of inequality'.

It is worth remembering, therefore, that 'there are no phenotypic (physically observable) or genetic differences within the human population that correspond with, or cause, cultural differentiation' (Smith 1989, 2) and it is through repetition 'that concepts like 'race' 'achieve a remarkable durability. In a sedimentary-like process the reinscription of social constructions can come to be widely accepted as unproblematic and as a natural given' (Dunn 2001, 292). Arguably, the processes through which peoples' everyday lives become racialised – through agents of socialisation such as the family and the media – work in similar ways across the UK.

Similarly, there is much evidence for the ways in which black and minority ethnic groups suffer due to the racism they experience on an everyday basis in

Scotland, as they do elsewhere in the UK. For example, in earlier work, Bowes, McCluskey and Sim (1990) documented the racism experienced by South Asian council tenants in Glasgow, Smith (1993) explored the ways in which racist behaviours and enactments permeate particular rural festivals in the Scottish borders, and Cant and Kelly (1995) sought to propose why there was a need for racial equality activity in Scotland by challenging a number of common-sense discourses about racism not being a Scottish problem. More recently, de Lima (2001, 2004) has documented the experiences of racism for those living in rural Scotland and Clayton (2005) sought to uncover the complex relationships between politics, sectarianism and racism. In my own work, I have highlighted the ways in which young Muslim men in Glasgow and Edinburgh experienced heightening levels of racism following the events of 11 September 2001 (Hopkins 2004, 2007a; Hopkins and Smith, forthcoming) as well as uncovering the everyday and institutional racism experienced by unaccompanied asylum-seeking children who are applying for asylum in Scotland (Hopkins and Hill 2006).

It is clear therefore, that racism is an everyday experience for many people in Scotland as it is for those living elsewhere in the UK. We know that the 'affective regulation of social belonging' (Noble 2005, 108) works to marginalise and exclude those who are perceived to be different as everyday behaviours and actions which are rude, undignified or threatening 'conspire to racialise, exclude and mark out differences on the basis of bodily markings, adornments and behaviours' (Hopkins and Smith, forthcoming). As I have shown, young Muslim men in Scotland often experience racism as a result of the ways in which bodily marking – their skin colour along with their dress and their choice to keep a beard – are interpreted and responded to by others (Hopkins 2004) So, 'in Scotland as in England, there is an all too familiar catalogue of insults, assaults, damage and harm, effected through both personal racism and political extremism, undermining health, welfare and wellbeing, and contributing to the separation and segregation of social life' (Hopkins and Smith, forthcoming).

These continuities and connections in the ways in which issues of race and racism permeate everyday life throughout Scotland as well as across the rest of the UK are crucial in the fight for equality, the battle against racism and the struggle for social justice. However, although these connections exist and it is crucial that we continue to explore and understand them, the uniqueness of the Scottish context also brings to light a number of discontinuities and disjunctures that highlight the ways in which race and racism in Scotland are different from elsewhere in the UK. It is the contention of this chapter that, since we already know much about the geographies of race and racism in the UK, it is those aspects that make Scotland unique that deserve particular attention. Such a claim is not new. Miles and Dunlop (1986, 24–5) noted that

... in Britain, the analysis of racialization, and of the role of racism in it, adopts almost exclusively a British perspective, thereby assuming that what has happened in England since 1945 has also happened in Scotland. Rarely, if ever, is any specific attention given to events in Scotland.

Moreover, as I argue in this chapter, such an approach is highly problematic as it ignores the influence of context, the significance of place and the importance of geography. For the remainder of this chapter, I therefore devote specific attention to the uniqueness of the Scottish context.

Discontinuities and Disjunctures

De Lima (2005, 143) argues that in Scotland racism has been 'portrayed as a predominantly 'English problem'. As a result of these assumptions about racism being an English problem, Rowena Arshad (1999, 221) has suggested that 'Scotland has avoided the realities of confronting racism as a door-step issue'. This is particularly concerning as the '... parallels between media and academic analyses' raise 'pertinent, and disquieting, questions about the ways in which selected academic discourses have fed into, and upon, populist soundbite understandings of race and ethnicity' (Alexander 2004, 544). These processes of displacement have worked in such a way to promote two related ideologies, one which sees Scotland as being 'white' and the other which sees Scotland as being free from any form of ethnic or racial tension. Although there are many black and minority ethnic communities living across Scotland, there are aspects of their experiences that differentiate them from their counterparts in England. I now therefore explore issues relating to the diversity, distribution and structure of the Scottish population as well as factors relating to Scottish politics, governance and national identities in order to explore some of the discontinuities and disjunctures between the Scottish and English contexts.

Diversity, Distribution and Structure

One of the important features of the Scottish context that differs from the situation in England relates to the diversity, distribution and structure of the black and minority ethnic population. The ethnic minority population of Scotland stands at just over 100,000, or 2 per cent of the population. The largest group are the Pakistanis, representing 31 per cent of the black and minority ethnic population, followed by the Chinese (16 per cent) and Indian (15 per cent) populations (Scottish Executive 2004). Clearly, this contrasts with the much larger populations of such groups for the UK as a whole where there were over 4.5 million people who identified themselves as belonging to black and minority ethnic groups (8 per cent of the total population), with 22 per cent of this population choosing the Indian group, followed by Pakistani (16 per cent), Black Caribbean (12 per cent) and Black African (10 per cent) (http://www.statistics.gov.uk). Not only is Scotland's ethnic minority community smaller than England's, the groups that constitute this population are also different from that found elsewhere in the UK. Most notably, the largest black and minority ethnic group in Scotland are the Pakistanis, with the Black Caribbean and African populations accounting for 2 per cent and 5 per cent respectively of the ethnic minority population. This contrast was observed some time ago now by Kearsley and Strivastava (1974,

110) who noted that 'Glasgow is exceptional among major British cities in that a very high proportion of its immigrant population is Indian and Pakistani, with the expected West Indian population notably absent'. So, the overall composition of Scotland's black and minority ethnic population differs from that in England. The case of unaccompanied asylum-seeking children in Scotland also highlights this insofar as the profile of this population is not simply a microcosm of the broader UK context and is instead more similar – in terms of age profile and countries of origin – to the profile of unaccompanied children found in the nations of Scandinavia (Hopkins and Hill 2006).

Scotland's black and minority ethnic population is concentrated in the main cities with 31 per cent residing in Glasgow and 18 per cent in Edinburgh, although 'a significant feature of the black and minority ethnic communities in Scotland is that they are scattered' (Cant and Kelly 1995, 10). In a sense, this distribution is also reflective of that elsewhere in the UK, with black and minority ethnic groups being concentrated in the main urban centres with smaller, more dispersed communities living in the rural areas (de Lima 2005). Furthermore, in Scotland – as in England, there are also neighbourhoods with high levels of ethnic residential clustering, most notably, Pollokshields in Glasgow, which has a minority ethnic population of nearly 50 per cent, the vast majority of which are Pakistani Muslims. So, although these observations point to continuities and connections with the situation elsewhere in the UK, the other issues of discontinuity and disjuncture relating to the diversity, distribution and structure of the population also act as a point of contrast.

As well as significant differences in diversity and distribution, the minority ethnic population of Scotland is also relatively middle-class compared with the same populations in England, and this relates to the ways in which the history of migration both into and out of Scotland is different from that of England, both in terms of the profile of new arrivals and their reasons for migration. Crucially, the migration of South Asian populations to Scotland was not 'as centrally related to the demands of the capitalist economy as in the instance of New Commonwealth migration to England in the same period and in the case of the Irish migration to Scotland in the 19th century' (Miles and Dunlop 1987, 125). These differences in context of migrants' arrival in Scotland has resulted in many of them being more middle-class than their counterparts in England. This is not to say that there is no poverty or deprivation amongst black and minority ethnic people in Scotland, however, in general, issues of deprivation, disadvantage and poverty are less salient in their lives compared with their counterparts south of the border. This is supported by Talib, a young Muslim man who had lived in England for many years before moving to Scotland:

> Talib: When I first came to Scotland I came in 96 and I came round this area just for a look around and … the first thing that I saw was how affluent people are, you know what I mean … that is like a major difference between Scotland and England in terms of like the Muslim community and the Pakistani community especially, you know what I mean. So in that way I don't think it's like England, and because there's a lot more affluence in Pollokshields, you haven't got to deal

> with same kind of problems and things that we've [poorer Pakistanis living in England] got to deal with, you know what I mean ...

This statement from Talib clearly demonstrates that issues of affluence and social class amongst Muslim communities in Scotland are a key point of contrast between Scotland and the rest of the UK. The majority of young Muslim men who participated in this project were middle-class, and although this may relate to how they were accessed during the project, Muslims in Scotland tend to live in more middle-class neighbourhoods compared with their counterparts in England. Related to the issue of class, there was also a sense that differences in the composition of the population influenced the likelihood of periods of urban unrest. This is highlighted by the views of Jamal and Latif during conversations about differences in levels of segregation and social class between Scottish and English Asians:

> Jamal: In Scotland you wouldn't get incidents like the Bradford riots, you know, that was a big thing, you wouldn't get anything like that in Scotland, you know

> Latif: ... in England there are a lot more fights between whites and blacks, there are a lot of fights that originally came from racism, whereas here [in Scotland] you hardly ever see anything like that, you know what I mean.

Scottish Politics and Governance

A second set of significant distinguishing features of the Scottish context relates to the particularities of Scottish politics and governance. An important argument here – proposed by Miles and Dunlop (1987, 119) – is that, in Scotland, there has been an absence of a 'racialisation of the political process in the period since 1945, rather than an absence of racism *per se*'. Therefore, there is a suggestion that political processes were racialised in England in a way that they were not in Scotland. This is often seen to be due to 'a political preoccupation with the religious divide between Catholic and Protestant Christians [which] displaced the racisms at the centre of English political affairs from Scottish affairs, producing what is now recognised as unwarranted complacency among Scottish decision-takers' (Hopkins and Smith, in press). Arguably this may be changing now with the Scottish Executive's promotion of the 'One Scotland, Many Cultures' campaign (see www.onescotland.com), and the increasing evidence that there is for racism to be taken seriously by all institutions in Scotland (Arshad 2003). However, the historical legacy of these processes are still likely to influence the ways in which issues of race and racism are taken into consideration, or not, within the Scottish political process.

As well as differences in the ways in which issues of race and racism come to enter the political arena, there are also two important discontinuities regarding the support for particular political parties in Scotland compared with England. The first point of contrast relates to the lack of success of the far right in Scotland. In Scotland, 'explicitly racist or fascist organisations made scarcely any impact during the 1970s' (Smith 1989, 153), and this has also continued through to the current times where the performance of the BNP (British National Party) at the Scottish Parliament elections in 2007 passed with little mention in the press or in political commentary with the party gaining little more than just over 4,000 votes in each of the eight Scottish regions. Although these numbers are by no means insignificant, Kabir, a young Scottish Muslim notes 'I think ... in Scotland, there's some things that so far haven't really been touched on in a big way, for example like, especially like the BNP – they haven't been very successful in Scotland'. Kabir therefore connects very positively with his Scottish heritage and sees organisations like the BNP as being associated with England rather than Scotland (Hopkins, forthcoming).

The second point of contrast relates to the ways in which 'political nationalism' (Miles and Dunlop 1986, 27) has been a stronger influence in Scotland compared with elsewhere in the UK (although Wales may act as an interesting point of similarity). In particular, the success of the Scottish National Party, bolstered by its promotion of 'civic nationalism' and the formation of 'Asians for Independence'[3] is an important point of discontinuity between Scotland and the rest of the UK. A number of young Muslim men who I interview stated that they would vote for the SNP. Faruk said 'It was the SNP I voted for', Omar stated 'yeah ... I voted SNP', Rehman recalled 'the last one I think I voted SNP' and Shafqat clarified that 'I voted for the SNP, yeah'. Furthermore, Bashir Ahmad – the initial founder Asians for Independence – was recently elected to the Scottish Parliament. He is Scotland's first Asian Member of the Scottish Parliament (MSP) and he was elected by standing for the Scottish National Party at the recent Scottish Parliament elections. Alongside the success of the SNP in Scotland, some young Muslim men also articulated a feeling of difference from certain aspects of England in terms of traditional arguments about Britain's north-south divide (e.g. Martin 1988). Here, there was an awareness of the ways in which Scotland is marginalised within the UK (along with northern England), and how Scotland – as Omar mentions – is not as 'well off' as England. Omar also mentioned how 'Labour is everywhere in Scotland' contrasting this with the Conservatives who win few seats in Scotland and are far more successful in the south of England. So, these differences in factors relating to politics, governance and power of Scotland compared with England also relate to differences in the economic circumstances of both places. For many, these differences also worked to the advantage of the SNP who gained increased support building upon the sense of difference experienced by many people in Scotland

A final point that distinguishes Scotland from the rest of the UK is its unique legislative framework. Matters devolved to the Scottish Parliament include health, education and training, local government, social work, housing, planning, tourism and economic development, some aspects of transport, police and fire services, the environment, sports and the arts and the natural and built environment. At the same time, Westminster responsibilities include, amongst others: immigration and nationality; defence and national security; social security; trade and industry; employment legislation; equal opportunities and energy. So, legislation relating to black and minority ethnic groups tend to be matters reserved to Westminster, yet there are also issues specific to the experiences of particular groups that are devolved to Scotland. An example here is the legislation relevant to asylum-seeking children in Scotland where there is legislation specific only to Scotland (for example, the Children (Scotland) Act 1995) as well as legislation that are common to the UK (e.g. immigration legislation). This may appear to be a minor point; however, the different legal, educational, health and social work contexts in Scotland may have an important influence on how issues of equality and diversity are embedded within institutional frameworks, how they are understood by employees and how they permeate into people's everyday experiences of social life.

Scottish National Identities

Connected closely with the second set of issues, the third main distinguishing feature of the Scottish situation is the distinctive ways in which black and minority ethnic groups relate to the construction and contestation of Scottish national identities. David McCrone (2002, 303) has noted that issues of 'what we might call 'ethnicity' and 'national identity' rarely connect with each other' in the academic and policy literature, however, recent research that has explored these issues has contributed much to understandings of the formation of 'Scottish Asian' and/ or 'Scottish Muslim' identities (Cassidy et al. 2006; Hopkins 2007b; Saeed et al. 1999). In particular, this work has shown how many ethnic minorities in Scotland identify positively with the nation, adopt hyphenated identities of various forms and see themselves as Scottish in many ways. This can be contrasted with the ambivalence sometimes associated amongst black and minority ethnic groups for the identities of 'Englishness' or 'Britishness'.

There could be a number of reasons for the ways in which Scottishness is often embraced by black and minority ethnic groups and this is an important issue for future research. However, an important point of differentiation between Scotland and England is, as mentioned earlier, the way in which 'political nationalism' (Miles and Dunlop 1986, 27) has shaped the Scottish context. Miles and Dunlop (1986) mention that the Scottish National Party increased its support base in the 1960s and 1970s. Even today, the SNP remains strong and has recently been the party with the most MSPs elected to the Scottish Parliament. As the most dominant voice of nationalism in Scotland, the SNP tends to promote a version of nationalism which sees everyone living in Scotland as having a valued contribution to make, regardless of race, ethnicity or place of birth (www.snp.org. uk). This can clearly be contrasted with the lack of any such vision of nationalism elsewhere in the UK, although the situation in Wales may act as a useful point of comparison. Instead, the voices we tend to hear about in promoting Englishness and Britishness often include the BNP's 'goal of a 'white' state' (Hopkins 2007a, 1127), Norman Tebbit's controversial 'cricket test', or John Major's reference to 'long shadows on cricket grounds' and 'warm beer'. These may appear as isolated comments or issues only relevant to political quarters; however they stand in stark contrast to the forms of nationalism, identity and community that tend to dominate within Scotland.

This contrast between the Scottish and English contexts is one of the reasons why I feel that black and minority ethnic groups in Scotland tend to identify strongly with Scotland and Scottishness. The vast majority of the young Muslim men who participated in focus groups and interviews in Glasgow and Edinburgh self-identified as Scottish Muslims, and they drew upon a range of markers of Scottishness in making such claims. Kiely et al. (2001) found that there are a number of markers that people use in this regard including: place of birth, ancestry, place of residence, length of residence, upbringing and education, name, accent, physical appearance, dress and commitment to place. The young Muslim men tended to draw upon a range of these markers in asserting their Scottish Muslim identities and this has been paralleled in work by Cassidy et al. (2006) and Saeed, Blain and Forbes (1999). The fact that black and minority ethnic groups tend to identify positively with Scotland and Scottishness is perhaps not exceptional, however, this contrasts with the sense of identities and affiliations of the same groups living elsewhere in the UK.

Conclusions

Overall then, this chapter has explored a number of important characteristics of the Scottish context in terms of race and racism to make a number of observations about the ways in which such issues are experienced differently in Scotland compared with elsewhere in the UK. In particular, I have argued that there are crucial discontinuities and disjunctures between the Scottish context and those found elsewhere in the UK that deserve careful consideration for researchers, policy makers and practitioners interested in understanding the experiences of black and minority ethnic groups. Specifically, I have proposed that the diversity, distribution and structure of minority ethnic groups as well as issues connected with Scottish politics and governance along with the construction and contestation of Scottish nation identities are crucial aspects of social and cultural life in Scotland that differentiate it from other places in the UK.

By conclusion, I would like to suggest that there are two issues in particular that these points of discontinuity and disjuncture raise. First, by highlighting the uniqueness of the Scottish context, I have emphasised that, although race and racism matter, so too does geography. Moreover, in identifying that Scotland is not simply a microcosm of the UK, it is important to emphasise the danger of generalising about the British context. Unfortunately, to date, this is a highly problematic gesture adopted by many academics who seek to generalise their studies of ethnic minority communities in London or Manchester to represent the experiences of all black and minority ethnic communities in Britain. Indeed, this is a matter for further investigation, as the divisions present within England between 'north' and 'south' may also have important influences over the construction and contestation of racial and ethnic identities and national affiliations. Overall then, the key argument of this chapter is that it is not possible to transpose experiences of race, racism and ethnicity in England and assume that they are the same as they would be in Scotland.

For a number of decades now, social geographers have studied ethnic residential clustering in cities (e.g. Jackson and Smith 1981; Johnston et al. 2002; Phillips 1998) and I recently noted that 'the measuring, mapping and monitoring of residential segregation continues apace, yet few studies have sought to explain or understand more fully the dynamics of ethnic residential clustering and the influence this has on the everyday lives of local residents' (Hopkins 2007c, 168). The ethnic composition of different communities and the dynamics present within these communities are likely to have important consequences for social interaction, community relations and experiences of everyday life. This leads me on to my second main conclusion. The differences in the diversity, distribution and structure of the black and minority ethnic population in Scotland is likely to have consequences for the everyday nature of social interaction in segregated as well as integrated settings. Building on my work with young Muslim men in Glasgow and Edinburgh, I noted the importance of locality and how the strong presence of white working-class young men as well as the relative absence of young black men is likely to influence how they construct their masculine identities (Hopkins 2006). In contrast, some young Muslims in England may construct their identities in the context of other young white men and young black men too and therefore have different social contexts through which to construct their identities. Future research could usefully explore further the uniqueness of the Scottish context and the influence this has on the ways in which race, racism and ethnicity permeate everyday life, influence interaction and structure people's experiences of and engagements with difference. In exploring these and other issues, researchers, policy-makers and practitioners will hopefully be able to better understand the difference that Scotland makes in terms of politics, race and nation.

Notes

1   The Scottish Executive is now called the Scottish Government
2   The first research project was about the geographies, politics and identities of young Muslim men living in Glasgow and Edinburgh in Scotland (see Hopkins 2004, 2006, 2007, a, b, c). This doctoral project was funded by the Institute of Geography at the University of Edinburgh and supervised by David Howard and Susan Smith. The second project was about the needs and experiences of unaccompanied asylum-seeking children in Scotland. This was funded by the Scottish Refugee Council and was conducted in collaboration with Malcolm Hill (see Hopkins and Hill 2006). Many thanks to the various people who participated in both projects.
3   Asians for Independence was founded in 1995 by Bashir Ahmad in order to encourage support for the SNP amongst Asian communities in Scotland. The SNP have also recently launched 'Young Asian Scots for Independence' in order to encourage political participation amongst young Asian Scots.

References

Alexander, C. (2004), 'Imagining the Asian Gang: Ethnicity, masculinity and youth after "the riots"', *Critical Social Policy* 24 4, 526–49.

Arshad, R. (1999), 'Racial Inclusion and the Struggle for Justice', in G. Hassan and C. Warhurst (eds), *A Different Future: A Modernisers Guide to Scotland* (Glasgow: Big Issue), 218–28.

Arshad, R. (2003), 'Race Relations in Scotland', in J. Crowther, I. Martin and M. Shaw (eds) *Renewing Democracy in Scotland* (Leicester: NIACE), 131–4.

Bowes, A., McCluskey, J. and Sim, D. (1990), 'Racism and Harassment of Asians in Glasgow', *Ethnic and Racial Studies* 13:1, 71–91.

Cant, B. and Kelly, E. (1995), 'Why is There a Need for Racial Equality Activity in Scotland?', *Scottish Affairs* 12, 9–26.

Cassidy, C., O'Connor, R. and Dorrer, N. (2006), *Young People's Experiences of Transition to Adulthood: A Study of Minority Ethnic and White Young People* (York: Joseph Rowntree Foundation).

Clayton, T. (2005), 'Diasporic Otherness': Racism, sectarianism and "national exteriority" in modern Scotland', *Social and Cultural Geography* 6:1, 99–116.

Dunn, K. (2001), 'Representations of Islam in the Politics of Mosque Development in Sydney', *Tijdschrift voor Economische en Sociale Geografie* 92:3, 291–308.

Hopkins, P.E. (2004), 'Young Muslim Men in Scotland: Inclusions and exclusions', *Children's Geographies* 2:2 257–72.

Hopkins, P.E. (2006), 'Youthful Muslim Masculinities: Gender and generational relations', *Transactions of the Institute of British Geographers* 31. 337–52.

Hopkins, P.E. (2007a), 'Global Events, National Politics, Local Lives: Young Muslim men in Scotland', *Environment and Planning A* 39:5, 1119–33.

Hopkins, P.E. (2007b), '"Blue Squares", "Proper" Muslims and Transnational Networks: Narratives of national and religious identities amongst young Muslim men living in Scotland', *Ethnicities* 7:1, 61–81.

Hopkins, P.E. (2007c), 'Young People, Masculinities, Religion and Race: New social geographies', *Progress in Human Geography* 31:2, 163–77.

Hopkins, P.E. (in press), 'Young, Scottish, Muslim and Male: A portrait of Kabir', in G. Jeffrey and J. Dyson (eds) *Telling Young Lives: Portraits in Political Geography* (Philadelphia: Temple University Press).

Hopkins, P.E. and Hill, M. (2006), *'This is a Good Place to Live and Think about the Future': The Needs and Experiences of Unaccompanied Asylum Seeking Children and Young People in Scotland* (Glasgow: Scottish Refugee Council).

Hopkins, P.E. and Smith, S.J. (in press), 'Scaling Segregation; Racialising fear', in R. Pain, and S.J. Smith (eds) *Fear: Critical Geopolitics and Everyday Life* (Aldershot: Ashgate).

Jackson, P. and Penrose, J. (1993), 'Introduction: Placing "Race" and '"Nation"', in P. Jackson and J. Penrose (eds) *Constructions of Race, Place and Nation* (London: UCL Press), 1–23.

Jackson, P. and Smith, S.J. (eds) (1981), *Social Interaction and Ethnic Segregation* (London: Croom Helm).

Johnston, R. Forrest, J. and Poulsen, M. (2002), 'Are there Ethnic Enclaves/Ghettos in English Cities?', *Urban Studies* 39:4, 591–618.

Kearsley, G. and Strivastava, S.R. (1974), 'The Spatial Evolution of Glasgow's Asian Community', *Scottish Geography Magazine* 90, 110–24.

Kelly, E. (2003), 'Integration, Assimilation and Social Inclusion: Questions of faith', *Policy Futures in Education* 1:4, 686–98.

Kiely, R., Bechhofer, F., Stewart, R. and McCrone, D. (2001), 'The Markers and Rules of Scottish National Identity', *Sociological Review* 49:1, 33–55.

Lima, P. de (2001), *'Needs not Numbers': An Exploration of Minority Ethnic Communities in Scotland* (Scotland: Commission for Racial Equality).

Lima, P. de (2003), 'Beyond Place: Ethnicity/race in the in the debate over social exclusion/inclusion in Scotland', *Policy Futures in Education* 1:4, 653–67.

Lima, P. de (2004), 'John O'Groats to Land's End: Racial equality in rural Britain', in N. Chakraborti and J. Garland (eds) *Rural Racism* (Devon: Willan Publishing), 36–60.

Lima, P. de (2005), 'An Inclusive Scotland? The Scottish Executive and Racial Inequality', in G. Mooney and G. Scott (eds) *Exploring Social Policy in the 'New' Scotland* (Bristol: Polity Press), 135–57.

Martin, R. (1988), 'The Political Economy of Britain's North-South Divide', *Transactions of the Institute of British Geographers* 13, 389–419.

McCrone, D. (2002), 'Who Do You Say You Are? Making Sense of National Identities in Modern Britain', *Ethnicities* 2:3, 301–20.

Miles, R. and Dunlop, A. (1986), 'The Racialization of Politics in Britain: Why Scotland is different', *Patterns of Prejudice* 20:1, 23–33.

Miles, R. and Dunlop, A. (1987), 'Racism in Britain: The Scottish dimension', in P. Jackson (ed.) *Race and Racism* (London: Unwin Hyman), 119–41.

Netto, G., Arshad, R., Lima, P. de, Diniz, F.A., McEwan, M., Patel, V. and Syed, R. (2001), *Audit of Research on Minority Ethnic Issues in Scotland from a 'Race' Perspective* (Edinburgh: Scottish Executive).

Noble, G. (2005), 'The Discomfort of Strangers: Racism, incivility and ontological security in a relaxed and comfortable nation', *Journal of Intercultural Studies* 26:1, 107–20.

Phillips, D. (1998), 'Black Minority Ethnic Concentration, Segregation and Dispersal in Britain', *Urban Studies* 35:10, 1681–702.

Saeed, A., Blain, N. and Forbes, D. (1999), 'New Ethnic and National Questions: Post-British identities among Glasgow Pakistani teenagers', *Ethnic and Racial Studies* 22:5, 821–44.

Scottish Executive, (2004), *Analysis of Ethnicity in the 2001 Census: Summary Report* (Edinburgh: Office of the Chief Statistician).

Smith, S.J. (1989), *The Politics of 'Race' and Residence* (Cambridge: Polity Press).

Smith, SJ. (1993), 'Bounding the Borders: Claiming space and making place in rural Scotland', *Transactions of the Institute of British Geographers* 18, 291–308.

Smith, S.J. (2005), 'Black: White', in P. Clokel and R. Johnson (eds) *Spaces of Geographical Thought* (London: Sage), 97–118.

Chapter 10

Managing 'Race' in a Divided Society: A Study of Race Relations Policy in Northern Ireland

Peter Geoghegan

Introduction

Sectarian division is such an integral, though unfortunate, part of the lived experience of Northern Ireland that it has long been considered a subject of mirth as well as conflict. A popular Belfast joke concerns a tourist walking through a no man's land between Protestant and Catholic neighbourhoods. The unsuspecting flaneur is grabbed by a masked man who asks 'Are you a Protestant or a Catholic?' The tourist answers: 'I'm a Jew.' Without a moment's hesitation the masked man enquires whether the tourist is 'A Catholic Jew or a Protestant Jew'. While the ridiculousness of trying to force a Jew into the Catholic/Protestant binary serves to mock this distinction it also attests to the power, and relative impermeability, of sectarian division. This joke has been recycled many times with different ethnic, religious and racial groups but the denial of the existence of 'other traditions' in Northern Ireland remains the salient point (Hainsworth 1998; McVeigh 1998).

Although sectarian tensions in Northern Ireland have not gone away since the signing of the Good Friday Agreement[1] in 1998, the ethnic conflict in the region has stabilised (Shirlow and Murtagh 2006). As efforts are made to regularise social and economic life post-conflict public policy has sought to emphasise the need for democracy and pluralism (Graham and Nash 2005). Over the last decade confidence in the peace process has grown and the region has experienced a minor economic resurgence. Northern Ireland, for years considered an exceptional case in the UK economy, is increasingly being repositioned as part of global economic restructuring (Shirlow 2006). Although previously an area of net out migration, the upturn in the economy has been accompanied by significant year on year increases in the number of people migrating to the region[2] (Jarman 2006; Mussano 2004). This increase in migration has coincided with an emerging recognition of groups and people not easily classified as Protestant and Catholic. Consequently issues of 'race',[3] and racialised difference, which were effectively marginalised and ignored during 'the troubles', have emerged as concerns animating government and civil society over the last decade (Connolly and Kennan 2002; Lentin and McVeigh 2006; McVeigh and Lentin 2002).

The focus of this chapter is on 'Race Relations' policy and practices in the specific, and quite unusual, governing structures of post-Agreement Northern Ireland. Previous research has highlighted the importance of both institutional structures and public policy discourses in either promoting or hindering 'Race Relations' (Anthias and Yuval-Davis 1992; Ben-Tovim et al. 1992). This chapter draws on interviews with civil servants and analyses of institutional structures, public policy and government statements to interrogate the nature, and limits, of 'Race Relations' policy in Northern Ireland. Tracing the emergence of 'Race Relations' policy, and institutions in which it is enacted, shall show how this policy has developed from a minor concern to a significant issue in the last decade. The analysis in this chapter demonstrates how the politics of 'Race Relations' is compromised by a complex, fragmented institutional structure which hinders its effective implementation. Although the Agreement sought to move society beyond the conflict the political settlement privileges sectarian political identities. 'Race Relations' in Northern Ireland are increasingly being tied in with a desire to construct the society as 'normal' but institutional structures and practices grounded in sectarian division, such as the reallocation of respondents in the Northern Ireland Census, seriously undermine this project.

'Race Relations' Policy in Northern Ireland

The history of 'Race Relations' policy in Northern Ireland is much briefer than in the rest of the UK. When the 1965 Race Relations Act was being drafted the Protestant dominated Stormont government requested that Northern Ireland be excluded from the Act on the grounds that religion rather than 'race' represented the most serious locus of discrimination (Dickey 1972). Although the public stance was that 'race' was not an issue worthy of legislative attention at the time of the drafting of the Act there seems to have been a concern among many Unionist politicians that Catholics could be defined as an ethnic group and use the legislation to seek redress against the government. For this reason the Stormont government, which notably did not lobby for specific legislation against discrimination on the grounds of religion, opted out of the Act (Dickey 1972, 135). When anti-discrimination legislation, such as the Fair Employment Acts,[4] was introduced in Northern Ireland, from the late 1960s on, it was only concerned with discrimination on the grounds of religion and political belief. While both Community Relations[5] and Fair Employment shared many common features with 'Race Relations' policy in the rest of the UK they were orientated around sectarian division and discrimination on grounds of religion and political belief rather than 'race' and racism.

A consensus on the need to implement some form of 'Race Relations' policy in Northern Ireland emerged in the 1990s (McVeigh 2002). The Standing Advisory Commission on Human Rights[6] recommended that 'Race Relations' legislation parallel to that in Britain be introduced (SACHR 1990). Eventually, in 1997, the Race Relations (NI) Order (RRO) was passed into law. McVeigh (2002) argues that civil society pressure led to the introduction of this legislation yet the most important factor seems to have been the election of a new Labour government in the UK in 1997. Unlike the Conservative government of the 1980s and early 1990s Labour was committed to supporting 'Race Relations' policy (Solomos 2003) and once in power it quickly moved to have the 'Race Relations' Act extended to cover Northern Ireland. The RRO mirrors the provisions of the 1976 Race Relations Act UK, outlawing discrimination on the grounds of colour, race, nationality or ethnic or national origin. Although 'Race Relations' legislation has been accused of essentialising socially constructed groups (Malik 1996), and possibly even contributing to racism (Gilroy 1987, 1990), its extension to Northern Ireland provided important legal protection for members of groups not easily classifiable as Catholic or Protestant (Connolly 2000).

The Agreement (1998) significantly changed the social and political environment for everyone in Northern Ireland. While a peaceful resolution of 'the troubles' was the primary aim of the Agreement it also provided important recognition for those beyond the 'two traditions'. During negotiations leading to the signing of the Agreement the British government pushed for the inclusion of a commitment to individual equality (McCrudden 1999). As a direct result an equality agenda based on 'the right to equal opportunity in all social and economic activity, regardless of class, creed, disability, gender or ethnicity' was included in the final settlement (Northern Ireland Office 1998, 17). Subsequently the Northern Ireland Act (1998) brought this commitment into political reality. Section 75(1) of this act commits a public body to ensuring equality of opportunity 'between persons of different religious belief, political opinion, racial group, age, marital status or sexual orientation.' The second clause of Section 75 states that 'a public authority shall in carrying out its functions relating to Northern Ireland have regard to the desirability of promoting good relations between persons of different religious belief, political opinion or racial group.' The commitment to racial equality and the promotion of good 'race' relations in Section 75, along with the Race Relations Order, are the key pillars of 'Race Relations' policy in post-Agreement Northern Ireland.

While to the outside world the terms Community Relations and Good Relations may appear virtually identical in the Northern Ireland context this shift in terminology marks a tectonic shift in identity politics. Community Relations has been heavily critiqued for its failure to address the economic and social basis of the conflict and its tendency to ignore those groups not perceived as Catholic or Protestant (McVeigh 2002). In the post-Agreement era there has been an emerging acceptance that the traditional sectarian binaries are unhelpfully reductive (Nash 2005). Increased cultural diversity, as well as the acknowledgement of the existence of groups and individuals who cannot, or do not want to be, defined in Catholic/ Protestant terms, has facilitated the emergence of the term Good Relations to include both Community Relations and 'Race Relations'. In less than a decade Northern Ireland has gone from a situation in which discrimination on the grounds of 'race' was not illegal to one in which 'Race Relations' policy is given a relatively central location in post-Agreement public policy.

'Race Relations' as Normalising Northern Ireland

The emergence of a concern for 'Race Relations' in public policy is closely tied in with the desire to 'normalise' social relations following years of sectarian conflict. During 'the troubles' the tendency to view Northern Ireland as 'a place apart', literally unrecognisable in comparison with other 'Western' contexts, was commonplace (Little 2003). Certainly many features of government and society such as the Diplock courts[7] and the presence of the army on the streets marked Northern Ireland as a state of exception removed from the 'normal' social space(s) of Britain and Ireland (Agamben 1998; Kearns 2006). Following the cessation of large scale armed conflict, and the tentative political progress since, there have been attempts to re-construct Northern Ireland as a 'normal' capitalist space (Graham and Nash 2005; Shirlow 2006; Shirlow and Murtagh 2006). The association of 'Race Relations' and 'normalisation' is evident in a speech delivered by the then Secretary of State Peter Hain on 21 September 2005:

> The vision we all share for Northern Ireland is of a normal civic society in which all individuals are treated as equals. Problems are resolved through dialogue and the state is impartial between contending claims. A Northern Ireland where the community or church you come from, your political opinion, race, gender, sexuality, age or disability makes no difference to where you are wanting to go.

In this speech an image of a 'normal' society based on equality and respect for diversity is constructed. A prominent feature of this normal space is Good Relations or the acceptance of difference based on 'political opinion, race, gender, sexuality, age or disability'. The presence of racialised diversity, and the promise of racial equality, is held up as evidence that Northern Ireland is moving beyond the dark days of sectarian division.

The contrast between a past of violence and intolerance and a potential future of normality characterised by the presence of racialised diversity emerged as a key theme in a number of interviews.[8] As one Community Relations policy maker noted:

> There have been examples in the past, in the height of the Troubles, Vietnamese people. When the first boat load was sent here, they wouldn't get off the boat, they said 'you must be joking, we are coming from war-torn Vietnam'. The first boat load went back to Liverpool, which I found hilariously funny. That's how bad it was … We are now, thankfully, I think, a growing society, becoming more normal, more multicultural. My view is that that is a sign of progress and peace. (Interview with Head of Policy, Community Relations Council, Belfast, 2005)

In the 1970s a relatively significant number of Vietnamese people came to Northern Ireland to escape the situation in their home country. Although many eventually settled there was initially some resistance to relocating to a region in conflict. In recent years Northern Ireland has moved away from this violent past to a better, more 'normal', 'now'. Cultural diversity and multiculturalism, represented by the presence of racialised 'Others', is interpreted as a sign that the violence of 'the troubles' has been left behind. The management of this diversity, through 'Race Relations' policy and practice, is supposedly something that takes place within the bounds of, and is defining of, a 'normal' society.

Practice of 'Race Relations' in Awkward Institutional Structures

The contemporary governing structure, within which 'Race Relations' policy is enacted, is the end result of a series of attempts to foster peace and political parity in Northern Ireland by changing its political institutions (Shirlow and Murtagh 2006). By the time of the Agreement it was widely acknowledged that the creation of power-sharing institutions which reflected the wishes of 'both communities' was the only way to secure a stable future (Thompson 2002). Following the Agreement Northern Ireland became a devolved constituent region of the UK and a devolved legislature, the Northern Ireland Assembly (NIA), was established. Alongside the NIA a power-sharing Executive, directly answerable to the Assembly, was created. The stated aim of the Executive is 'not only to administer certain government departments but also to promote cultural and political equality' (Shirlow and Murtagh 2006, 36). Under the terms of the Agreement economic and social issues were devolved to a new Executive which is made up of eleven devolved departments. However, the Agreement did not devolve all powers to the Executive. Instead constitutional and security issues remain under the control of the Northern Ireland Office (NIO) which is directly answerable to the Home Office and the Secretary of State for Northern Ireland rather than the Executive.

'Race Relations' policy is being enacted within both the devolved and non-devolved administrations. Although all devolved departments are subject to the equality and good relations duties contained in Section 75 and the Race Relations (NI) Order the main responsibility for 'Race Relations' policy rests with the key ministry of the Office of the First Minister and Deputy First Minister (OFMDFM).[9] In 2000 the Race Unit, the first governmental institution specifically focused on 'Race Relations' in Northern Ireland, was formed in OFMDFM 'to reduce social exclusion amongst minority ethnic people' (OFMDFM 2000, 2). Since renamed the Racial Equality Unit (REU) this institution has produced *A Racial Equality Strategy for Northern Ireland* (OFMDFM 2006) as well as liaising with minority ethnic representative groups and statutory bodies. Working alongside the REU in the Equality Directorate of OFMDFM is the Community Relations Unit (CRU). Although the CRU is primarily concerned with relations between Catholics and Protestants *A Shared Future* commits the CRU to addressing both sectarianism and racism (OFMDFM 2005, 5). While the REU, and to a lesser extent the CRU, play integral roles in the production of devolved policy on 'Race Relations' the non-devolved administration, the NIO, is not subject to devolved policy, instead its institutions are subject to Home Office 'Race Relations' policy. Alongside these government institutions a number of non-governmental public bodies, or 'quangos', such as the Community Relations Council, the Equality Commission and the Northern

Ireland Human Rights Commission, have particular, though often overlapping, remits around 'Race Relations'.

The political settlement in Northern Ireland has produced a particularly complex institutional structure (Lentin and McVeigh 2006). Although this structure was designed to manage Nationalist/Unionist relations it encompasses 'Race Relations' policy and practice. The awkward reality of trying to engage with 'Race Relations' policy in this institutional landscape is clearly expressed by a member of the Community Safety Unit:

> The problem that we face in Northern Ireland is that for community safety to work in a kind of cohesive way, it needs to have a landscape that isn't completely fragmented, in the way that it is here. Issues like race hate crime really do need a joined up approach, because it is a complex issue. And it's about prevention, support, protection. So all of these things have to be connected. Northern Ireland is a very bad place for things to be connected, we have so many silos, so many government departments, so many councils. I used to work in Liverpool, I don't know how much you know about Merseyside but it has got roughly the same population as Northern Ireland, 1.6 million people. But if you compare the structures on Merseyside: there are five councils, not twenty-six. (Interview with Community Safety Unit, February 2006)

The Community Safety Unit (CSU) is a department within the Criminal Justice Directorate of the non-devolved NIO. Alongside a commitment to tackling anti-social behaviour and sectarian hate crime the CSU is charged with trying to reduce racist attacks and manifestations of racism. However, the lack of effective connections between the vast array of institutions that make up the post-Agreement governing structure makes it particularly difficult to fulfil this remit on 'race hate crime'. The fragmentation evident in the large number of District Councils,[10] government bodies and other political institutions involved in different aspects of 'Race Relations' policy negatively impacts on attempts to address issues like racist attacks and racially motivated crime.

Reflecting on attempts to tackle social exclusion through planning policy Geraint Ellis (2001, 407) opines that it is 'somewhat ironic that the political climate of the peace process that has allowed these issues to emerge on the policy agenda has also necessitated an administrative structure that frustrates the process of implementation.' Although Ellis is commenting on the fact that responsibility for planning in Northern Ireland does not fall within a single department his point is equally valid in considerations of 'Race Relations' policy. Here the set-up is so complex, and institutional name changes so frequent, that one interviewee who held a very senior position in the Equality Directorate in OFMDFM was unable to remember the title of one of the units in the directorate. The peace process, and political changes accompanying it, provided an opportunity for 'Race Relations' policy to emerge in Northern Ireland for the first time but these initiatives have been stymied by the high level of fragmentation, and confusion, within the institutional structures bequeathed by the Agreement.

The division between the devolved and non-devolved administrations is particularly problematic, as a senior civil servant explains:

There is a Chinese Wall, if you like, between the NIO and the devolved administration in practical matters. If I access my computer, the devolved administration is all on one computer network, the NIO is a separate one ... I can log onto a dial system which will give me a telephone number of anybody here (devolved administration) but not there (NIO). (Interview with Department of Social Development, Belfast, January 2006)

The existence of distinct computer systems and telephone directories attests to the sense of disconnection, of a 'Chinese Wall' dividing the two administrations. This separation is not just cumbersome and inconvenient; it has important implications for 'Race Relations' policy. While individuals from devolved and non-devolved departments might work together on bodies like the Racial Equality Forum, the separate telephone networks are symptomatic of the lack of opportunities for everyday interaction between 'Race Relations' practitioners working in the two administrations. Consequently policy initiatives and advances made in one administration may not be carried over into the other, hindering attempts to promote best practice across the board.

Equality is a devolved issue in Northern Ireland but policing and immigration, two areas which have huge implications for equality issues in general and the situation of racialised groups in particular, remain outside the direct control of the devolved administration. Policing, a key nexus in the management of 'race', has often been implicated in processes of racialisation (Keith 1993; Jackson 1993; 1994; Solomos 2003). In Northern Ireland criminal justice is a reserved matter; it is not yet devolved but potentially could be in the future. Consequently the Police Service of Northern Ireland (PSNI) is located within the institutional structures of the NIO. This produces an anomalous situation whereby *A Racial Equality Strategy for Northern Ireland* refers to 'Race Relations' in the PSNI (OFMDFM 2006) but devolved policy has no official jurisdiction over police practice. Although policing may be devolved as the political situation improves, at the present time the devolved administrations 'Race Relations' structures and policies are effectively separated from policing policy and practice.

Immigration and border control also play crucial roles in processes of racialisation (Kemp 2004). As immigration is an excepted issue, can never be devolved, Northern Ireland is bound to enforce increasingly tight Home Office restrictions. The UK wide 'Managed Migration ' policy (Home Office 2005) was produced without policy makers in institutions such as the Racial Equality Unit having any influence over the content of the policy or the option to opt out of it. The lack of devolved control over asylum policy was most graphically illustrated when, from 2000 on, asylum applicants were locked up alongside dissident paramilitaries in Maghaberry Prison, Co. Derry (Tennant 2001). As this procedure was in accordance with Home Office policy for dealing with asylum seekers in the absence of other secure accommodation, the Assembly, despite expressing cross-party condemnation of the practice, was powerless to intervene. It was only following negative media attention that the NIO established temporary holding centres on alternative sites in Belfast (Lentin and McVeigh 2006). This example illustrates that while OFMDFM is making policy commitments to racial equality Home Office policy sometimes aids processes of racialisation and discrimination.

Although the paradoxical relationship between commitments to 'Race Relations' and practices which encourage racialisation has become the hallmark of the actions of many contemporary states (Lentin 2001; Solomos 2003) what makes the Northern Ireland context distinct is that the devolved government will never have the opportunity to select its own immigration policy.

The Agreement and the Reproduction of Sectarian Identities

The public administration of 'Race Relations' in Northern Ireland is undermined by a fundamental tension between a discourse of Good Relations and normalisation stressing equality and social diversity and a set of structures and practices which privilege sectarian identities. Although the Agreement includes a commitment to diversity beyond the 'two traditions' the text itself is a product of sectarian division and, in many important respects, continues to reproduce this bifurcation (Shirlow and Murtagh 2006). The privileging of sectarian identities is particularly evident in the workings of the political institutions bequeathed by the Agreement (Little 2003). In accordance with the consociational[11] principles of the Agreement, the Northern Ireland Assembly (NIA) demands that all elected representatives declare themselves Nationalist, Unionist or Other. In order for a bill to become law it requires the support of 60 per cent of one ethno-national bloc and 40 per cent of the other. This institutionalisation of sectarian division is replicated in the devolved Executive; here the eleven departments are a direct reflection of the perceived numerical division between Catholics and Protestants (Ellis 2001). As Northern Ireland is crudely assumed to be divided between Protestants and Catholics in a 55:45 ratio the eleven departments are, in practice, split into six Unionist controlled and five headed by Nationalists.[12] Clearly a significant portion of the fragmentation and confusion outlined previously is a direct result of this attempt to create institutional structures which recognise sectarian identities above all others.

The peace process helped create a climate of relative stability in which new discourses of diversity and difference, such as 'Race Relations' policy, could emerge but institutional structures and practices continue to reproduce the divisions the Agreement ostensibly sought to ameliorate. A particularly striking example of the reproduction of sectarian identities comes from the 2001 Northern Ireland Census. The census is both an important public policy tool and a political practice through which putatively 'natural' category constructions are reified (Hannah 2001). In a divided society the census can often assume even greater significance as the decline in numerical strength of one group is often mirrored by the demographic growth of another (Anderson et al. 2004). Sectarian head counting has been integral to the governing of Northern Ireland since its foundation; in fact the state was specifically constructed to ensure a 66 per cent Protestant and 33 per cent Catholic demographic balance (Anderson and Shuttleworth 1998). The religious question on the Northern Ireland Census is used to compute the Catholic/Protestant demographic breakdown, helping to ensure compliance with Fair Employment legislation (Gallagher 1992). The census is also a highly politicised process. There is an assumed high correlation between religious affiliation and political identities with numerical strength often used as a proxy measure for support for alternative Nationalist/Unionist political visions (Anderson et al. 2004).

Although the Northern Ireland Census has been used to quantify the 'two traditions', from 1971 on respondents have shown much less willingness to commit themselves to a religious identity (McNair 2006). The figure for 'No religion/not-stated' responses to the religious question in the census rose from 9.4 per cent in 1971 to 11 per cent in 1991. In the 2001 census 13.9 per cent of respondents (see Table 10.1), when asked if they were 'Protestant', 'Roman Catholic' or 'Other', chose the final option. When this lack of self-allocation to one religious group or another threatened to undermine the ability of the census to give a definite figure on the Catholic/Protestant breakdown[13] the Northern Ireland Statistical Research Agency (NISRA) took the decision to reallocate the 'No Religion' group based on 'communal background' (Hadden 2003). This occurred firstly on the basis of a question asking which religion the respondent was brought up in. This reduced the figure from 13.9 per cent to 7.5 per cent. Then a computerised matching process known as 'donor imputation' was used to assign a religious community to the remaining 'No Religion' responses on the basis of a range of items, including geographical proximity to someone who did state their religion or communal background (Hadden 2003). As a result a total of 127,000 respondents who refused a religious identity were reassigned as Protestants and 59,000 as Catholics. After reallocation the percentage of Catholics in Northern Ireland stood at 43.8 per cent, Protestants made up 53.1 per cent and None/Non-stated 2.7 per cent (see Table 10.1).

Table 10.1 Reallocation of 'non-religious' into sectarian identities in Northern Ireland Census 2001

	Recorded	**Reallocated**
Catholics	678,500 40.3%	737,450 43.8%
Protestants	767,900 45.8%	895,400 53.1%
None/not stated	233,900 13.9%	45,850 2.7%

Source: Hadden (2003).

The social practice of this the first post-Agreement census undermines the expressed desire to normalise social relations outlined previously. The process of reallocating all those who expressed a clear preference for avoiding religious identities 'institutionalises the Northern Irish truism that there are only Catholic and Protestant atheists' (Alexander 2002, n.p.). The dangers inherent in this process are clearly flagged by Wilson:

> [T]he communal registration process has militated against the emergence of a strong political centre that might engender stability in the institutions. Indeed, on the contrary, it has reinforced 'groupist' stereotyping characteristic of media reporting of Northern Ireland, where actual Protestant and Catholic individuals are constantly hovered up into ethnonationalist 'communities' belying the pluralism of real social life. (Wilson 2003, 15; cited in Shirlow and Murtagh 2006, 34)

The practice of communal registration, and active reallocation, is evidence that, in some locations, there is a continuation, and *strengthening*, of sectarianism occurring within the very same institutional structures and practices which are supposed to help Northern Ireland move beyond sectarian division. Despite the commitment to Good Relations between *all groups*, non-sectarian groups and people, including those groups not classifiable as Catholic or Protestants, are often side lined. The reallocation of the 'No Religion' responses illustrates that 'Race Relations', far from shifting the dominance of sectarian narratives and worldviews, is effectively denied, and undermined, by institutional structures and practices grounded in, and reproductive of, sectarian division.

Conclusion

Over the last decade 'Race Relations' policy in Northern Ireland has changed remarkably. In this time Northern Ireland has gone from a situation in which discrimination on the grounds of 'race' was not illegal to one in which 'Race Relations' policies and practices are being produced in a swath of government institutions. These new policy developments have, generally, been very positive for many racialised groups (McVeigh and Lentin 2002). The changing political climate, as well as social and economic transitions, has precipitated the emergence of a discourse on diversity beyond the staid 'two traditions' binary. As part of this process the Community Relations model of negotiating difference, specifically based on Catholic/Protestant divisions, has been expanded to include a concern for 'Race Relations' as part of the new Good Relations agenda. Good Relations signifies a movement beyond a singular focus on sectarianism as the locus of anti-discriminatory practices and a heightened awareness of issues of 'race', racism and 'Race Relations'. For many politicians and policy makers the presence of cultural diversity is to be welcomed as a sign of progress towards a 'normal' society. In this chapter it was clearly evidenced that 'Race Relations', and the management of 'race', is conceived of as part of this process of 'normalisation'.

Despite many positive developments the public administration of 'Race Relations' still faces a number of significant challenges. Although the Agreement created a favourable climate for such progressive political developments it also ushered in a fragmented state structure which stands in the way of successful implementation of such policies (Lentin and McVeigh 2006). The current situation, under which racial equality is devolved, criminal justice reserved and immigration excepted, stymies effective 'Race Relations' with attendant negative effects for racialised groups. The Northern Ireland Assembly may never have full control in all these areas but a more clearly connected approach to 'Race Relations' between devolved and non-devolved institutions must be pursued to prevent these issues being marginalised further.

Sectarian division remains a pervasive feature of the 'lifeworlds' of many in Northern Ireland (Shirlow and Murtagh 2006). Sectarian cleavages are not confined to the private sphere; they also structure political institutions and their practices. The fragmentation and diffusion of powers described in this chapter are direct products of a political settlement which sought to temper sectarian division by privileging the recognition of Nationalist and Unionist political identities. The analysis of the process of reallocating census data into sectarian categories highlights the contradiction between a commitment to Good Relations and a set of structures and social practices which reproduce sectarian division. Rather than construct 'Race Relations' as vehicle for re-imagining a 'normal' Northern Ireland policy makers and politicians need to acknowledge the resilience of engrained sectarian ways of thinking and doing within post-Agreement political institutions and practices. Only then will it be possible to attend to the quite specific challenges of 'Race Relations' and the precarious situation facing many racialised groups living in a violent, divided society that is slowly moving out of conflict.

Acknowledgements

I would like to thank Jane Jacobs, Jan Penrose and Karen Keaveney for their varied but always helpful feedback on my thesis in general and earlier versions of this chapter in particular. I would also like to thank Owen McEldowney for his advice on some sections of this chapter.

Notes

1   The official name of the settlement reached in April 1998 is *the Agreement Reached in the Multi-Party Negotiations.* As with so much in Northern Ireland the title of this settlement has become the source of some debate and consternation, here I refer to it as the Good Friday Agreement and henceforth as the Agreement.

2   The bulk of new migrants come from 'A8' countries, i.e. the eight Eastern European countries signed the Accession Treaty and joined the EU in May 2004. According to Jarman (2006) from 2003 to January 2006 there were 12020 applications for National Insurance numbers from Polish nationals, 4,987 from Lithuanians, 3,605 from Portuguese, 3,469 from Slovakians, 2,486 from Indians, 1,524 from Filipinos, 1,358 from Latvians, 1,338 from Czechs, 1,317 from Chinese and 867 from Ukrainians. The migrants are employed in the following industries, arranged in order of greatest number: admin./management, manufacturing, construction, hospitality, food processing, agriculture, health, retail, transport and entertainment (from Jarman, 2006).

3   The term 'race', as well as related concepts such as racial, was once understood to imply the existence of naturally discrete, and discernible, groups in human populations. But it is now widely accepted that 'race' is a socially constructed category such that it is common now to talk of 'racialised' groups and processes of 'racialisation'. However,

the debate surrounding this facet of difference in contemporary society continues to be couched in the terms 'race' and 'racial'. Mindful of the need to use the term but also of its social construction I am placing the term 'race' in quotes to stress that 'race' has no ontological status as a categorisation and that the distinctions made by the term have no biological basis. In doing this I follow a long tradition, e.g. Bonnett (1996), Jackson and Penrose (1993) and Miles (1982). In this chapter I use the term 'Race Relations' to refer to those policies and practices aimed at effecting racialised groups and peoples.

4   The Fair Employment Act 1976 (modified in 1989) is concerned with workplace discrimination on the grounds of religious belief or political opinion. Under the Act tribunals were set up ensure that workforces accurately reflected the religious backgrounds of the local area and did not display political symbols (see Osborne and Shuttleworth (2004) for a contemporary discussion on the effectiveness of Fair Employment legislation).

5   The Community Relations Act (Northern Ireland) established a Community Relations Commission in Northern Ireland at the end of 1969 (Griffiths 1972). This Commission was similar in structure to the Community Relations Commission created in Britain at the time under the 1968 Race Relations Act but instead of promoting good 'race' relations it was charged with improving relations between Catholics and Protestants (Griffiths, 1972; Solomos, 2003). While the Race Relations Act 1976 replaced the Community Relations Councils, and the Community Relations Commission, in Britain with the Commission for Racial Equality, the infrastructure of Community Relations has remained in place in Northern Ireland.

6   The Standing Advisory Commission on Human Rights (SACHR) was a body formed under the Northern Ireland Constitution Act 1973 to advise the direct rule government, and principally the Secretary of State for Northern Ireland, on issues of discrimination and human rights in Northern Ireland. The body was eventually replaced by the Northern Ireland Human Rights Commission (NIHRC) following commitments made in the Agreement.

7   The Diplock courts were established by the British government in Northern Ireland in 1972 following recommendations in a report by Lord Diplock. These controversial courts attempted to overcome jury intimidation by suspending trial by jury, instead of which a single judge adjudicated.

8   This chapter arises from my doctoral research on the politics and policy of accommodating, and managing, ethnicised and racialised diversity in post-Agreement Northern Ireland. This project involved ethnographic fieldwork, analysis of anti-racist visual culture and over forty semi-structured interviews with 'Race Relations' policy makers, civil servants working in Racial Equality Units and Community Relations, minority ethnic representative groups and anti-racist groups conducted in Belfast and Lisburn between December 2004 and August 2006.

9   OFMDFM is here described as a 'key ministry' as it has a very wide range of responsibilities, including designing a Programme for Government which sets budgets and priorities for all departments. It is also unique in that it is not run by a single Nationalist or Unionist minister but rather jointly by two junior ministers, one from each ethno-national bloc.

10 A review of public administration was launched by the then Secretary of State for Northern Ireland Peter Hain in 2005. As well as making proposals for streamlining measures in areas such as education this review recommended that the number of District Councils be reduced to seven. At present these recommendations have not

been adopted but it is expected that the number of district councils shall be radically reduced in the coming years.

11 Consociationalism is a form of government involving power-sharing and group representation to manage divided and conflictual societies by promoting accommodation between political elites (Tonge 2005). In a consociational settlement institutions are seen as key to the creation and maintenance of political stability (Lijphart 1977). The importance of guaranteeing Nationalist and Unionist representation and participation in the political process, and its institutions, has been one of the founding principles of the Agreement (McGarry and O'Leary 2004).

12 The D'Hondt system is used for to allocate ministerial positions in Northern Ireland. Although the Agreement does not have written into it that ministerial portfolios should be divided in a 6:5 ratio between Unionists and Nationalists, the use of this system means that in practice this division of ministerial labour occurs.

13 In the lead-up to the 2001 census, the first census since the Agreement and the first in which Sinn Féin urged Republican participation; there was a growing media discourse on changing demographics, particularly the possibility of a Catholic majority in favour of a United Ireland (Anderson et al. 2004). This led to the 2001 census becoming even more politicised than in previous years. In the run up to the census many Unionist politicians claimed it would show that the Protestant numerical superiority while Nationalists suggested it would show population changes which would see a Catholic majority in Northern Ireland in the coming decades (Anderson et al. 2004). In this highly politicised context having almost 14 per cent of respondents avoid a religious identity caused a serious problem which was 'resolved' by the reallocation.

References:

Agamben, G. (1998), *Homo Sacer: Sovereign Power and Bare Life* (Stanford, CA: Stanford University Press).

Alexander, N. (2002), 'Count me Out – the Northern Ireland Census', *The Vacuum* 6.

Anderson, J. and Shuttleworth, I. (1998), 'Sectarian Demography, Territoriality and Political Development in Northern Ireland', *Political Geography* 17:2, 187–208.

Anderson, J., Shuttleworth, I., Lloyd, C. and McEldowney, O. (2004), *End of Award Report: Political Demography: The Northern Ireland Census, Discourse and Territoriality* (published online 2004), <http://www.esrcsocietytoday.ac.uk/> (home page), accessed 4 July 2007.

Anthias, F. and Yuval-Davis, N. (1992), *Racialised Boundaries: Race, Nation, Gender, Colour and Class and the Anti-racist Struggle* (London: Routledge).

Ben-Tovim, G., Gabriel, J., Law, I. and Stredder, K. (1992), 'A Political Analysis of Local Struggles for Racial Equality', in P. Braham, A. Rattansi and R. Skellington (eds) *Racism and Antiracism: Inequalities, Opportunities and Policies* (London: Sage), 201–17.

Bonnett, A. (1996), 'Constructions of "Race", Place and Discipline: Geographies of "racial" identity and racism', *Ethnic and Racial Studies* 19:4, 864–83.

Connolly, P. and Kennan, M. (2002), *Tackling Racial Inequalities in Northern Ireland: Structures and Strategies* (Belfast: Northern Ireland Statistics and Research Agency).

Dickey, A. (1972), 'Anti-incitement Legislation in Britain and Northern Ireland', *New Community* 1:2, 133–8.

Ellis, G. (2001), 'Social Exclusion, Equality and the Good Friday Peace Agreement: The implications for land use planning', *Polity and Politics* 29:4, 393–411.

Gallagher, T. (1992), 'Fair Employment and Religion in Northern Ireland: An overview', *International Journal of Manpower* 13:6, 245–64.

Gilroy, P. (1987), *There Ain't no Black in the Union Jack: The Cultural Politics of Race and Nation* (London: Macmillan).

Gilroy, P. (1990), 'The End of Antiracism', *New Community* 17:1, 71–84.

Graham, B. and Nash, C. (2005), 'A Shared Future: Territoriality, pluralism and public policy in Northern Ireland', *Political Geography* 25:2, 253–78.

Griffiths, H. (1972), 'The Northern Ireland Community Relations Commission', *New Community* 1:2, 128–32.

Hadden, T. (2003), 'Putting People into Boxes', *Fortnight Magazine* 411, 6.

Hainsworth, P. (1998), 'Politics, Racism and Ethnicity in Northern Ireland', in P. Hainsworth (ed.) *Divided Society: Ethnic Minorities and Racism in Northern Ireland* (London: Pluto Press), 33–51.

Hannah, M.G. (2001), 'Sampling and the Politics of Representation in US Census 2000', *Environment and Planning D: Society and Space* 19:5, 515–34.

Home Office (2005), *Managed Migration: Working for Britain* (London: Home Office).

Jackson, P. (1993), 'Policing Difference: 'Race' and crime in Metropolitan Toronto', in P. Jackson and J. Penrose (eds) *Constructions of Race, Place and Nation* (London: UCL Press), 181–200.

Jackson, P. (1994), 'Constructions of Criminality: Police–community relations in Toronto', *Antipode* 26:3, 216–35.

Jackson, P. and Penrose, J. (1993), 'Introduction: Placing "race" and "nation"', in P. Jackson and J. Penrose (eds) *Constructions of Race, Place and Nation* (London: UCL Press), 1–23.

Jarman, N. (2006), 'Diversity, Economy and Policy: New patterns of migration to Northern Ireland', *Shared Space: A Research Journal on Peace, Conflict and Community Relations in Northern Ireland* 2 (Belfast: Community Relations Council), 45–61.

Kearns, G. (2006), 'Bare Life, Political Violence and the Territorial Structure of Britain and Ireland', in D. Gregory and A. Pred (eds) *Violent Geographies: Fear, Terror and Political Violence* (New York: Routledge), 9–34.

Keith, M. (1993), *Race, Riots and Policing: Lore and Disorder in a Multi-racist Society* (London: UCL Press).

Kemp, A. (2004), 'Labour Migration and Racialisation: Labour market mechanisms and labour migration control policies', *Social Identities* 10:2, 267–92.

Lentin, R. (2001), 'Responding to the Racialisation of Irishness: Disavowed multiculturalism and its discontents', *Sociological Research Online* 5:4, <http://www.socresonline.org.uk/5/4/lentin.html>, accessed 4 March 2007.

Lentin, R. and McVeigh, R. (2006), *After Optimism? Ireland, Racism and Globalisation* (Dublin: Metro Eireann Publications).

Lijphart, A. (1977), *Democracy in Plural Societies: A Comparative Exploration* (New Haven: Yale University Press).

Little, A. (2003), 'Multiculturalism, Diversity and Liberal Egalitarianism in Northern Ireland', *Irish Political Studies* 18:2, 23–39.

Northern Ireland Office (1998), *The Agreement Reached in the Multi-Party Negotiations* (Belfast: Northern Ireland Office).

Malik, K. (1996), *The Meaning of Race: Race, History and Culture in Western Society* (New York: New York University Press).

McCrudden, C. (1999), 'Mainstreaming Equality in the Governance of Northern Ireland' *Fordham International Law Review* 22, 16–19.

McGarry, J. and O'Leary, B. (2004), *The Northern Ireland Conflict: Consociational Engagements* (Oxford: University Press).

McNair, D. (2006), 'Social and Spatial Segregation: Ethno-national separation and mixing in Belfast' (unpublished PhD thesis: Queen's University, Belfast).

McVeigh, R. (1998), '"There's no Racism Cause There are no Black People Here": Racism and anti-racism in Northern Ireland', in P. Hainsworth (ed.) *Divided Society: Ethnic Minorities and Racism in Northern Ireland* (London: Pluto Press), 11–32.

McVeigh, R. (2002), 'Between Reconciliation and Pacification: The British state and community relations in the North of Ireland', *Community Development Journal* 37:1, 47–59.

McVeigh, R. and Lentin, R. (2002), 'Situated Racisms: A theoretical introduction', in R. Lentin and R. McVeigh (eds) *Racism and Anti-racism in Ireland* (Belfast: Beyond the Pale), 1–48.

Miles, R. (1982), *Racism and Migrant Labour: A Critical Text* (London: Routledge and Kegan Paul).

Mussano, S. (2004), 'Citizenship Education Policies in Northern Ireland and the Recognition of Ethnic and Racial Diversity in the Wake of New Immigration', *Migration Letters* 1:1, 2–10.

Nash, C. (2005), 'Equity, Diversity and Interdependence: Cultural policy in Northern Ireland', *Antipode* 37:2, 272–300.

Office of the First Minister and Deputy First Minister (2000), *Internal Memo: Proposal for a Race Equality Branch within OFMDFM* (Belfast: New TSN Unit).

Office of the First Minister and Deputy First Minister (2005), *A Shared Future: Policy and Strategic Framework for Good Relations in Northern Ireland* (Belfast: Office of the First Minister and Deputy First Minister).

Office of the First Minister and Deputy First Minister (2006), *A Racial Equality for Northern Ireland* (Belfast: Office of the First Minister and Deputy First Minister).

Osborne, B. and Shuttleworth, I. (2004), *Fair Employment in Northern Ireland: A Generation On* (Belfast: Blackstaff Books).

Shirlow, P. (2006), 'Belfast: The "post-conflict" city', *Space and Polity* 10:2, 99–107.

Shirlow, P. and Murtagh, B. (2006), *Belfast: Segregation, Violence and the City* (London: Pluto Press).

Solomos, J. (2003), *Race and Racism in Britain* (Basingstoke: Palgrave Macmillan).

Standing Advisory Commission on Human Rights (1990), *Second Report: Religious and Political Discrimination and Equality of Opportunity in Northern Ireland* (London: HMSO).

Tennant, V. (2001), *Sanctuary in a Cell* (Belfast: Law Centre (NI)).

Thompson, S. (2002), 'Parity of Esteem and the Politics of Recognition', *Contemporary Political Theory* 1:2, 203–20.

Tonge, J. (2005), *The New Northern Irish Politics?* (Basingstoke: Palgrave Macmillan).

Wilson, R. (2003), *Northern Ireland: What's Going Wrong?* (London: Constitutional Unit).

Chapter 11

Race and Immigration in Contemporary Ireland

Una Crowley, Mary Gilmartin and Rob Kitchin

Introduction

Since the 1990s the Republic of Ireland has become, for the first time in modern history, a country of net immigration. In the period from 1995 to 2004, 486,300 people moved to Ireland. In the same period, 263,800 people emigrated, resulting in net immigration of 222,500 (see Table 11.1). Since then, between May 2004 and December 2006 around 300,000 PPSNs (Personal Public Service Numbers) were issued to workers from accession states (O'Brien 2007).

The Central Statistics Office reported, in September 2004, that the population of Ireland had exceeded four million for the first time since 1871 and between 1996 and 2006, the overall population of the state increased by 16.8 per cent from 3.62 million to 4.23 million (CSO 2007). Reasons for this recent growth in immigration to Ireland are complex, but they include Ireland's economic strength (the 'Celtic Tiger' era), the Northern Ireland ceasefires, and EU enlargement in 2004 (see Kitchin and Bartley 2007).

Growing numbers of immigrants is not in itself an unusual phenomenon in a wealthy Western country. Yet Ireland is different from many Western countries because of its long experience of emigration. Throughout the nineteenth and twentieth centuries, Ireland was a net exporter of people. Millions of Irish people – thousands within each generation – fled poverty and political and social repression to seek work and new lives abroad, thus creating a large, global Irish diaspora. The extent of emigration led geographer Jim MacLaughlin, writing just over ten years ago, to describe Ireland as an 'emigrant nursery' (MacLaughlin 1994). Despite recent economic growth in Ireland, emigration from the country continues, and Irish politicians and the Catholic Church continue to be active in lobbying on behalf of the thousands of undocumented Irish immigrants currently living in the United States. Irish identity has thus traditionally been associated with the act of migration. As writer Polly Devlin commented, 'emigration was a big sad Irish word in every sense ... We were all poised on the point of eternal emigration' (in Logue 2000, 42).

In this chapter, we reflect on the impact of these recent changes in migration patterns to and from Ireland. We are primarily interested in the relationship between these changes and the construction, or indeed reconstruction, of Irish

Table 11.1 Emigration from and immigration to Ireland and net migration rates, 1995–2004

Emigration from Ireland											
To	*1995*	*1996*	*1997*	*1998*	*1999*	*2000*	*2001*	*2002*	*2003*	*2004*	
UK	13,300	14,100	12,900	8,500	11,200	7,200	7,800	7,400	6,300	4,900	93,600
EU	5,100	5,100	4,100	4,300	5,500	5,500	5,600	4,800	4,300	3,400	47,700
USA	8,200	5,200	4,100	4,300	5,300	4,000	3,400	4,800	2,500	2,800	44,600
ROW	6,600	6,800	7,900	4,100	9,500	10,000	9,500	8,500	7,600	7,400	77,900
	33,200	31,200	29,000	21,200	31,500	26,700	26,300	25,500	20,700	18,500	263,800
Immigration to Ireland											
Nationality	*1995*	*1996*	*1997*	*1998*	*1999*	*2000*	*2001*	*2002*	*2003*	*2004*	
Irish	17,600	17,700	20,500	23,200	26,700	24,800	26,300	27,000	17,500	16,900	218,200
UK	5,800	8,300	8,200	8,300	8,200	8,400	9,000	7,400	6,900	5,900	76,400
EU	3,200	5,000	5,500	5,800	6,900	8,200	6,500	8,100	6,900	10,600	66,700
USA	1,500	4,000	4,200	2,200	2,500	2,500	3,700	2,700	1,600	1,800	26,700
ROW	3,100	4,200	5,500	4,500	4,500	8,600	13,600	21,700	17,700	14,900	98,300
	31,200	39,200	43,900	44,000	48,800	52,500	59,100	66,900	50,600	50,100	486,300
Net migration											
	1995	*1996*	*1997*	*1998*	*1999*	*2000*	*2001*	*2002*	*2003*	*2004*	
	−2,000	8,000	14,900	22,800	17,300	25,800	32,800	41,400	29,900	31,600	222,500

Sources: CSO (2000, 2004).

identity. In particular, we discuss the changing racialisation of Irish identity as a consequence of and response to immigration. We thus chart the change in migration patterns and reflect on the state's attempts to define and place new immigrants through a variety of political responses. In particular, we discuss the 2004 Citizenship Referendum, its racist overtones, and its fundamental rescripting of the nature of Irish citizenship and identity designed to limit immigration of non-white peoples. We conclude with a discussion of the impact of European accession on race and immigration in the Irish context.

From Emigration to Immigration

For the two centuries prior to the present period of immigration, Ireland had been a net exporter of people. Migration from Ireland has peaked periodically, usually connected to difficult economic, political and social periods in the country. Periods of significant out-migration include the years immediately following the Great Famine, the 1950s and the 1980s. For example, it is estimated that over 600,000 people emigrated from Ireland (twenty-six counties) in the period from 1851 to 1855 (Miller 1985). By 1961, there were over 750,000 people of Irish birth living in Britain (Ferriter 2004). Between 1987 and 1996, the Central Statistics Office estimates that over 430,000 people emigrated from Ireland, peaking in 1989 when over 70,000 people – 2 per cent of the population – left the country. The process continues today, with roughly 20,000 people – the majority under 25 years of age – emigrating annually (CSO 2006).

However, while emigration from Ireland continues, its significance has been occluded by recent changes in patterns of immigration to Ireland. In 1996, Ireland experienced net immigration, and since that year levels of immigration and of net immigration have been steadily increasing producing the most sustained period of net immigration to Ireland since independence. This was not the first time: for a short period in the 1970s, Ireland also experienced net immigration, but this was quickly reversed with the advent of a global economic downturn and the escalation of violence in Northern Ireland. Just as in the 1970s, the majority of immigrants to Ireland in the contemporary period are returning Irish and their families, or citizens of other EU countries, most noticeably the UK. For the period from 1995 to 2004, 45 per cent of immigrants to Ireland were Irish, and 30 per cent were from the EU (over half of these were from the UK). For these immigrants, entry to Ireland is unproblematic, and they have unrestricted access to the Irish labour market. This is not the case for immigrants with other than EU citizenship. The main entry routes to Ireland for such immigrants were as work permit holders, as asylum seekers or refugees, or as working visa/work authorisation holders. Work permits, working visas and work authorisations represent Ireland's labour migration strategies. Working visas and work authorisations (WV/WA) were issued to highly skilled workers required for the information technology, medical and construction sectors. The numbers of immigrants admitted to Ireland under this scheme were low (see Table 11.2). Work permits were issued to less skilled workers needed in the service sectors, agriculture, catering and industry. Prior to

EU enlargement in May 2004, the majority of work permits were issued to citizens of the EU-10, particularly Poland, Lithuania and Latvia. Since May 2004, the number of work permits issued annually has fallen (see Table 11.3).

Table 11.2 Total working visas/work authorisations issued 2000–2005

Year	Working visas	Work authorisations	Total
2000	991	392	1,383
2001	2,667	1,082	3,749
2002	1,753	857	2,610
2003	791	367	1,158
2004	1,098	346	1,444
2005	2,307	278	2,585

Source: Department of Enterprise, Trade and Employment (personal communication).

Table 11.3 Total work permits issued 1999–2006

Year	New permits	Renewals	Group permits	Total	Refused
1999	4,328	1,653	269	6,250	Not known
2000	15,434	2,271	301	18,006	Not known
2001	29,594	6,485	357	36,436	Not known
2002	23,326	16,562	433	40,321	1,310
2003	21,965	25,039	547	47,551	1,838
2004	10,020	23,246	801	34,067	1,486
2005	7,354	18,970	812	27,136	1,215
2006	6,289	14,258	848	21,395	1,123

Source: Department of Enterprise, Trade and Employment. http://www.entemp.ie/labour/workpermits/statistics.htm, accessed October 2004, March 2005, March 2007.

The schemes clearly made assumptions about the desirability and treatment of different labour migrants. The WV/WA was a fast-track programme where the migrant worker applies for a visa/authorisation through the Irish embassy in their country. In contrast, employers applied for and held work permits on behalf of their employees, leaving the employee tied to that site of work and creating large power differentials that have clearly been exploited in a number of cases. Work permit holders have significantly less rights to state services and family reunification than WV/WA holders. These distinctions have been maintained in new employment permits legislation, introduced with effect from February 2007

(DETE 2007). Figures for asylum applications have risen from a low of 39 in 1992 to an annual level of over 4,000, peaking at 11,634 in 2002 (see Table 11.4).

Table 11.4 Asylum applications in Ireland 1992–2006

Year	No. of applications	Top five countries of origin
1992	39	Not available
1993	91	Not available
1994	362	Not available
1995	424	Not available
1996	1,179	Not available
1997	3,883	Not available
1998	4,626	Not available
1999	7,724	Not available
2000	10,938	Nigeria, Romania, Czech Republic, Moldova, DR Congo
2001	10,325	Nigeria, Romania, Moldova, Ukraine, Russia
2002	11,634	Nigeria, Romania, Moldova, Zimbabwe, Ukraine
2003	7,900	Nigeria, Romania, DR Congo, Moldova, Czech Republic
2004	4,766	Nigeria, Romania, Somalia, China, Sudan
2005	4,323	Nigeria, Romania, Somalia, Sudan, Iran
2006	4,314	Nigeria, Sudan, Romania, Iran, Iraq

Source: ORAC (2006).

Other routes of entry for migrants from outside the EU are as holders of student visas (especially Chinese), working holiday visas (issued to people under 35 from Australia, New Zealand, Canada and Hong Kong) or holiday visas. The preliminary results of the 2006 Census indicate that over 10 per cent of the resident population of Ireland classifies their nationality as other than Irish, and over 14 per cent were born outside Ireland (CSO 2007).

What is striking about these recent migration streams to Ireland is their heterogeneity. Official figures from the CSO prior to Census 2006 classify all migrants from outside the EU and the US as ROW (Rest of World). Hidden in that catch-all classification is the range of nationalities immigrating to Ireland on an annual basis. For example, in 2006 work permits were issued to citizens of over 100 countries, with significant numbers issued to citizens of the Philippines, India, Ukraine, South Africa, Romania, Brazil and China (DETE 2006), and the majority of asylum application were received from citizens of Nigeria, Sudan, Romania, Iran and Iraq (ORAC 2006). In 2005, working visas and work authorisations were issued to citizens of over 50 countries, particularly India and the Philippines. It is important to highlight that heterogeneity, in this context, refers entirely to nationality. The official collection of information on immigrants to Ireland predominantly focuses on nationality, and gives little information on racial diversity in the country. Census 2002 collected no data about racial identification in Ireland. The only direct question on ethnicity asked respondents to identify if they were members of the Irish Traveller community. The most recent Census, in 2006, included a new question entitled ethnic and cultural background. Despite its title, the focus of the question was on race, even though the racial categories provided were confusing and inadequate (O'Toole 2006).

Table 11.5   Ethnic and cultural background of Irish residents, 2006

Category	%
White Irish	87.4
White Irish Traveller	0.5
Any other White background	6.9
Black or Black Irish: African	1.0
Black or Black Irish: Any other Black background	0.1
Asian or Asian Irish: Chinese	0.4
Asian or Asian Irish: Any other Asian background	0.9
Other including mixed background	1.1
Not stated	1.7
	100.0

Source: Central Statistics Office (2007).

As a consequence, there are no official statistics on the ethnic composition of the resident population of Ireland. In addition, the state's failure to gather even the most basic data on the racial composition of its population means that assertions about the relationship between 'race' and immigration are difficult to substantiate.

Responding to Immigration

Many public debates about immigration in Ireland are framed in terms of the Irish experience of emigration. The National Action Plan against Racism makes this explicit, claiming that 'the Irish have been a migrant people for a very long time', subject to racism both at home and abroad (NPAR 2005). There is significant evidence to support the claim of racism directed against Irish emigrants, particularly in the United States and the United Kingdom, where the Irish were described as simian and wild, and associated with dirt, disease, poverty and savagery (see, for example, Engels 1958; Miller 1985; Ignatiev 1995; Curtis 1997;

Hickman and Walter 1997; Roediger 1999). Reminders of the Irish emigrant experience periodically resurface in political utterances about immigration: in the words of the President, Mary McAleese: 'we hope our distilled wisdom and experience will enable us to ensure rapid and easy melding of our new citizens into Irish life. Of all people on the planet we have no excuse for getting it wrong' (McAleese 2007).

Mindful of this experience, but aware of significant changes in the pace and scale of immigration to Ireland, people have struggled to find an adequate vocabulary to describe and debate these changes. Many public figures are careful to speak about immigration in neutral language, and those who breach these unwritten codes are often publicly censured.[1] The six main political parties endorsed an anti-racism protocol prior to the 2002 elections (Fanning and Mutwarasibo 2007, 442). Politicians insisted that they would not 'play the race card', and made political capital from claims that others did. Concurrently, policy initiatives and legislation directed against discrimination and racism were introduced. These included the ratification of the UN Convention on the Elimination of All Forms of Racial Discrimination (CERD) in 2000, the Employment Equality Act (1998) and the Equal Status Act (2000) (McVeigh and Lentin 2002, 6; Tannam 2002, 197). The government also established two bodies with responsibility for monitoring these acts and providing guidance on issues of equality, race and interculturalism. The National Consultative Committee on Racism and Interculturalism (NCCRI), established in 1997, acts in a policy advisory role to the government and develops anti-racist programmes. The Equality Authority was established in 1999, and its role is to promote and defend equality rights. The NCCRI and the Equality Authority together act as the national focal points for RAXEN, the European Racism and Xenophobia Network. Membership of the EU has been instrumental in providing the impetus for legislative change with an anti-racist and anti-discriminatory agenda. The National Action Plan against Racism (NPAR), published in January 2005, provides the most explicit statement about the relationship between immigration and racism:

> Inward migration is not the cause of racism. Cultural diversity and racism existed in Ireland before the recent increase in people coming to live in Ireland ... [R]acism is an issue we must tackle regardless of inward migration. (NPAR 2005)

Despite these initiatives and pronouncements, it is clear that early debates about immigration to Ireland were racialised. From the 1990s onwards, public debates about immigration focused primarily on refugees and asylum seekers, who represented a small proportion of the overall numbers of immigrants. Refugees and asylum seekers were often racialised as 'black', even though no accurate figures were publicly available to legitimate this claim. As a consequence, there has been a widespread belief that Ireland is being 'overrun' by black immigrants (Cullen 2000), generally understood as refugees and asylum seekers. Fianna Fáil TD Noel O'Flynn voiced these sentiments in a speech in 2002, when he said that 'the asylum seeker crisis was out of control' and the country was being held hostage by 'spongers, wasters and conmen.' Though denying his remarks were racist, O'Flynn claimed that putting large amounts of refugees from different ethnic backgrounds together was a 'powder keg waiting to explode' (in Spendiff 2002). It is important to acknowledge, however, that for many would-be immigrants from countries with substantial non-white populations, asylum represents one of the main entry routes to Ireland (see Table 11.4). Working visas, work authorisations and work permits are predominantly issued to citizens of countries with substantial white populations, or to citizens of countries like the Philippines, which shares a Catholic heritage with the Republic of Ireland. In addition, citizens of such countries have been specifically targeted to fill job vacancies in Ireland (Loyal 2003). In this way, public discourses helped to create a semantic association of black people with asylum seekers, and of asylum seekers with immigrants (White 2002, 104). This was further exacerbated by a seeming reluctance on the part of some government officials and media to separate issues of asylum and immigration, preferring instead to conflate the two.

In this way, asylum seekers – implicitly understood as black – became increasingly problematised. In response to the growing numbers of asylum seekers, the government introduced a range of punitive measures, including the Immigration (Trafficking) Bill of 1999 and the amendment of the Refugee Act in 2000, which together served to make it more difficult to successfully claim asylum in Ireland and to increase deportation rates. In terms of the treatment of asylum seekers, the government introduced a system of direct provision in April 2000, which limited support to basic accommodation, meals, and weekly cash allowances of IR£15 for adults and IR£7.50 for children (Fanning 2002, 103).[2] Asylum seekers were also dispersed outside Dublin to centres of direct provision, often local hostels and hotels commandeered for the purpose, and often in the face of widespread local opposition because of a perceived connection between asylum seekers, crime and disease. Asylum seekers are regularly portrayed by politicians in negative terms. John O'Donoghue, a previous Minister of Justice, Equality and Law Reform, described asylum seekers as 'illegal immigrants and as exploiters of the Irish welfare system' (in Fanning 2002, 103). More recently, the recently deposed Minister of Justice, Michael McDowell, described the accounts of asylum seekers as 'cock-and-bull' stories, claiming that he 'would prefer to interview these people at the airport' (Holland 2005).

The problematisation of asylum seekers has led to a variety of moral panics: about the 'invasion' of (black) asylum seekers, and about the consequent abuse of Irish citizenship legislation and social welfare benefits (see Luibhéid 2004). This gives legitimacy to practical actions that are both controlling and excluding: controlling racial and ethnic diversity while at the same time managing and facilitating white migration. This has been obvious in the state enforcement of immigration policy through deportation. Two groups of people are generally deported: those whose asylum applications were refused, and those whose parents' applications for asylum were refused (though the deportees were, themselves, Irish citizens by birth). Deportations of people who have entered Ireland on holiday or work visas or permits and overstayed or violated the conditions of those visas are extremely rare. These actions have, in part, been successful because they are building on already present xenophobic and racist sentiments (see Cullen 2002;

McVeigh 2002; Rolston and Shannon 2002). Policy discourse has thrived upon these anxieties. In this way, coercive state policies and everyday discriminatory practices in relation to the perceived invasion of unwanted black migrants are legitimised, allowing the state to prevent 'black' immigration while at the same time making it easy to import low-cost, 'white' labour.

The 2004 Citizenship Referendum

The most striking political response to immigration, and the response with the most fundamental and lasting impact on Irish identity, was the 2004 Citizenship Referendum. In this referendum, the Irish electorate voted by a margin of four to one for a change in the definition of Irish citizenship. Prior to the referendum, any child born on the island of Ireland had an automatic right to Irish citizenship – a right that was enshrined in the Irish Constitution. As a consequence of the referendum, the right to citizenship by birth was removed from the Constitution. Irish citizenship is now primarily defined by descent (see Lentin 2007).

The referendum was championed by the ruling coalition partners, Fianna Fáil and the Progressive Democrats (PDs), and it was supported by the second largest party, Fine Gael. It was first announced in March 2004 by Michael McDowell, the then PD Minister for Justice, Equality and Law Reform. McDowell described the referendum as 'both rational and necessary'. There was, he wrote, 'a steady stream of people coming to Ireland, both legally and illegally, so as to ensure that their children avail of our present law so as to secure the entitlement to Irish citizenship' (McDowell 2004b). His remarks were echoed by Fianna Fáil, the largest political party in Ireland. The party urged the electorate to support the referendum: their campaign posters read 'Vote Yes for Common Sense Citizenship'. For McDowell and other advocates of the referendum, Irish law had created a 'loophole' that permitted undeserving and undesirable people to gain access to Irish citizenship. Closing this loophole by changing the basis on which citizenship could be granted was, they claimed, the commonsense thing to do. The campaign in favour of the referendum created the illusion of hordes of pregnant women – so-called citizenship tourists – arriving in Ireland late in their pregnancy to give birth, putting themselves and the integrity of Irish citizenship at risk (see Garner 2005; Luibhéid 2004; Lentin 2004, 2005). For example, Michael McDowell commented that pregnant women traveling to Ireland were 'putting themselves at risk (or being put under pressure by their partners or others to take that risk) by undertaking hazardous journeys' (McDowell 2004a). His remarks were supported by Declan Keane, the Master of Holles Street, one of Dublin's main maternity hospitals, who claimed that:

> when these women are arriving late from Nigeria ... and often arriving, as I say, unwell, with no idea of when their first menstrual period was, no idea of their dates, some of them with complex medical disorders ... Some issues we can't [deal with]. We've seen a massive increase in HIV from some of these countries because in some of them HIV is almost endemic. (RTÉ 2003)

These kinds of claims, bolstered by misleading figures and statistics about births to 'non-national' women in Irish maternity hospitals, helped to create a moral panic around citizenship tourism. The introduction of 'commonsense citizenship' was the state's attempt to assuage the panic it had instigated.

The appeals to commonsense in the lobbying of voters had a number of powerful discursive effects. First, by appealing to commonsense, the 'Yes' campaign sought to stunt accusations of racism or xenophobia. Voting 'Yes' was not about discriminating against immigrants. Rather, its aim was to protect and benefit Ireland through the long-term preservation of culture and safeguarding of the economy. Second, 'commonsense citizenship' worked to focus on the present – the here and now – casting as irrelevant Ireland's own history of emigration and anti-Irish racism. The referendum was about the future, not the past: the past is, after all, a foreign country. It was therefore commonsense for people to vote to shape and protect their future in ways that ignored earlier generations' experiences as immaterial to the contemporary context. Third, commonsense understandings of immigration worked to undermine the legitimacy of a range of immigrants – asylum seekers, refugees, pregnant women – by questioning their authenticity and by generalising their motivations and experiences. The discursive construction and denigration of asylum seekers and 'economic migrants' as bogus, spongers or economic parasites cast doubt on their right to stay in Ireland and claim citizenship for themselves and their children. Fourth, 'commonsense citizenship' worked to (re)define and fix notions of Irishness. On the one hand, there was an appeal to a national, shared culture, but on the other Irishness was defined by blood ties and a rooted legacy in Ireland. To be Irish, one had to have grown up in Ireland or the Irish diaspora, and therefore be assimilated to the 'Irish way of life'. Commonsense therefore cast culture and identity in essentialist terms – as having inherent characteristics – rather than seeing Irishness as something constructed or performed; diverse, contingent, relational and constantly in the process of formation. This essentialist notion of Irishness therefore worked to create an exclusive, universal, rational category, difficult to challenge due to its commonsensical nature. In doing so, it sought to unite anyone who considered themselves Irish through an appeal to a common cultural and genetic heritage.

The attempts by the Citizenship Referendum to fix notions of Irishness mirror earlier attempts to construct a notion of Ireland as a monocultural, homogenous community, where 'masses of cultural expression – alternative realities, virtually alternative countries – were ignored while the Free State/Republic fetishised "Irishness"' (Longley 2001, 9). This version of Irishness emphasised Catholicism, the Irish language, the importance of rural Ireland, and exclusionary cultural organisations such as the GAA (Gaelic Athletic Association), and was underpinned by an assumed whiteness. In doing so, it denied the reality of the Irish State as multicultural, with citizens and residents with different ethnicities, religions and racial identities, with variances along lines of gender, class, sexuality and so on. Recent academic work has introduced complexity into discourses of Irish identity (see, for example, Brown 1985; Crowley 2005; Kiberd 1995; Kitchin and Lysaght 2004; Cullen 2000; Garner 2003; Longley and Kiberd 2001; Loyal 2003; Ó Gráda 2006; Rolston and Shannon 2002). Other critical work highlights the extent of dissent from the prevailing orthodoxy, suggesting that many Irish citizens disrupted and challenged organisational and institutional attempts at personal and national regulation (see Ferriter 2004 for a broad-ranging account of many of these acts of resistance). This tradition of dissent was similarly evident in the run up to the referendum. Opposition to the referendum was coordinated by CARR (Campaign against a Racist Referendum), and its supporters included political parties (Labour, Greens and Sinn Féin), trade unions, and special interest groups such as the Union of Students in Ireland, the National Youth Council of Ireland, the National Women's Council of Ireland, the National Traveller Women's Forum and the National Lesbian and Gay Federation. Despite this broad coalition of opposition, the calls for 'commonsense citizenship' were overwhelming accepted. In the wake of the referendum, debates about race and immigration have increasingly shifted focus, from asylum seekers and refugees to economic migrants and 'non-nationals.' The term 'non-national' is frequently used by media and politicians to categorise immigrants, and the discursive denial of nationality again serves to create a distinction between legitimate residents and those whose presence in Ireland is in some way less legitimate.

Conclusion

In May 2004, immediately prior to the Citizenship Referendum, ten new members were admitted into the EU. Ireland was one of just three existing members of the EU (along with the UK and Sweden) who allowed unrestricted movement and access to employment for citizens of the new member states. The consequences of this decision are abundantly clear in the preliminary demographic reports from Census 2006 (see Table 11.6), with Poles and Lithuanians as the third and fourth largest national groups currently resident in the country, a very significant change since Census 2002.

New migrants from the EU-10 are predominantly young and male, and their increasing presence in Ireland is raising new questions in public discourse. Issues of welfare abuse, fraud and citizenship tourism were raised in relation to asylum seekers, but now the concerns are with displacement and violence.[3] Concerns with displacement surface in relation to employment practices, particularly with regard to the replacement of Irish workers by migrant workers receiving lower wages and less benefits. The most high profile expression of this concern related to the passenger and freight shipping company, Irish Ferries. When the company offered redundancy payments to its Irish workers, who they wanted to replace with Eastern European workers, a protest march organised by trade unions attracted up to 100,000 people onto the streets of Dublin. Trade union leaders insisted their concern was with workers' rights, claiming 'there is a threshold of decency below which the Irish people will not accept anybody being dragged, no matter where they come from' (David Begg, quoted in Dooley 2005). However, the leader of the Irish Labour Party, Pat Rabbitte, told a different story, when he insisted that 'displacement is going on in the meat factories and it is going on in the hospitality industry and it is going on in the building industry ... there are 40 million or so

Table 11.6   Persons resident and present in Ireland classified by nationality, 2006

Nationality	Number ('000s)	%
Irish	3,706.7	88.84
UK	112.5	2.70
Polish	63.3	1.52
Lithuanian	24.6	0.59
Other EU	75.4	1.81
Other European	24.4	0.58
Nigerian	16.3	0.39
Other African	19.0	0.46
Chinese	11.2	0.27
Other Asian	35.8	0.86
USA	12.5	0.30
Other American	8.6	0.21
Other	16.3	0.39
No nationality	1.3	0.03
Not stated	44.3	1.06
	4,172.2	100.00

Source: Central Statistics Office (2007).

Poles after all, so it is an issue we have to have a look at' (Collins 2006). Opinion polls taken shortly after Rabbitte's statement showed strong support for his views, indicating that 70 per cent of people in Ireland believed that no more foreign workers should be admitted to the country (*Irish Times* 2006). Discourses around violence focus on the alcohol abuse, speeding and road accidents, assaults and murder among EU-10 immigrant communities in particular. This shifting focus serves, in many ways, to deflect attention from issues of racism and discrimination faced by immigrants on a daily basis, experiences that are charted in reports on racist incidents from a variety of state and state-sponsored bodies (CSO 2004; NCCRI 2006).

The shifting focus also deflects from broader and deeper concerns about Irish identity, as migrants come to represent 'all that is newly unfamiliar' (Pred 2000, 31) in a rapidly-changing society. Ireland's new-found confidence and cosmopolitanism is a mask hiding deeply etched and historically rooted anxieties and insecurities. In its new prosperity, 'commonsense' Ireland has been quick to sideline its own emigrant past. Instead, it has embraced the neoliberal, right-wing and racist rhetoric of Europe and North America. It wants immigrant workers, but it does not want them to stay if the economy experiences a downturn. It will tolerate a minimum of asylum seekers, but only if they are 'genuine' and contribute in 'positive' ways to Irish society. It uses racist rhetoric while at the same time insisting that it is not racist. It expects its citizens to be able to move to and work in any part of the world, but not vice versa. These paradoxes are troubling, but the articulation of 'commonsense citizenship' serves to dispel disquiet in the interests of the common good of the state and its existing citizens. It also mirrors similar experiences in other EU countries such as Denmark, the UK and the Netherlands. The difference, in the case of Ireland, is a long, sustained and substantial history of out-migration, which continues today in the bodies of undocumented Irish in the US. 'Take away the immigrants and their children, and the exiles and theirs', journalist Fintan O'Toole wrote, 'and we have no Irish nation, no Irish culture, no Irish identity' (O'Toole 2004). Through an acknowledgement of Ireland's emigrant past, the personal circumstances of immigrants, the racialised construction of the immigration 'problem', and the recognition of racist practices, the contradictions at the heart of 'commonsense citizenship' are exposed and possibilities for other, less restrictive understandings of citizenship and belonging are made apparent.

Notes

1   For example, Fianna Fáil junior minister Conor Lenihan was under pressure to resign when he described Turkish immigrants as 'kebabs'. The pressure dissipated quite quickly (O'Toole 2005), and Lenihan is now the country's first Minister for Integration.
2   The current weekly rates are €19.10 per adult (£12.90 or $25.50) and €9.50 per child (£6.40 or $12.70).
3   The Habitual Residency clause, introduced in May 2004, means that in general people have to be resident in Ireland for two years before they have any entitlement to social welfare payments.

References

Bartley, B. and Kitchin, R. (eds) (2007), *Understanding Contemporary Ireland* (London: Pluto Press).

Brown, T. (1985), *Ireland: A Social and Cultural History 1922–1985* (London: Fontana Press).

Central Statistics Office (2004), *Quarterly National Household Survey: Crime and Victimisation*, < http://www.cso.ie/qnhs/documents/qnhscrimeandvictimisation.pdf>, accessed March 2007.

Central Statistics Office (2006), *Population and migration estimates: April 2006*, <http://www.cso.ie/releasespublications/documents/population/current/popmig.pdf>, accessed April 2007.

Central Statistics Office (2007), *Census 2006: Principal Demographic Results*, <http://www.cso.ie/census/documents/Final%20Principal%20Demographic%20Results%202006.pdf>, accessed April 2007.

Collins, S. (2006), 'Rabbitte Calls for Rethink of Policy on Immigrants', *Irish Times*, 3 January.

Conway, B. (2006), 'Who Do We Think We Are? Immigration and the Discursive Construction of National Identity in an Irish Daily Mainstream Newspaper,

1996–2004', *Translocations* 1:1, <http://www.imrstr.dcu.ie/firstissue/conway.shtml>, accessed March 2007.

Coulter, C. and Coleman, S. (eds) (2003), *The End of Irish History: Critical Reflections on the Celtic Tiger* (Manchester: Manchester University Press).

Crowley, U. (2005), 'Liberal Rule through Non-liberal Means: The attempted settlement of Irish Travellers (1955–1975)', *Irish Geography* 38:2, 128–50.

Cullen, P. (2000), *Refugees and Asylum-Seekers in Ireland* (Cork: Cork University Press).

Curtis, L.P. (1997), *Apes and Angels: The Irishman in Victorian Caricature*, 2 edn (Washington, DC and London: Smithsonian Institute).

DETE (2006), *Permits issued by Nationality, 2006*, <http://www.entemp.ie/labour/workpermits/statistics.htm>, accessed March 2007.

DETE (2007), *New Employment Permit Schemes*, <http://www.entemp.ie/labour/workpermits/index.htm>, accessed March 2007.

Dooley, C. (2005), 'Further Effort to End Ferry Row as Thousands March in Protest', *Irish Times*, 10 December.

Engels, F. (1958), *The Condition of the Working Class in England*, trans. W.O. Henderson and W.H. Chaloner (Oxford: Basil Blackwell).

Fanning, B. (2002), *Racism and Social Change in the Republic of Ireland* (Manchester: Manchester University Press).

Fanning, B. and Mutwarasibo, F. (2007), 'Nationals/non-nationals: Immigration, citizenship and politics in the Republic of Ireland', *Ethnic and Racial Studies* 30:3, 439–60.

Ferriter, D. (2004), *The Transformation of Ireland 1900–2000* (London: Profile Books).

Garner, S. (2003), *Racism in the Irish Experience* (London: Pluto Press).

Garner, S. (2005), 'Guests of the Nation', *Irish Review* 33, 78–84.

Hickman, M. and Walter, B. (1997), *Discrimination and the Irish community in Britain* (London: Commission for Racial Equality).

Holland, K. (2005), '€50,200 Spent to Deport One Person Last Year', *Irish Times*, 19 May.

Ignatiev, N. (1995), *How the Irish Became White* (London: Routledge).

Irish Times (2006), 'Growing Anxiety over Migrant Labour', *Irish Times*, 23 January.

Kiberd, D. (1995), *Inventing Ireland: The Literature of the Modern Nation* (London: Random House).

Kitchin, R. and Lysaght, K. (2004), 'Sexual Citizenship in Belfast, Northern Ireland', *Gender, Place and Culture* 11, 83–103.

Kitchin, R. and Bartley, B. (2007), 'Ireland in the Twenty-first Century', in Bartley and Kitchin (eds).

Lentin, R. (2004), 'Strangers and Strollers: Feminist reflections on researching migrant m/others', *Women's Studies International Forum* 27:4, 301–14.

Lentin, R. (2005), 'Black Bodies and "Headless Hookers": Alternative global narratives for 21st century Ireland', *Irish Review* 33, 1–12.

Lentin, R. (2007), 'Ireland: Racial State and Crisis Racism', *Journal of Ethnic and Racial Studies* 30:4, 610–27.

Lentin, R. and McVeigh, R. (eds) (2002), *Racism and Anti-racism in Ireland* (Belfast: Beyond the Pale Publications).

Logue, P. (ed.) (2000), *Being Irish: Personal Reflections on Irish Identity Today* (Dublin: Oak Tree Press).

Longley, E. (2001), 'Multi-culturalism and Northern Ireland: Making Differences Fruitful', in Longley and Kiberd (eds).

Longley, E. and Kiberd, D. (eds) (2001), *Multi-culturalism: The View from the Two Irelands* (Cork: Cork University Press).

Loyal, S. (2003), 'Welcome to the Celtic Tiger: Racism, immigration and the state', in Coulter and Coleman (eds).

Luibhéid, E. (2004), 'Childbearing against the State? Asylum Seeker Women in the Irish Republic', *Women's Studies International Forum* 27, 335–49.

MacLaughlin, J. (1994), *Ireland: The Emigrant Nursery and the World Economy* (Cork: Cork University Press).

McAleese, M. (2007), 'The Changing Faces of Ireland – Migration and Multiculturalism', address to the British Council, 14 March, <http://www.ireland.com/focus/2007/mcaleese/index.html>, accessed March 2007.

McDowell, M. (2004a), 'Referendum on Citizenship of New Babies not Racist', *Sunday Independent*, 14 March.

McDowell, M. (2004b), 'We Must be Able to Manage Migration in a Sensible Fashion', *Irish Times*, 24 April.

McVeigh, R. (2002), 'Nick, Nack, Paddywhack: Anti-Irish racism and the racialisation of Irishness', in Lentin and McVeigh (eds).

McVeigh, R. and Lentin, R. (2002), 'Introduction: Situated racisms', in Lentin and McVeigh (eds).

Miller, K. (1985), *Emigrants and Exiles: Ireland and the Irish Exodus to North America* (Oxford: Oxford University Press).

NPAR (National Action Plan against Racism) (2005), *Planning for Diversity: The National Action Plan against Racism*, <http://host2.equinox.ie/diversity/procontent/Publications/upload/File/NPAR.pdf>, accessed March 2007.

NCCRI (2006), *Report on Incidents Relating to Racism in Ireland*, <http://www.nccri.ie/incidents-reports.html />, accessed March 2007.

O'Brien, C. (2007), 'Immigrant Influx is Largest Ever as 200,000 Register', *Irish Times*, 2 January.

Ó Gráda, C. (2006), *Jewish Ireland in the Age of Joyce: A Socioeconomic History* (Princeton: Princeton University Press).

ORAC (Office of the Refugee Applications Commissioner) (2006), *Statistics*, <http://www.orac.ie/pages/Stats/statistics.htm>, accessed March 2007.

O'Toole, F. (2004), 'They Made Us What We Are ...', *Irish Times*, 25 May.

O'Toole, F. (2005), 'Tolerance of Racism Exposed', *Irish Times*, 24 May.

O'Toole, F. (2006), 'Skin Colour Query Sours the Census', *Irish Times*, 4 April.

Pred, A. (2000), *Even in Sweden: Racisms, Racialized Spaces and the Popular Geographical Imagination* (Berkeley: University of California Press).

Roediger, D. (1999), *The Wages of Whiteness: Race and the Making of the American Working Class*, revised edition (London: Verso).

Rolston, B. and Shannon, M. (2002), *Encounters: How Racism Came to Ireland* (Belfast: Beyond the Pale Publications).

RTÉ (2003), *The Marian Finucane Show*, broadcast on RTÉ Radio 1, 16 October.

Spendiff, S. (2002), 'TD Calls for Action on "Sponger Immigrants"', *Irish News*, 28 January.

Tannam, M. (2002), 'Questioning Irish Anti-racism', in Lentin and McVeigh (eds).

White, E. (2002), 'The New Irish Storytelling: Media, representations and racialised identities', in Lentin and McVeigh (eds).

Chapter 12

The 'New Geography' of Ethnicity in England and Wales?

Michael Poulsen and Ron Johnston

Introduction

There has recently been considerable debate over how the geography of ethnicity has changed in Britain since 1991 during a period when the number of immigrants moving to many developed countries has reached near record levels, and birth rates have fallen. Immigrants and their descendants have provided both a replacement population of unskilled and low-skilled workers, needed to support economic development in Britain, while also making major contributions in some middle-class occupations. Most of them have chosen to reside within either gateway cities (i.e. the international nodes) or the old industrial towns where cheap housing is available and communities of their coethnics have been established. At the same time a sizeable number of the host population has been moving out of the major urban areas into smaller urban places and rural areas, in many cases reflecting longer post-retirement periods of good health. With higher levels of natural increase among the immigrant populations a number of cities are trending towards them becoming the majority there. Of concern to policy makers and the media continues to be the growth in the immigrant population living in ethnically-segregated areas and only slowly economically and socially assimilating with the white population.

How nations handle these changes is a source of much media reporting. Amongst the issues covered in the British media is a 'fear' that its multicultural society is becoming increasingly segregated, with a significant percentage of the ethnic minority population living parallel lives to their hosts. It is argued that an increasing percentage of immigrants are residing in neighbourhoods that are non-white, their children are attending schools that are even more ethnically segregated (Johnston et al. 2006), and the degree of interaction between the non-white and white population is relatively slight. These concerns were highlighted by Trevor Phillips's 2005 claim that Britain was 'sleepwalking towards segregation' – a phrase that has become almost a commonplace in media discussions of multicultural issues,[1] despite Phillips' later partial withdrawal of the analogy.[2] Ted Cantle's earlier report on the disturbances in the northern towns in 2001 concluded that the problem was the development of 'self-segregation' and 'parallel lives', and the Commission for Racial Equality's valedictory booklet (produced to mark its incorporation into the wider Commission on Equality and Human Rights) claims on its opening page that 'Segregation – residentially, socially and in the workplace – is growing' (CRE 2007, 1).

But are such claims true? Studies using the traditional indices of segregation – the indices of dissimilarity, isolation, and diversity – argue that this is not the case, claiming instead that segregation levels are slowly but steadily declining and diversity increasing across urban areas (Simpson 2004, 2007; Vertovec 2007). They contend that the major issue is increased socio-economic segregation/stratification that all ethnic and other minority groups face. But are those conclusions flawed because they reflect the use of a inappropriate methodology?

What is in dispute is not the key role of socio-economic segregation/ stratification but the methodology used to measure changes in residential segregation. Two competing methodologies exist; the traditional indices approach and the typology approach. The traditional indices are averaging approaches that discount changes in the population living in segregated areas by changes in the population living in mixed and other areas. The typology approach separates out the population living in segregated areas from those in mixed areas, and from those in exclusively white areas (for details see Johnston et al. 2005); in other words, it identifies how many people live in particular types of residential area defined by their ethnic composition. Both methodologies produce very similar results when applied to the extreme levels of segregation, as with African-Americans in US cities. But in the British context, with lower levels of segregation and higher levels of residential mixing amongst ethnic groups, they produce conflicting results. (Poulsen and Johnston (2006), for example, showed the traditional indices based on 2001 data for London and Bradford indicate a decline in levels of segregation, whereas the typology measures suggest an increase.) Use of the traditional indices of segregation is not appropriate in the British and other multicultural contexts, because they can generate misleading conclusions. The aim of this chapter, therefore, is to apply the typology approach to measure changes in levels of segregation and mix in the major ethnic urban areas in England and Wales between 1991 and 2001, updating an earlier analysis (Johnston et al. 2001).

The importance of using a more appropriate methodology to measure segregation in Britain was recently highlighted by the use made of the State of English Cities report by the Interim Statement of the Commission on Integration and Cohesion (2007a). The former states that (Office of the Deputy Prime Minister 2006, 22):

> segregation by ethnic group has declined during the past decade. The level of residential segregation fell slightly between 1991 and 2001 in 48 out of 56 cities. It increased in only 8 cities between 1991 and 2001. But this was by very small amounts in 6. In only 2 cases was the increase significant.

What that summary does not mention is that the analyses were conducted on data at the ward scale, based on a single-number index – the Index of Segregation – calculated across a set of cities the majority of which had no significant levels of segregation in either 1991 or 2001 because they had only small minority ethnic group populations. Nor does the report identify what constitutes a segregated area, or measure the change in the number or percentage of persons living in such areas. Instead it applies an index that combines data on the ethnic population who live across a wide range of different types of areas, and produces an average value across a relatively large area (many wards in urban England have populations in excess of 10,000).

But that is not where it ends. The Commission in Integration and Cohesion (CIC) interim statement (2007a, 20) took the State of English Cities findings and, faced with confusion over the measurement of segregation, concluded that:

> the main message here seemed to be that how people feel about their areas is more important than statistics – that whatever the data suggests, it is people's perceptions that will drive cohesion. Or, as one respondent put it, 'focus first on what people feel not facts.'... But above all it is our view at this point that it is social segregation more than residential segregation which is the main barrier to integration and cohesion.

Residential segregation, they conclude is, almost a 'red-herring' (2007a, 27). This deliberate break in coupling residential with social segregation is based on both the confusion over the measurement of segregation and an unstated premise that these ethnic communities are permanent. As such the CIC's interim statement and final report (Commission on Integration and Cohesion 2007a, 2007b) contend that residential segregation in itself is not an issue. This argument is in line with a multicultural policy which places the continued development of residentially-distinct ethnic communities above that of spatial assimilation. The interim CIC statement goes on to argue that the 'efforts to reduce [social] segregation must therefore focus on ensuring people are given the opportunity to meet and mix socially and in informal ways' (21) '... we want to think more about how to build integration and cohesion at a neighbourhood level by using community pride' (2007a ,26). The vision, as stated in the final report, is 'a much greater sense of shared futures and mutual interdependence .. [not] reasserting their group identities, or by focusing on the differences between them and others' (2007b, 46–7).

In this chapter we use the typology approach to examine two fundamental propositions using our typology methodology.[3] Our concern is not whether residential segregation is permanent or problematic, but rather that an appropriate methodology be used to measure levels of residential segregation in Britain, and what its use reveals. The first proposition to be evaluated is that segregation levels amongst the ethnic groups in England and Wales (as defined by the Office of National Statistics (2005), based on a self-reported ethnic identity census question) are declining as the defined minority groups (most of which are predominantly non-white, thereby failing to reflect the massive recent migration from eastern European countries) increasingly move into mixed areas (i.e. the conclusion of the State of English Cities report). The second proposition is that the changes in the ethnic composition of urban places between 1991 and 2001 is so complex and varied that the only conclusion we might draw is that everywhere is different. To address these propositions four sequential research questions are examined:

1) Is the ethnic population living in segregated areas increasing or declining? Segregated areas are defined as those in which 70 per cent or more of the population is non-white – Types IV, V and VI areas in the typology (Figure 12.1).

2) Do increases in the number of persons in the ethnic groups living in mixed areas exceed the increases of those living in segregated areas? Mixed areas are defined as white majority mixed areas which have between 20 and 50 per cent of their population non-white (Type II), and minority dominant mixed areas with between 50 and 70 per cent of their population non-white (Type III areas).

3) Can we, using regression, accurately model the changes in the percentage living in the exclusive white areas (Type I), the mixed areas (Types II and III) and the ethnically segregated neighbourhoods (Types IV,V and VI) across Britain?

4) Has there been a shift in favour of the percentage of the ethnic population living in mixed areas (Types II and III) compared with the percentage living in segregated areas (Types IV,V and VI)?

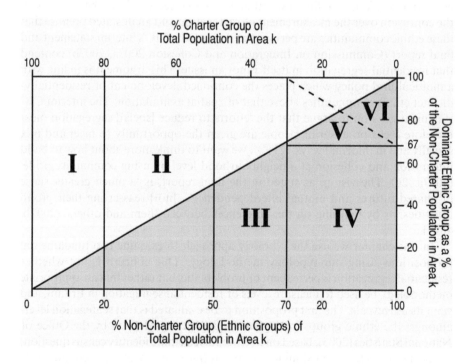

Figure 12.1 The Poulsen, Johnston and Forrest typology

To answer these questions the data were taken from the 1991 and 2001 census data at the smallest geographical scale – for England and Wales only because of the lack of comparability with Scottish data. They covered 110,148 enumeration districts (EDs) for 1991 and the 175,435 output areas (OAs) for 2001. In doing so we accept that there will be a limited degree of error in the conducting comparative analyses across two data sets that do not share common boundaries. We adopted the Simpson-Akinwale (2004) method to create a comparable ethnic data set – because the 2001 census used a larger number of categories than 1991,[4] and used Office of National Statistics enumeration districts to output areas lookup table to identify which enumeration districts and output areas were located within each urban place and for Greater London, which borough, so that the boundaries of these entities were common at each census. We then undertook the typology analyses on the enumeration district and output area data separately, and aggregated the distribution of the population across the six area types defined in Figure 12.1 four different ways, producing the following conclusions.[5] For the total population the number of members of the defined ethnic groups living in segregated areas increased over the decade. Furthermore those rates are substantial. Also increasing is the percentage of the ethnic population living in mixed residential areas, particularly those that are white dominated. The absolute number involved is approximately five times the increase in numbers in segregated areas, hence the declines in the traditional indices despite the increase in the number living in segregated areas. In a series of regression analyses modelling change in the percentage of the population living in the three types of area we accurately model change in each type using a single model. Next we compare the percentage increase by ethnic groups in mixed areas compared to segregated areas. The expectation was that the rate of increase in mixed areas would outstrip that in segregated areas but we observed the converse. The rates of increase in segregated areas were almost twice those in mixed areas, resulting in a shift towards segregated areas. As a result segregated areas had a larger percentage of the total population and the population of each ethnic group in 2001 than they did in 1991. What follows in this chapter are the incremental steps in the analyses that allowed us to establish these findings.

The Results

Table 12.1 presents data on changes in the ethnic composition of the population of England and Wales between 1991 and 2001. The largest change was the addition of 600,000 persons to the category other. Many of these persons in 1991 almost certainly identified themselves as either white, Black Caribbean, Black African or Asian, with the remainder either post-1991 immigrants or the descendants of earlier immigrants born between 1991 and 2001. Of the six major groups examined, the largest numerical increase was for the white population (421,000). The Black African (270,000) and the Pakistani populations (257,000) had similar increases, while the Indian and Bangladeshi populations increased by 147,000 and 118,000 respectively. Only the British Black Caribbean population exhibited a smaller increase, of 61,000.

Table 12.2 displays the distribution of the total population across the five different types of areas established by the typology (Types V and VI are combined because of the absence of substantial numbers living in the latter).[6] The first and

Table 12.1 Ethnic population of England and Wales 1991 and 2001

	1991	2001	Change	% change
White	47,099,453	47,520,921	421,468	0.9
Black Caribbean	502,728	564,061	61,333	10.9
Black African	209,471	479,460	269,989	56.3
Black Other	177,409	95,854	−81,555	−85.1
Indian	889,859	1,036,577	146,718	14.2
Pakistani	457,846	714,727	256,881	35.9
Bangladeshi	162,676	280,873	118,197	42.1
Chinese	146,322	226,933	80,611	35.5
Other Asian	194,816	219,376	24,560	11.2
Other	284,150	902,579	618,429	68.5
Mixed: White and Black Caribbean		237,468		
Mixed: White and Black African		78,871		
Mixed: White and Asian		188,995		
Mixed: Other Mixed		155,645		
Asian or Asian British: Other Asian		241,600		
Total persons	50,124,730	52,041,361	1,916,631	3.7

Table 12.2 Typology classification 1991 and 2001

1991	Type I	Type II	Type III	Type IV	Type V+VI	Total
Total	45,666,921	3,321,027	684,097	150,918	288,466	50,111,429
White	44,453,275	2,265,955	283,042	30,751	53,392	47,086,415
Indian	321,050	270,739	142,027	38,525	117,466	889,807
Pakistani	112,205	151,886	90,804	36,129	66,800	457,824
Bangladeshi	41,777	54,135	29,515	14,450	22,593	162,470
Black_Caribbean	188,343	228,257	60,940	15,249	9,914	502,703
Black_African	68,476	108,318	25,278	4,340	3,010	209,422
Other	481,795	241,737	52,491	11,474	15,291	802,788

Table 12.2 cont'd

% 1991	Type I	Type II	Type III	Type IV	Type V+VI
Total	91.1	6.6	1.4	0.3	0.6
White	94.4	4.8	0.6	0.1	0.1
Indian	36.1	30.4	16.0	4.3	13.2
Pakistani	24.5	33.2	19.8	7.9	14.6
Bangladeshi	25.7	33.3	18.2	8.9	13.9
Black_Caribbean	37.5	45.4	12.1	3.0	2.0
Black_African	32.7	51.7	12.1	2.1	1.4
Other	60.0	30.1	6.5	1.4	1.9

2001	Type I	Type II	Type III	Type IV	Type V+VI	Total
Total	45,335,082	4,598,286	1,309,199	329,645	454,837	52,027,049
White	43,712,488	3,105,010	540,072	66,354	83,146	47,507,070
Indian	345,712	305,242	182,460	72,496	130,564	1,036,474
Pakistani	154,768	204,852	146,909	70,851	137,251	714,631
Bangladeshi	58,025	83,359	58,586	34,783	45,793	280,546
Black_Caribbean	163,971	243,806	114,023	29,929	12,305	564,034
Black_African	108,397	222,908	116,585	20,175	11,291	479,356
Other	791,721	433,109	150,564	35,057	34,487	1,444,938

% 2001	Type I	Type II	Type III	Type IV	Type V+VI
Total	87.1	8.8	2.5	0.6	0.9
White	92.0	6.5	1.1	0.1	0.2
Indian	33.4	29.5	17.6	7.0	12.6
Pakistani	21.7	28.7	20.6	9.9	19.2
Bangladeshi	20.7	29.7	20.9	12.4	16.3
Black_Caribbean	29.1	43.2	20.2	5.3	2.2
Black_African	22.6	46.5	24.3	4.2	2.4
Other	54.8	30.0	10.4	2.4	2.4

second panels show that in 1991 approximately 91 per cent of the total population and 94 per cent of the white population lived in Type I areas. By contrast the ethnic groups had between only 24 and 37 per cent of their populations in such areas with approximately another 35 per cent in Type II areas, plus 15 per cent in Type III areas, and 20 per cent in segregated type areas (Type IV and Type V+VI). In each case, therefore, they had a majority living in the two types of mixed areas in 1991 (II and III). The third and fourth panels of Table 12.2 display similar distributions in 2001.

Table 12.3 shows the net change that occurred in the total population across the five types of areas between 1991 and 2001. For England and Wales approximately

332,000 fewer were living in Type I areas in 2001 than in 1991. Increases of 1.3 million persons occurred in Type II areas (mixed white majority areas), 625,000 in Type III areas (mixed non-white majority areas), and 345,000 in segregated areas (Types IV and V+VI). In 2001 784,000 persons lived in segregated areas (1.5 per cent), and 5.9 million in mixed areas (11.3 per cent), which was a 78 per cent increase over the number living in segregated areas in 1991. So a partial answer to the first two research questions is: (i) the total number of persons living in segregated areas did increase between 1991 and 2001; however, (ii) the total increase in numbers living in mixed areas was five times greater.

Table 12.3   Change in distribution of the total population between 1991 and 2001

	Type I	Type II	Type III	Type IV	Type V+VI
Total	−331,839	1,277,259	625,102	178,727	166,371
Percent	−4.0	2.2	1.2	0.3	0.3

When we divide the total population into two groups, those living in the set of 525 major urban places and rural/smaller urban places (in this classification Greater London refers to the built-up area not the 32 boroughs of the Greater London Authority), a different set of relationships is established. The bottom panel of Table 12.4 indicates that 1.3 million fewer persons lived in Type I areas in major urban places in 2001 than in 1991. By contrast there was a net increase of 970,000 persons in Type I areas in the rural/small urban places. These persons, who were overwhelmingly white, are almost certainly those who moved out of the major cities – whether this is a feature of lifestyle change or white flight can only be determined by future qualitative research into individual experiences. The second feature is that the growth in the population in Type II, III, IV, and V+VI areas was almost exclusively within the major urban places. The net increase in mixed areas (Types II and III) was 1.8 million, while that in the segregated areas was 342,000.

Turning to the 30 principal ethnic urban places in 1991 (Table 12.5), these accounted for 21 per cent of the population of England and Wales living in Type I areas, 89 per cent in Type II areas, 92 per cent in Type III areas, 99 per cent in Type IV areas, and 99 per cent in Type V+VI areas: virtually all of the segregated neighbourhoods were concentrated in this small number of places. The largest number of persons lived in the Greater London built-up area (7.7 million), Birmingham (964,000), Leicester (560,000), Leeds (432,000) and Manchester (404,000). Across the 30 individual urban places (Table 12.2) it was rare in 1991 for the percentage living in Type I areas to exceed 90 per cent: this occurred only in Burnley, Dudley and Southampton. By contrast in Batley, Birmingham, Bradford, Greater London, Leicester, Nelson, and Slough less than 70 per cent did so. Similarly there were major differences in proportions living in the Type II and III mixed areas: Greater London, High Wycombe, Nelson and Slough all had more than 20 per cent living in Type II areas, while Keighley, Leicester, Nelson,

Table 12.4   Net change in the total population: major urban places (MUP) and rural/small urban places (R/SUP)

1991	Type I	Type II	Type III	Type IV	Type V+VI	Total
MUP	26,762,581	3,251,308	671,306	149,493	284,675	31,119,363
R/SUP	18,904,340	69,719	12,791	1,425	3,791	18,992,066

% 1991	Type I	Type II	Type III	Type IV	Type V+VI
MUP	86.0	10.4	2.2	0.5	0.9
R/SUP	99.5	0.4	0.1	0.0	0.0

2001	Type I	Type II	Type III	Type IV	Type V+VI	Total
MUP	25,455,926	4,474,239	1,296,363	329,645	446,666	32,002,839
R/SUP	19,879,156	124,047	12,836	0	8,171	20,024,210

% 2001	1991	Type II	Type III	Type IV	Type V+VI
MUP	79.5	14.0	4.1	1.0	1.4
R/SUP	99.3	0.6	0.1	0.0	0.0

Change 1991 to 2001	Type I	Type II	Type III	Type IV	Type V+VI
MUP	–1,306,655	1,222,931	625,057	180,152	161,991
R/SUP	974,816	54,328	45	–1,425	4,380

% Change 1991 to 2001					
MUP	–6.5	3.5	1.9	0.5	0.5
R/SUP	–0.3	0.3	0.0	0.0	0.0

and Slough had more than 10 per cent living in Type III areas. And, possibly more importantly, four places had segregation levels exceeding 10 per cent (Types IV and V+VI areas); Batley, Blackburn, Bradford, Dewsbury and Leicester.

When we examine the net changes that took place in population numbers in these 30 major ethnic urban places between 1991 and 2001 all but two lost population from their Type I areas: Dudley and Preston were the exceptions. Numerically the largest losses were from Greater London (954,000), Leicester (172,000) and Birmingham (113,000). In terms of percentages, there was in excess of a 20 per cent decline in Dewsbury, Luton and Slough. Conversely the population in Type II areas increased in all but Blackburn, Burnley, Leicester and Nelson. The largest increases exceeding 10 per cent were in Coventry, Dewsbury, Luton and West Bromwich. Again Type III areas increased their population in all but six places – Bradford, Chadderton, Dudley, Keighley, Leicester and West Bromwich – while at the same time there were increases in the population living in segregated areas exceeding 5 per cent in Batley, Birmingham, Blackburn,

Bradford, Keighley and Leicester. Finally, only in Batley, Birmingham, Bradford, Chadderton, Keighley, Nelson, Oldbury/Smethwick, and Oldham did the increase in the population number living in segregated areas exceeded the increase in the population living in mixed areas.

Table 12.6 presents data on those 30 places in 1991 and 2001 ranked according to the percentage of the largest ethnic minority group there living in segregated areas (Types IV and Types V+VI), with the number of persons involved also indicated. While the urban places are primarily the same at both dates the rankings changed dramatically, while the numbers and percentage living in segregated areas in most places increased markedly for most of the dominant ethnic groups within individual urban areas. One of the most extreme changes was in Keighley, where in 1991 approximately 18 per cent of its Pakistani population (855) lived in segregated areas, whereas in 2001, 51 per cent (2868) did. In general the larger the ethnic group in a place in 1991 the larger the increase in its percentage living in segregated areas. However, Greater London as a single entity has no groups in this table, but it does have segregated areas when we examine the individual boroughs.

Within Greater London built-up area, 35 of the 66 boroughs experienced decline in the population living in Type I areas. Increases in the population in Type II areas were a feature of all but 7 boroughs; similarly an increase in the population in Type III areas, where they existed in 1991, occurred in all but two boroughs. All boroughs with a population living in segregated areas in 1991 had increases in the number of persons living in those areas; that is in Barking and Dagenham, Brent, Camden, Croydon, Ealing, Greenwich, Hackney, Harrow, Hillingdon, Islington, Lambeth, Lewisham, Newham, Redbridge, Southwark, Tower Hamlets, Waltham Forest, Westminster, and Woking/Byfleet. Increases in excess of 10,000 persons occurred in Brent, Ealing, Hounslow, Newham, Redbridge, and Tower Hamlets. And in all those boroughs with the exception of Brent and Newham the increase in the population living in mixed areas exceeded that of the increase in segregated areas. Table 12.7 presents the changes in segregation levels. For the top three boroughs (Ealing, Newham, and Tower Hamlets) in 1991 segregation levels were high for the dominant ethnic groups. Elsewhere, segregation levels were low, especially when compared to the levels across the other major urban areas. But again almost all boroughs with segregated areas in 1991 have seen a marked increase in the populations within the different groups living in those segregated areas by 2001, bringing them closer to the segregation levels recorded for the other major urban areas.

What these analyses suggest has occurred is that as the ethnic minority populations increased in most places, many Type I areas in 1991, especially those adjoining Type II areas (see Johnston et al. 2006) became Type II areas in 2001. That transition process continued on up through each of the different pairs of area types. As such, some Type II areas in 1991 became Type III areas in 2001, some Type III areas became Type IV areas, and some Type IV areas in 1991 became Type V+VI areas in 2001.

This sequence of changes suggests that a generalised model of how places are currently transitioning over time, which we can posit as the basis of further

Table 12.5 Distribution of the total population in 1991 across the different types of areas for the 30 major urban places

	GOR	Type I	Type II	Type III	Type IV	Type V+VI	Total
Batley	Yorkshire	21,694	7,023	2,010	0	4,779	35,506
Birmingham	West Midlands	666,909	125,751	71,514	58,163	41,741	964,078
Blackburn	North West	75,039	13,663	6,506	4,667	5,905	105,780
Bolton	North West	137,334	22,850	6,070	0	4,067	170,321
Bradford	Yorkshire	206,449	37,395	26,181	4,234	25,674	299,933
Burnley	North West	69,023	5,369	417	601	1,064	76,474
Chadderton	North West	37,258	2,043	1,193	0	3,236	43,730
Coventry	West Midlands	257,584	29,170	8,498	2,560	2,103	299,915
Derby	East Midlands	182,595	18,849	7,965	1,357	1,434	212,200
Dewsbury	Yorkshire	36,107	5,797	1,243	4,416	864	48,427
Dudley	West Midlands	178,311	9,844	1,820	0	1,017	190,992
Greater London	London	5,039,658	2,228,502	325,568	43,890	75,660	7,713,278
High Wycombe	South East	52,986	17,504	2,975	0	1,016	74,481
Huddersfield	Yorkshire	101,492	21,271	6,702	355	1,875	131,695
Keighley	Yorkshire	35,537	3,271	4,951	0	1,349	45,108
Leeds	Yorkshire	385,729	32,918	7,209	4,445	2,662	432,963
Leicester	East Midlands	333,793	89,246	55,896	1,836	80,126	560,897
Luton	East of England	122,902	31,963	8,469	5,810	3,831	172,975
Manchester	North West	322,063	66,169	11,665	1,121	3,326	404,344
Nelson	North West	16,342	8,972	3,011	0	793	29,118
Oldbury/ Smethwick	West Midlands	105,789	21,836	10,745	1,415	2,120	141,905
Oldham	North West	77,726	10,953	2,167	884	5,613	97,343
Preston	North West	156,338	15,249	3,370	0	2,217	177,174
Rochdale	North West	74,300	9,706	3,629	3,100	3,630	94,365
Shipley	Yorkshire	26,340	1,500	1,457	0	458	29,755
Slough	South East	59,825	33,422	11,450	3,311	1,321	109,329
Southampton	South East	206,036	2,628	2,830	380	351	212,225
Walsall	West Midlands	138,429	26,009	6,076	3,328	986	174,828
West Bromwich	West Midlands	112,010	13,123	7,215	2,149	1,007	135,504
Wolverhampton	West Midlands	183,698	44,229	20,821	803	3,829	253,380

Table 12.6 Rank order by place of percentage in a group living segregated areas

	Place 1991	Group	Population in segregated area	% in segregated area	Place 2001	Group	Population in segregated area	% in segregated area	% change
1	Dewsbury	Indian	2587	80.9	Birmingham	Bangladeshi	11613	72.3	14.7
2	Chadderton	Bangladeshi	310	76.5	Bradford	Pakistani	32270	69.3	22.2
3	Batley	Indian	3227	61.9	Leicester	Bangladeshi	94	67.8	16.4
4	Oldbury/Smethwick	Bangladeshi	671	61.2	Batley	Indian	4777	67.0	5.0
5	Birmingham	Bangladeshi	7486	57.6	Birmingham	Pakistani	41482	62.4	6.2
6	Birmingham	Pakistani	37051	56.2	Dewsbury	Indian	2631	61.6	-19.3
7	Blackburn	Indian	5441	52.2	Blackburn	Indian	8671	60.1	7.9
8	Leicester	Bangladeshi	674	51.5	Leicester	Pakistani	2108	59.6	10.5
9	Bradford	Bangladeshi	1128	50.7	Blackburn	Pakistani	5014	58.8	18.2
10	Luton	Bangladeshi	2238	49.7	Leicester	Indian	43112	57.3	9.2
11	Leicester	Pakistani	2572	49.2	Bradford	Bangladeshi	1621	55.7	4.9
12	Oldham	Pakistani	4096	48.3	Keighley	Pakistani	2868	51.4	33.2
13	Leicester	Indian	57966	48.1	Blackburn	Bangladeshi	196	51.0	15.8
14	Bradford	Pakistani	18358	47.0	Chadderton	Pakistani	1462	48.2	37.4
15	Blackburn	Pakistani	3170	40.6	Luton	Bangladeshi	3336	47.9	-1.9
16	Birmingham	Indian	19043	37.9	Oldham	Bangladeshi	570	47.5	47.5

Table 12.6 cont'd

Place 1991	Group	Population in segregated area	% in segregated area	Place 2001	Group	Population in segregated area	% in segregated area	% change
17 Dewsbury	Pakistani	1993	37.8	Chadderton	Bangladeshi	416	47.4	-29.1
18 Rochdale	Pakistani	3846	35.9	Leeds	Bangladeshi	997	45.4	12.2
19 Luton	Pakistani	3804	35.7	West Bromwich	Bangladeshi	976	45.1	42.3
20 Blackburn	Bangladeshi	93	35.2	Coventry	Pakistani	8194	43.7	18.4
21 Leicester	Black African	362	34.0	Birmingham	Indian	23470	43.1	5.2
22 Leeds	Bangladeshi	197	33.2	Derby	Pakistani	3182	42.9	17.5
23 Leeds	Black Caribbean	1355	32.0	Luton	Pakistani	6380	39.9	4.2
24 Derby	Bangladeshi	40	26.5	Coventry	Bangladeshi	1041	37.9	17.1
25 Chadderton	Black Caribbean	70	25.9	Derby	Bangladeshi	66	37.7	11.2
26 Bradford	Indian	2859	25.5	Batley	Pakistani	1444	37.4	14.8
27 Derby	Pakistani	1411	25.5	Rochdale	Bangladeshi	746	37.4	17.4
28 Coventry	Pakistani	972	25.2	Blackburn	Black African	40	36.7	31.0
29 Birmingham	Black Caribbean	10066	23.7	Rochdale	Pakistani	3746	35.3	-0.6
30 Keighley	Bangladeshi	144	23.0	Oldham	Indian	338	35.2	19.1

research. Figure 12.2 sets out four major processes and how they relate to the typology model: (i) settlement, (ii) spatial assimilation, (iii) the transition of areas (also termed displacement and succession), and (iv) out migration. As depicted, once the structure of the different areas has been established in a city, most unskilled immigrants probably move initially into either segregated (Types IV and V+VI) or majority mixed areas (Type III). Skilled migrants by contrast probably move into Type I and Type II areas. Within two generations the descendants of many unskilled migrants move out of those segregated and majority mixed areas into Type I and II areas. In doing so they displace the resident white population by occupying existing housing, and unless that spatial assimilation process stops the area will transition over time to a higher level type of area (i.e. from I towards V) through the process of natural increase. In this way spatial assimilation can be a temporary change. In some instances the white population in impacted areas responds by moving out of the urban area although occasionally the white population will move back into an area occupied by the ethnic population through the process of gentrification. The final aspect of this model is that the transition of areas primarily takes place within areas that have similar socio-economic levels, which means that the potential for mixed areas developing over time is declining, and as a result segregation levels will continue to increase. Countering that trend would be a slow down in the natural increase in the minority population, migration of the ethnic population to another urban area, an increased level of movement up by individuals to areas at higher socio-economic levels, or the transition of existing enclaves into ethnoburbs (Li 2006) as their economic status begins to increase.

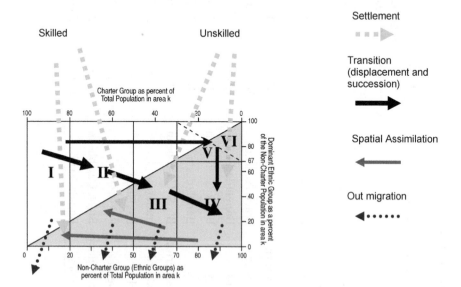

Figure 12.2 Dynamics of ethnic concentration over time

To develop this research further we explore whether these changes between 1991 and 2001 can be successfully modelled using regression analysis. Three analyses were conducted for both the 30 major urban places and the 66 boroughs of the Greater London built-up area. The aim was to explore in sequence: the causes of the increase in the total population in Type II areas; the decline in Type I areas; and the increase in segregated areas (the combined Type IV and Type V+VI areas). In the first analysis the change in the percentage of the population living in Type II areas is the dependent variable; in the second it is the change in the percentage of the population living in Type I areas; and in the third it is the change in the percentage living in segregated areas (Types IV and V+VI). In each analysis the expectation is that the size of place and their ethnic character will be key determinants of that change (as argued successfully in a recent international comparative study (Johnston et al. 2007)). The size of the place is presented as the log of the total population in 1991. The ethnic character of each place is represented by two sets of variables: the percentage change in the Indian, Pakistani, Bangladeshi, Black Caribbean and Black African populations; and three dummy variables which describe the general character of ethnicity in a place. To acquire these dummy variables the places were grouped using the SPSS K-Means algorithm on the basis of the percentage of their populations in each of those five ethnic groups in 1991 plus the percentage living in segregated areas in 1991. Three groups were extracted: for the 30 major urban places the three dummy variables are places that are predominantly Indian, Pakistani, and white in their population composition, respectively. For the 66 boroughs in Greater London the groups were boroughs that were predominantly Bangladeshi, Indian and white. Only two of the dummy variables were used in the regression analyses as the three form a closed number set so the regression results therefore need to be interpreted relative to the missing dummy variable – the white population for the 30 major urban places and the Bangladeshi population in Greater London.

The regression analyses confirm what we know about the dynamics of changing ethnic settlement patterns. The upper panel in Table 12.8 gives the results of the analyses for the 30 major urban places. In the first two columns are the results from modelling the proportional increase in Type II areas which, perhaps not surprisingly, is directly linked to the decline in the population in Type I areas. The larger the decline in the percentage in Type I areas the greater the increase in Type II areas. That growth is also linked to places that did not have a large percentage change in their ethnic group composition. However, places that were characterised as Indian, as opposed to those that were primarily white, had higher levels of increase.

In the second model we want to know why the population in Type I areas decreased more in some places than others. Interpretation of the third and fourth columns is that the decline in Type I areas occurred more in places characterised by their Indian population, and where there was an increase in the Pakistani and Bangladeshi population. And in the third model (the fifth and sixth columns) where the dependent variable is the percentage increase in the population living in segregated areas, the interpretation is that it increased most in places where there has been a marked increase in the Black African and Pakistani populations.

Table 12.7 Rank order by borough of percentage in a group living segregated areas

	Borough 1991	Group	Population in segregated area	% in segregated area	Borough 2001	Group	Population in segregated area	% in segregated area	% change
1	Wandsworth	Bangladeshi	5277	58.1	Newham	Indian	19760	66.7	24.1
2	Ealing	Indian	23541	53.2	Newham	Pakistani	9933	57.9	26.1
3	Ealing	Pakistani	3238	43.6	Wandsworth	Bangladeshi	9657	57.7	-0.3
4	Newham	Indian	11755	42.5	Southwark	Pakistani	2636	56.1	55.5
5	Newham	Bangladeshi	3155	40.9	Ealing	Indian	26701	55.0	1.7
6	Tower Hamlets	Pakistani	394	33.7	Bromley	Pakistani	896	52.9	52.9
7	Newham	Pakistani	3967	31.8	Newham	Bangladeshi	9354	51.0	10.1
8	Tower Hamlets	Bangladeshi	7337	25.4	Ealing	Pakistani	4459	42.5	-1.1
9	Hounslow	Pakistani	872	16.6	Tower Hamlets	Bangladeshi	19030	37.9	12.5
10	Hounslow	Indian	4811	16.5	Southwark	Bangladeshi	2388	36.0	36.0
11	Tower Hamlets	Indian	273	16.3	Hounslow	Indian	13246	35.6	19.1
12	Newham	Black African	1849	16.1	Bromley	Black Caribbean	2353	34.2	22.5
13	Newham	Black Caribbean	2403	15.7	Brent	Indian	15589	30.9	18.0
14	Ealing	Bangladeshi	121	15.2	Redbridge	Pakistani	4024	29.1	22.2
15	Ealing	Black African	636	14.9	Hounslow	Pakistani	2517	27.9	11.3
16	Ealing	Black Caribbean	1628	13.6	Tower Hamlets	Pakistani	306	23.9	-9.8
17	Brent	Indian	5391	12.9	Newham	Black Caribbean	4166	21.8	6.1
18	Bromley	Black Caribbean	418	11.8	Bromley	Black African	970	21.5	12.3
19	Brent	Black Caribbean	2765	11.7	Newham	Black African	6488	21.0	4.8

Table 12.7 cont'd

Borough 1991	Group	Population in segregated area	% in segregated area	Borough 2001	Group	Population in segregated area	% in segregated area	% change
20 Brent	Black African	1071	10.9	Redbridge	Bangladeshi	786	20.7	16.8
21 Brent	Bangladeshi	78	10.2	Ealing	Bangladeshi	242	20.6	5.4
22 Brent	Pakistani	701	9.6	Ealing	Black African	2276	19.2	4.3
23 Bromley	Black African	119	9.2	Tower Hamlets	Indian	567	18.8	2.5
24 Hounslow	Bangladeshi	56	8.4	Brent	Pakistani	1737	18.3	8.7
25 Redbridge	Pakistani	437	6.9	Brent	Black African	3791	18.1	7.3
26 Hounslow	Black African	149	6.7	Brent	Black Caribbean	4294	17.5	5.9
27 Newham	White	7447	6.1	Redbridge	Indian	5810	17.3	13.9
28 Tower Hamlets	Black African	241	5.5	Brent	Bangladeshi	182	16.2	6.0
29 Tower Hamlets	Black Caribbean	299	4.3	Newham	White	15302	15.8	9.7
30 Westminster	Bangladeshi	153	4.2	Redbridge	Black Caribbean	1325	15.2	13.5

Table 12.8   Modelling the percentage change of Type II (model 1), Type I (model 2) and the segregated areas (Types IV and V+VI) (model 3)

	Model 1		Model 2		Model 3	
	B	**Sig**	**B**	**Sig**	**B**	**Sig**
30 major urban places						
(Constant)	4.448	0.573	−12.183	0.074	−13.680	0.274
Log91tot	−0.445	0.757	2.245	0.072	2.197	0.335
PopT1cptp	−1.134	0.000				
Tsegcptp	−0.122	0.654				
Indiancptp	−3.000	0.001	−0.230	0.734	−2.033	0.118
Pakiscptp	−1.821	0.005	1.235	0.001	−1.368	0.028
Banglcptp	−1.338	0.020	1.035	0.012	−0.581	0.419
Black_Ccptp	0.889	0.691	3.215	0.071	−3.196	0.325
Black_Afrcptp	−3.595	0.023	−1.543	0.142	−7.467	0.001
Indian_Char	5.796	0.045	5.366	0.022	5.450	0.194
Pakist_Char	0.693	0.689	−0.654	0.673	0.431	0.882
R-square =	0.87		0.718		0.697	
Greater London Urban Area						
(Constant)	−7.521	0.351	36.198	0.064	3.563	0.306
Log91tot	−1.058	0.154	−5.025	0.004	−0.577	0.061
PopT1cptp	−1.196	0.000				
Tsegcptp	−0.929	0.009				
Indiancptp	−1.773	0.000	−4.650	0.000	−0.297	0.090
Pakiscptp	−2.518	0.008	−0.990	0.646	0.713	0.069
Banglcptp	0.400	0.582	−2.541	0.082	0.946	0.001
Black_Ccptp	−0.533	0.361	−3.154	0.022	0.035	0.883
Black_Afrcptp	−2.547	0.000	−2.430	0.000	0.303	0.004
Indian_Char	11.237	0.111	−4.288	0.794	5.598	0.062
White_Char	13.735	0.041	−13.526	0.404	−1.196	0.681
R-square =	0.948		0.715		0.896	

Notes

1   Where Log91tot = the log of the total population; PopT1cptp =change in the per cent living in Type I areas;

2   Tsegcptc = change in the per cent living in segregated areas; Indiancptcp = change in the per cent Indian; Pakiscptp = change in the per cent Pakistani; Banglcptp = change in the per cent Bangladeshi; Black_Ccptp = change in the per cent Black Caribbean;

Table 12.8 cont'd

Black_Afrcptp = change in the per cent Black African; Indian_Char = dummy variable for places characterised as Indian, Pakist_Char = dummy variable for places characterised as Pakistani; White_Char = dummy variable for places characterised as white.

3   Dependent variables. Per cent change in Type II areas, Type I areas and Segregated Areas.

4   All percentages based on the total population in 1991.

For Greater London, essentially the same interpretation applies. In the first model in the lower panel in Table 12.8, Type II areas increase where there has been a decline in the percentage living in Type I areas. These boroughs were away from where the percentage in the ethnic groups was increasing, especially the Indian, Pakistani and Black African populations, and from those boroughs experiencing increased percentages of their population living in segregated areas. The decrease in population living in Type I areas (model 2) occurred primarily in the larger boroughs outside the inner and middle suburbs, again away from where the ethnic populations were displaying large percentage increases, especially the Indians, Black Caribbeans and Black Africans, whereas the areas where segregation increased are the boroughs with major increases in their Bangladeshi and Black African populations.

Conclusion

We began this chapter by identifying two very different methodologies for determining changes in residential segregation in an urban area, which when applied to British urban areas produce conflicting information. We argued that it is important in a multicultural society to use a methodology that distinguishes between segregated and mixed residential areas, rather than use simple single-number indices which fail to make that distinction. Policy makers need accurate information on levels of segregation and mix to inform their decisions, not simple averages which conceal the complex structures.

We have used a typology methodology which we consider more suitable for analyses of segregation levels in Britain. The analyses were conducted on enumeration district and output area data, and we accepted a small degree of error in the results due to the fact that these data do not share common boundaries. The key findings from those analyses with regard segregated areas include that at the national level, in 1991, 18 per cent of the Indian population lived in segregated areas (according to our threefold typology used here), compared to 22 per cent of the Pakistani 23 per cent of the Bangladeshi, 5 per cent of the Black Caribbean, 3 per cent of the Black African, and 0.1 per cent of the white population. In total 0.9 percent (440,000) of the total population lived in segregated areas. In no case did the majority of any group live in the segregated areas. On the question of whether segregation levels are increasing in England and Wales, we established that the number of persons living in segregated areas increased by 345,000, bringing the total in 2001 to 784,000 or 1.5 per cent of the total population. That increase of 78 per cent varied across the country's urban places, however; larger increases characterised places with large Pakistani and Bangladeshi populations. In the major urban places with large ethnic minority populations, and the similar Greater London boroughs, segregation levels increased markedly between 1991 and 2001. By contrast most of the growth in ethnically-mixed residential areas occurred in parts of Greater London characterised by relatively large Indian populations. And although more people (approximately 1.5 million) moved into mixed areas, the rate of increase of the ethnic population in the mixed areas was less than the percentage increase in the segregated areas. Those rates were almost two to one in favour of the segregated areas. In addition there has been a net loss of the white population from the major urban places to rural and small urban places, totalling almost 1 million persons.

Finally this study updates our 2001 paper in three ways. First it reconfirms that residential segregation remains an important urban issue within a limited set of British cities and towns. Secondly, it suggests that increasing levels of residential segregation are likely to continue because mixed areas are primarily concentrated within the lowest socio-economic strata. Many of the areas that were classified as mixed in 2001 appear to be simply transitory and by 2011 or 2021 could well be classified under the typology as segregated areas. The fear by many policy makers that a policy that privileges the continued development of ethnic communities over spatial assimilation is wrong needs to be further researched. It is seen to led to (i) parallel lives, and (ii) limitations on the life chances of the children in those areas. Certainly qualitative research is needed on these topics to evaluate the experiences of individuals. Thirdly, despite the increasing levels of ethnic diversity in all cities, this study reconfirms the fact that most of the white population have very little potential for neighbourhood social contact with the ethnic population – or, almost certainly, for their children to have contact with members of minority groups at their local schools.

While there are many aspects of the CIC report that are very good, unfortunately they have been misled into dismissing residential segregation as important by information derived from the use of simple single index measures of segregation and studies conducted at the wrong geographical scale. Public policy that does not focus on facilitating movement out of residentially segregated areas to higher socio-economic areas of mix, or the socio-economic development of segregated areas, misses two of the key prerequisites for integration and social cohesion and as a result is unlikely to be successful.

Notes

1   As in an article in *The Observer*, 27 May 2007: http://education.guardian.co.uk/ print/0,,329932205-115999,00.html. Phillips was Chair of the Commission for Racial Equality.
2   Opening address Royal Geographical Society's Conference 2006. London.

3   The original of this chapter was presented as a paper at the Royal Geographical Society/Institute of British Geographers annual conference in August 2005, and subsequently formed the basis of Trevor Phillips' statement on 'Sleepwalking Towards Segregation'.
4   Six major groups were used: white, Pakistani, Bangladeshi, Black Caribbean and Black African. The same approach as used the State of English Cities report. Experiments on five alternative methods produced essentially the same results when the data was classified by the typology.
5   The sum of the population across the six area types were acquired for: (i) England and Wales, (ii) Major Urban Areas (525) versus Rural and Small Urban Areas, (iii) 30 major ethnic urban places, and (iv) Greater London's 66 boroughs. The 525 major urban areas were the 524 urban places at the 2001 census with a six digit code plus the 66 boroughs of the Greater London built up area (the GOR), and top 30 urban places were the subset which had population living in Type V+VI areas in 2001. Those 30 major ethnic urban places accounted in 2001 for more than ninety per cent of persons living in Type II to Type V+VI areas across England and Wales.
6   The variations between the totals in Tables 12.1 and 12.2 are because observations with fewer than 20 persons were excluded from the analyses.

References

Commission on Integration and Cohesion (2007a), *Our Interim Statement*, <http://www.integrationandcohesion.org.uk>.
Commission on Integration and Cohesion (2007b), *Our Shared Future*, <http://www.integrationandcohesion.org.uk>.
CRE (2007), *A Lot Done, a Lot to Do: Our vision for an integrated Britain* (London: Commission for Racial Equality).
Johnston, R. and Poulsen, M. (2007), 'London's Changing Ethnic Geography', *Geography Review* 21, 21–5.
Johnston, R., Forrest, J. and Poulsen. M. (2001), 'Are there Ethnic Enclaves/Ghettos in English Cities?', *Urban Studies*, 39, 591–618.
Johnston, R., Poulsen, M. and Forrest, J. (2005), 'On the Measurement and Meaning of Residential Segregation: A response to Simpson', *Urban Studies*, 42, 1221–7.
Johnston, R., Poulsen, M. and Forrest, J. (2006), 'Ethnic Residential Segregation and Assimilation in British Towns and Cities: A comparison of those claiming single and dual ethnic identities', *Migration Letters* 3, 11–30.
Johnston, R., Burgess, S., Wilson, D. and Harris, A. (2006), 'School and Residential Ethnic Segregation: An analysis of variations across England's local education authorities', *Regional Studies* 40:9, 973–1004.
Johnston, R.J., Poulsen, M. and Forrest, J. (2007), 'The Geography of Ethnic Residential Segregation: A Comparative Study of Five Countries', *Annals Association of American Geographers* 97,4.
Li, W. (ed.) (2006), *From Urban Enclave to Ethnic Suburb: New Asian communities in Pacific Rim countries* (Honolulu: University of Hawai'i Press).
Office of National Statistics (2005), *A Guide to Comparing 1991 and 2001 Census ethnic group data*, www.statistics.gov.uk/.
Office of the Deputy Prime Minister. (2006), *State of English Cities: Urban Research Summary 21*, <http://www.odpm.gov.uk/>.

Poulsen, M. and Johnston, R. (2006), 'Commentary: Ethnic residential segregation in England: Getting the right message across', *Environment and Planning A* 38. 2195–9.

Simpson, L. (2004), 'Statistics of Racial Segregation: Measures, evidence and policy', *Urban Studies* 41, 661–81.

Simpson, L. (2007), 'Ghettos of the Mind: The empirical behaviour of indices of segregation and diversity', *Journal of the Royal Statistical Society A* 170, 405–24.

Simpson, L. and Akinwale, B. (2004), *Quantifying Stability and Change in Ethnic Group* (Manchester: Cathie Marsh Centre for Census and Survey Research, University of Manchester).

Vertovec, S. (2007), 'Super-diversity and its Implications', *Ethnic and Racial Studies* 30, 1024–54.

Chapter 13

The Problem with Segregation: Exploring the Racialisation of Space in Northern Pennine Towns

Deborah Phillips

The twenty-first century has seen a reawakening of discourses around residential segregation and social integration in Britain. It is a racialised discourse, which has little to say about the segregation of people's lives on the basis of social class or lifestyle. It focuses instead on ethnic separation and particularly on the settlement patterns of minority ethnic groups (as opposed to majority whites groups) and those cast as cultural and religious 'others'. Following the urban disturbances in several Pennine towns in 2001, the political and policy discourse on ethnic segregation has become integral to wider debates about national identity, unity and citizenship in multicultural Britain. The persistence of minority ethnic clustering, particularly within poorer inner-city areas, tends to be read as evidence of the failure of migrants to follow normative paths towards social and spatial integration. Explanations as to why such groups continue to cluster are contested, although it is evident that poverty, a lack of structural integration of disadvantaged groups, racism, religious and cultural factors, social capital, and a legacy of racial discrimination in the housing market all play a role (Phillips 1998, 2006). Recent political and policy debates about solutions to the 'problem' of segregation, however, have largely been constructed in cultural terms. Ethnic segregation is seen to be indicative of minority ethnic isolationism, weak social integration and the failure of multiculturalism to unite the nation through common values and identity.

Britain is not alone in expressing anxiety about the apparent entrenchment of minority ethnic segregation and the development of so-called 'parallel societies' along ethnic lines. The EU Agency for Fundamental Human Rights (FRA 2007) notes that many northern European Union member-states are expressing growing disquiet about ethnic segregation, and its apparent intensification, and Muslim populations find themselves at the epicentre of concern (EUMC 2006). Politicised discourses on segregation have brought calls for interventions to promote social and spatial integration in many countries. Arguments for targeted policy initiatives, which include de-concentration through urban renewal and migrant dispersal, are often framed in terms of a desire to avoid the development of United States style ghettos; an international benchmark of a failing and divided society.

New Geographies of Race and Racism

This chapter begins by critically examining the politicised discourse on segregation in Britain and its policy implications. It explores the construction of minority ethnic clustering as a 'problem' and argues that this discourse embodies anomalies and contradictions. The chapter goes on to investigate everyday understandings of 'race' and place amongst British Asian and white young people living in three northern, former textile towns; Bradford, Oldham and Rochdale. These localities have in many ways become emblematic of the failings of multicultural Britain in the public imagination. Bradford and Oldham experienced racialised disturbances in 2001 and the legacy of a historical racialisation of space in all of these towns lives on through the choices young people make about where to live. We nevertheless see a trend towards diminishing ethnic segregation and the desire, on the part of many young people, for further social and spatial mixing.

Constructing Ethnic Segregation as a Problem

The problematisation of ethnic segregation is not new, although the object of concern has been constructed in different ways at different points in time. The nineteenth and early twentieth centuries, for example, witnessed national expressions of anxiety about the integration and settlement of Irish and 'alien' Jewish 'others' because of the presumed threat they posed to the social integrity and economic health of the nation (Gainer 1972). By the 1950s and 1960s, however, political attention had been realigned towards the apparent dangers associated with the concentration and segregation of newly arriving colonial migrants from India, Pakistan and the Caribbean. Policy interventions during this era included the institutionalisation of a discriminatory programme of black minority ethnic tenant dispersal by Birmingham City Council in the 1970s (Henderson and Karn 1987). Then, in 2001, the focus of political concern shifted once again; this time from apprehensions about 'race' and integration to the potentially divisive effects of cultural and religious difference. The urban unrest experienced by some northern cities highlighted tensions between British Asians (most of whom were Muslims) and whites in these particular places and raised wider anxieties about the link between residential segregation, social integration and the cohesion of different faith communities in multicultural Britain.

Official reports into the northern disturbances, and the media coverage at the time, were replete with the spatial vocabulary of polarisation, segregation, ghettoisation, territorialisation and no-go areas. The dominant political discourse was one of rising levels of ethnic residential segregation and poorly integrated minority ethnic groups, often with strong transnational connections. Separate lives, it was claimed, were being sustained by the 'self-segregation' of British Muslims, a theme captured in the national discourse on 'parallel lives' (Phillips 2006). The government's response to the 'problem' was to advocate policy initiatives to 'break down segregation' and promote community cohesion (Community Cohesion Panel 2004). After the London bombings in 2005, the justification for such policy interventions was widened to draw in narratives about national security, which have represented segregated, introspective British Muslim communities as both social and spatially isolated and an incubator for so-called 'home-grown terrorists' and radicalism.

At first glance, the argument for breaking down segregation may seem compelling, especially given the spectre of United States-style ghettos alluded to by the chairperson of the former Commission for Racial Equality, Trevor Phillips (2005). Americans have a long history of tackling ghettoisation through programmes of population deconcentration (Goetz 2003). However, a more critical look at the evidence for ethnic segregation in Britain, its likely causes and the link between residential segregation and social integration casts some doubt on the direction of recent integration and cohesion debates and raises questions about appropriate policy interventions.

Measuring Ethnic Segregation: The Evidence

Politicians and policy makers have been keen to acquire 'objective', quantitative indicators of ethnic segregation. Such measures, however, embody a degree of subjectivity and uncertainty that is rarely acknowledged in the policy domain. The assertion that levels of ethnic segregation are rising, for example, has been contested. Both Simpson (2007) and Dorling (2007) contend that this is a 'myth', and that we should be more concerned about the growing separation of the social classes in Britain than ethnic segregation. Simpson argues that the indices of segregation for all ethnic minority groups identified by the census in fact fell between 1991 and 2001. Others (e.g Poulsen in this volume) have, however, disputed this, and have used different indices of geographical separation to support the argument that ethnic segregation is intensifying. The controversy over the relative merits of particular statistical indicators (as illustrated in the exchange between Simpson (2004) and Johnston et al. (2005), and in Wong's (2004) analysis of the segregation 'index war') suggests that we should be treating claims about growing ethnic segregation with caution.

Arguments over appropriate indicators reflect the multi-dimensionality of segregation as a construct. Political and policy anxieties about minority ethnic segregation are particularly grounded in concerns about 'exposure' or 'isolation', as evidenced by the 'parallel lives' debate and notions of 'self-segregation'. The quantitative evidence would seem, however, to point to a more complex picture; one that cannot simply be equated with minority ethnic withdrawal from mixing. British Asian concentrations, for example do seem to be increasing in many inner-city areas, but this, Simpson (2007) argues is as much a feature of demographic growth 'in situ' through natural increase, immigration and white out-migration (white flight), as 'self-segregation' and withdrawal. Furthermore, ethnic clusters are not necessarily static phenomena, but will incorporate transient populations that move through the urban structure with time. It is clear that increasing numbers of British Asians, including Muslims, are dispersing away from established inner-city clusters to relocate in more ethnically mixed neighbourhoods (Simpson 2004; Phillips 2006). As indicated later in the chapter, however, this mobility can be hindered by fears of racism and racist harassment, and the sustained growth of mixed neighbourhoods can be impeded by 'white flight'.

There are also conceptual challenges to measuring changing levels of segregation, which tend to be sidestepped in political discourses and policy debates. The reconfiguration of ethnic and 'race' categorizations over the decades and shifting conceptualisations of ethnic and racial difference complicate comparisons over time. The social construction of ethnic groups in the census politicises the process and the object of measurement (and the dynamics of 'concern'), bringing invisibility for some minority groups and an overly zealous attention to other populations (typically, in the past, non-whites[1]). The growing number of mixed-heritage households and people who feel they have hybrid or multiple ethnic identities complicates the analysis further. Finally, since there is no objective definition of 'intense' or 'acceptable' segregation, uncertainty clouds policy-related judgements about when a group's level of clustering might become a cause for 'concern'. This raises questions about what ethnic segregation means and whether it is appropriate to construct it as a problem.

Interpreting Ethnic Segregation

The reasons for persistent ethnic segregation have been widely debated in the academic and policy domain. An extensive literature points to the role of social, cultural and structural factors in producing and sustaining minority ethnic clusters (Huttman 1991; van Kempen and Ozuekren 1998; Phillips 1998; Goldberg 1998; Johnston et al. 2002), although the significance of minority choice versus constraint, and the power of individual agency versus institutional discrimination to shape ethnic geographies remain contentious. Qualitative research, such as that presented later in the chapter, indicates that minority ethnic clustering is unlikely to be a product of one set of forces, but rather reflects the intersection of a range of factors related to 'race', social class, culture, religiosity, lifestyle etc.

Contemporary political discourses on community relations in Britain are nevertheless apt to represent the minority ethnic segregation of some, but not all groups[2], as a sign of failure; the result of minority ethnic groups' reluctance to adapt to mainstream society and/or the outcome of the misguided precepts of multiculturalism. These recent discourses rest on normative assumptions about the social and spatial trajectories of minority ethnic groups and, in the case of British Muslim clustering, are primarily underpinned by worries about the incorporation of cultural and religious difference. In the process, the segregation of those marked as religious, cultural and ethnic 'others' tends to be problematised, whilst the segregation of white Christians and Jews in the suburbs or the protective, gated enclaves of our cities draws little attention. This is simply viewed as normal in multicultural Britain.

The construction of British Muslim segregation as a particular problem deserving of policy attention is encapsulated in the discourse on 'self-segregation' that followed the 2001 disturbances (see Phillips 2006) and Trevor Phillips's assertion that Britain is 'sleepwalking' towards United States style ghettos. The use of the emotive language of the 'ghetto' evokes images of poverty, exclusion and polarisation. Evidence suggests that British Muslims are indeed overrepresented in some of the poorest neighbourhoods of our cities (Beckford et al. 2006; Peach

2006). Nationally, one third of Muslims live in the bottom decile of deprived areas, a pattern broadly replicated in our case study localities. It is therefore perhaps surprising that Trevor Phillips' speech gave less weight to the need for poverty alleviation than the imperative of social and cultural integration on the part of self-segregating minorities. The ensuing public debate focussed on improving 'community relations' (*Hansard* 28 November 2005) and endorsed the nation's community cohesion agenda.

The government's approach to tackling social and spatial divisions in Britain's cities through the community cohesion agenda has been widely criticised for its desegregationist ethos (Phillips 2006), assimilationist tone (Lewis and Neal 2005), the primacy of the local neighbourhood as a setting for interaction (Robinson 2005; McGhee 2005), the de-racialisation of inequality (Worley 2005; Harrison et al. 2005) and the romanticising of community (Phillips et al. 2008). The cohesion agenda is rooted in the premise that greater intercultural contact at the neighbourhood scale will bring social integration and help to foster a sense of common identity and belonging. This presumption is, however, contentious. On the one hand, there is some evidence to support the 'contact hypothesis' (Vertovec 2007). For example, Sanjek's (1998) study of a diverse neighbourhood in New York revealed how a shared 'politics of place', or common concerns about local 'quality of life' issues, helped to build alliances between settled and new groups. On the other hand, the relationship between social integration and spatial mixing at the neighbourhood level is unclear. Contact does not always bring positive interaction and there is a risk that increased contact may reinforce perceptions of 'difference'. Past evidence suggest that unwelcome incursions into a neighbourhood by those labelled 'outsiders' may bring isolation for the newcomers and can lead to racist harassment (Chahal and Julienne 1999; Amin 2002; Phillips et al. 2007). Furthermore, the strength of local bonds to, and within, a neighbourhood clearly varies between different groups of people. For some, especially professional people, other spheres of interaction (e.g. work and virtual spaces) are more important than residential spaces for expressions of self, the forging of friendships and the making of identity. National and transnational connections and post-national forms of citizenship may also render the neighbourhood relatively insignificant for some minority ethnic families, whose communities of association may well extend beyond their places of residence (e.g. Soysal 1994; Kaplan and Holloway 2001; Kennett and Forrest 2006).

Every local authority has been enjoined to devise strategies to promote cohesion in Britain's fractured communities. The task is focussed on the social and spatial divisions associated with ethnic difference. The possibilities for enhancing intercultural engagement at the local scale are, however, likely to depend on a number of contextual factors, which reflect the specificity of place and the shape of ethnic relations there. As Amin (2002) has argued, local histories and politics of place, local micro-cultures, employment conditions, new migration streams etc. will all help to mould opportunities for interaction and possible spaces of engagement.

In the next section, we explore the potential for growing social and spatial interaction at the neighbourhood scale in the northern towns of Oldham, Rochdale and Bradford from the perspective of British Asian and white young people living there. These particular localities have a special place in the national discourse on citizenship and belonging in twenty-first century Britain. All have sizable British Muslim populations of Pakistani, Kashmiri and Bangladeshi origin, which are characterised by relatively high levels of unemployment, deprivation and spatial clustering. Following the 2001 disturbances, Oldham and Bradford were explicitly labelled as cities troubled by Muslim 'self-segregation' and ghettoisation, and Rochdale has, by association, been tarred with the same brush.

The findings presented here derive from research conducted between 2001 and 2006 as part of two projects exploring the housing and neighbourhood decisions made by different ethnic groups, and the way in which geographies of ethnic segregation in each town were interpreted and imagined (see Phillips 2006; Phillips et al. 2007; Simpson et al. 2007). Qualitative data were collected through household interviews and focus groups with young people of Pakistani/Kashmiri and Bangladeshi origin and whites aged between 18 and 25. The focus groups were constructed on the basis of gender, ethnicity and current living arrangements, and most of the minority ethnic participants lived in inner-city neighbourhoods associated with their ethnic group.

Racialised Spaces in Three Northern Towns

Ludi Simpson has produced statistical evidence to suggest that, contrary to popular perceptions, ethnic residential clustering in Oldham, Rochdale and Bradford is not growing. In fact, in line with the rest of England and Wales, segregation decreased in Oldham and Rochdale between 1991 and 2001, and showed no change in Bradford (Simpson and Gavalas 2005; Simpson 2004). For example, the Index of Dissimilarity for electoral wards in Oldham decreased from 65 to 63 between 1991 and 2001, and from 60 to 58 in Rochdale.

There is nevertheless a clear racialisation of space within these towns, which lives on in people's imaginations and influences decisions about where to live. Young focus group participants and interviewees indicated that certain areas were strongly associated with particular minority ethnic groups or labelled as 'white areas', although there was a recognition that some neighbourhoods were gradually becoming more ethnically mixed. There was certainly evidence of a growing out-migration of British Asians from the inner-city areas most closely associated with their group. There were, however, also perceived barriers to mobility. Some young people were particularly reticent about the prospect of moving into 'white' areas; as one Pakistani woman in Rochdale commented, 'We were scared to move where we didn't know what it was like'. This feeling was closely linked to worries about a sense of isolation (especially for women) and fear of racist harassment.

Our research with young people in Oldham, Rochdale and Bradford revealed a complex picture of housing and neighbourhood aspirations and preferences, which varied with social class, gender, ethnic group and individual biographies. Nevertheless, both white and Asian young people specified similar attributes when describing a desirable neighbourhood; security, safety, low crime rates, good community and neighbours, and a pleasant environment. It was clear that many young people from all ethnic groups valued their attachments, or the idea of attachments, to a local 'community'. However, the research provided little evidence of a desire for the social isolation implicit in the dominant political discourse on Asian 'self-segregation' and withdrawal. Whilst many young British Asian interviewees felt that their parents would think of 'community' in terms of people from the same ethnic/religious background, these young people conceptualised it more broadly in terms of 'good' or 'tolerant' people from any ethnic background. In fact, many went one step further in asserting a preference for living in an ethnically mixed community because, in the words of one Muslim woman, 'you learn from each other and it can only make us richer'.

It may be argued that the greatest challenge to greater residential mixing lies neither in 'cultural difference', nor in the strength of introspective communities, but in the constraints on spatial mobility and settlement experienced by minority ethnic groups. The continuing racialisation of space means that some people found it difficult to envisage crossing the established boundaries (real and imagined) of ethnic space. This section explores how these boundaries are perpetuated by national discourses, local myths and the legacy of racism.

Discourses and Myths

National discourses and local myths (often perpetuated by the media) continue to place minority ethnic groups in specific, usually marginal spaces of the city. It might be argued, for example, that references to 'ghettos' and to the process of 'ghettoisation' ultimately serve to reinforce racialised social and spatial divisions. The idea of 'ghettoised' inner areas stands in contrast, in the public imagination, to the 'whiteness' of many other parts of the city, drawing racially-coded distinctions between the urban and the rural, and the 'dark' inner-city and the white suburbs (cf. Bonnett 2002).

One focus group of young Muslim men living in Rochdale felt that the stigmatisation of the 'Asian' areas in this town was holding them back. They argued that they needed to 'get out [of their area] ... in order to get on' because their neighbourhood was '... almost like a ghetto'. Such views were not universally held, although there was a general feeling amongst both Asian men and women that ethnic mixing at the neighbourhood level could help ethnic groups to learn more about each other. The sentiments expressed by one British Pakistani woman in Oldham when talking about ideal neighbourhoods were not uncommon:

> Why can't we have a mixed balance of Asians and whites and appreciate each others' cultures?

Both whites and Asians, however, said they might feel 'out of place' in the urban spaces traditionally associated with other ethnic groups. The comments of a British Muslim woman, who had considered moving to a rural, white area, illustrate the obstacles to crossing racialised boundaries:

I think there's places, even in Rochdale, like [village] ... and you go there, and people just look at you funny .... At one stage we were looking for a house there ... but just the thought of moving there, and people staring at you and everything .... They look at you like you're an alien or something.

National discourses on Muslim citizenship and belonging in Britain, as well as constructions of religious 'otherness', have touched the young British Asians living in the three research localities. In so doing, they have played a role in shaping their experiences, and readings, of these places. Muslim women in Oldham, for example, said that they felt 'in the firing line' when walking in public places, especially if veiled, because of 'identifying yourself as a Muslim and all the negativity'. One woman commented on the stereotyped reaction of white people, which she directly attributed to media coverage, saying:

... its all the publicity that Islam has had cause they don't know you as a person, they just know that you're a Muslim ... they just see 'right Muslim woman: trouble.

Bradford women also discussed how the national vilification of Islamic states and British Muslims made them 'feel black' and 'feel different', whereas before they saw themselves as 'just a Bradfordian'.

Local myths about the difficulty of building bridges between ethnic communities abound. The discourse on 'parallel lives' surfaced in the course of accounts of how one ethnic group did not want to talk to, or mix with, another group. Both Asian and white focus group participants invoked similar narratives of avoidance by the other group. Myths were, however, often challenged in the course of discussions, with participants qualifying their descriptions of social distances between ethnic groups with examples of friendships/neighbourliness across them. A white focus group participant in Oldham, for example, started out by complaining that Asians '... just wanted to be in their own little community, and I felt like they didn't want to speak to me ...', but then went on to recount a number of positive contacts with her Asian neighbours.

The Legacy of Racism

Young people's perceived neighbourhood options and views on the possibilities of mixing were in part shaped, in all three localities, by decades of institutional racism and other exclusions within the local housing market (Ratcliffe et al. 2001; Commission for Racial Equality 1990; Phillips et al. 2007). This legacy of discrimination has undoubtedly reinforced minority ethnic attachments to particular inner-city neighbourhoods and will impede the pace of change. Many young women, both Asian and white, aspired to live within easy access of their families, and some young Muslim men (particularly elder sons) talked of remaining 'close by' because of their family obligations. Although private transport is loosening constraints on mobility, there was still a strong preference by many Asian women in particular to live 'not too far away' from parents and other family members in the established ethnic areas. As more British Asian women find work outside the home, they are increasing looking to the family for childcare. As one young Bangladeshi woman living independently in Oldham explained:

> You've got your own separate houses, but you still spend a lot of time with the main family, like your parents or the grandparents or whoever it is .... They help with the cooking and the children when we are at work.

Such close attachments may well help to perpetuate spatial clustering, even when there is an expressed desire for greater mixing.

The legacy of racism also lives on in the reputation of areas, and the belief (both real and imagined) that certain urban and rural spaces might be closed to minority ethnic groups, especially British Muslims. White working class areas were still avoided by many British Asians in all three localities, with white social estates that were seen as 'rough' and 'unwelcoming to Asians' coming top of the list. One Pakistani woman in Rochdale summed up the legacy of exclusion from social housing noting, 'I think traditionally our parents were afraid of council estates, so that's made us afraid'. She observed that patterns were changing, but felt that 'ideally nobody wants to move there'.

Anxieties about racism and racist harassment present a real barrier to spatial mobility and to achieving, and sustaining, ethnic mixing. Young Bangladeshi men in Rochdale contended that their housing options were constrained by the risk of harassment, explaining that they wanted 'cheap house prices, but if it means that you have to compromise on getting abused and having all these problems ... we won't touch that'. A Pakistani woman in Rochdale commented:

> I think, being Asian, we've always got the fear that if we move into a white area where there are no Asians, we are going to get racially harassed

In Bradford, nearly one in five of those interviewed said that they had experienced harassment in their current neighbourhood, even though they had opted to live in areas they perceived to be safe. One young Bangladeshi man recalled an incident from his childhood which made a lasting impression:

> I went to [a social housing estate] and this women turns round and 'says you black bastard what are you doing here?' ... and I still remember that, every time I go into [that area]....so I've got fear

Harassment has long served to re-inscribe the whiteness of social housing estates in Bradford as elsewhere (cf. Bowes et al. 1998). Middle-class areas of private housing were generally seen as less threatening, although some still viewed them with caution. As one young person said, in middle class areas 'you could be walking on the street and nobody's going to call you a black git are they?' Many felt that such areas may well be superficially convivial, although there was some scepticism about the likelihood of sustained positive encounters.

Crossing Boundaries: Sustaining Ethnic Mixing

The racialisation of space has hindered movement across neighbourhood boundaries in Oldham, Rochdale and Bradford, but it has not halted it. Whilst particular areas were seen as out of bounds by many, there appeared to be a growing confidence amongst some young British Asians that they could venture into 'white' areas, although concerns about safety were always paramount. The impetus for mobility comes mainly from a desire for better private housing or affordable social housing, although some young Asians also sought a more socially integrated lifestyle for themselves and their children.

The legacy of ethnic exclusion and segmentation within the social housing markets of Oldham, Rochdale and Bradford has posed particular challenges to young British Asians wishing to cross the boundaries of these particular racialised spaces, as well as to policy makers concerned with implementing the community cohesion agenda. Social housing providers have taken a number of steps to erode barriers to integration within this sector in recent years (see Ratcliffe et al. 2001; Blackaby 2004). Supported tenancies schemes implemented in Oldham and Rochdale, which assist minority ethnic households wanting to move into traditionally white social housing areas, have received national acclaim. Both Asian and white young people in our focus groups praised the Community Induction Project operating in Rochdale, which has facilitated the social transformation of a particular housing estate. One Pakistani woman described this estate as now having 'a fair mix – you haven't got too much of one [group] or too much of the other'. Another group in Oldham saw the potential for creating new intercultural spaces, free from a history of ethnic tension, through the construction of new-build housing. Referring to a specific new development, one Asian woman observed:

> … there's no set nationality or culture settled there, so all sorts of people are buying houses there. Nobody's asking 'what kind of area is it? who lives there?', because nobody lives there. It is [a] completely new area … we have got a mix of Pakistanis, whites, Bangladeshis, Indians, Chinese, black people all speaking when you walk around. It's a good mix.

Evidence suggests, however, that sustaining such mixing, especially in areas with a history of ethnic tension, may require long-term policy interventions. This demands both skills and resources, and good, but fragile, relations can be easily disrupted by events (e.g. security alerts) that reignite deeper suspicions and resentments.

The potential for achieving sustained ethnic mixing in the private housing areas of Oldham, Rochdale and Bradford is uncertain. British Asian households are gaining a foothold in high status, previously white neighbourhoods in all these localities; areas like Chadderton in Rochdale and Heaton in Bradford are notable for their increasing ethnic diversity. Our research also uncovered an expressed desire for greater social and spatial mixing amongst many Asian and white young people in these towns. However, there may be limits to the direction and progress of residential mixing.

First, spatial integration is mainly a one-way process of outwards movement of people of Asian origin. Young Asians were sceptical about the likelihood of whites moving into 'their areas', now or in the future. One Bangladeshi woman thought there was 'no way' that white people would move into her area in inner Rochdale, and a white focus group later endorsed her view.

Second, can ethnic mixing at the neighbourhood level be sustained? Again Asian youngsters were sceptical, recounting narratives of 'white flight' in all localities. In Rochdale, for example, a Bangladeshi man described how

> ... when we first came to this area there were many English, but now they have all gone
> ...We can tolerate them but they can't tolerate us! [Laughs]

Another young man thought that white out-migration was 'going to be the trend, whether we'd want that or not'. Focus group participants in Oldham and Bradford told a similar story, observing that 'You see "for sale" signs everywhere' once Asians start moving into an area, concluding that '... Slowly, slowly – all the white people move out ... They feel invaded'.

An analysis of migration data suggests that so-called 'white flight' from 'Asian' areas is probably relatively small (Simpson et al. 2007), although white people may stop moving into an area that they perceive to be changing. There was, nevertheless, a perception amongst both Asian and white youngsters that 'white flight' was helping to sustain racialised divisions in these towns.

Conclusions

The dominant discourse on ethnic segregation as a problem associated with cultural difference can be challenged. Alternative demographic interpretations of statistical data on ethnic segregation, which emphasise the natural growth of youthful populations, suggest that 'commentators have been misled by the numbers game, which see large numbers of Black and Asian people as a threat to integration' (Simpson 2006, 22). Qualitative research in towns such as Oldham, Rochdale and Bradford, which epitomise the so-called 'crisis of community cohesion' (Flint and Robinson 2008), indicate that there is a discernable trend towards British Asian spatial dispersal and, importantly, a desire for greater ethnic mixing at the neighbourhood scale. Although mobility amongst older community members is relatively low, younger generations are laying claim to new urban spaces. A suburban presence reflects the growing class differentiation within the British Asian population and new affordable housing options are slowly emerging in the previously exclusionary social rented sector. Whereas initiatives to widen minority ethnic group access to 'white' social housing estates failed twenty years ago (Phillips 1987), largely because of weak institutional support, they now seem to be achieving a measure of success.

There is thus room for cautious optimism. Some of the barriers to minority ethnic residential mobility are being eroded and there seems to be a genuine desire, on the part of some young people at least, for greater intercultural engagement. The relationship between spatial segregation and social integration is still, however, unclear. Furthermore, national discourses on segregation and integration, together with evidence on the ground in places like Oldham, Rochdale and Bradford, places the responsibility for, and the costs of, social and spatial integration largely on the shoulders of minority ethnic groups. The assimilationist and de-segregationist tone of the community cohesion agenda put the onus on minority ethnic group change. As McGhee (2005) has argued, community cohesion seeks to 'replace past attachments to tradition, culture and faith with future oriented allegiances based on civic engagement and active citizenship'. The de-racialised language of 'community', however, diverts attention from the legacy of structural inequalities, such as poverty and inequality of housing access, that helps to keep minority ethnic groups in areas dubbed as 'ghettos'. In addition, whilst national discourses continue to represent minority ethnic groups, especially British Muslims, as cultural and religious 'others', the costs of crossing the boundaries of racialised spaces will be high; feelings of isolation, of being 'out of place' in white areas, the experience of harassment and 'white flight' are all signs of the continuing marginalisation and outsider status of Britain's minority ethnic citizens.

Notes

1   Much policy-related research, for example, has a tendency to racialise constructions of 'ethnicity', which is seen as something ascribed to 'others', particularly people who were 'not white'.
2   The residential segregation of the Jewish minority in twenty-first century Britain provides a clear exception.

References

Amin, A. (2002), 'Ethnicity and the Multicultural City: Living with diversity', *Environment and Planning A* 34, 959–80.
Beckford, J., Gale, R., Owen, D., Peach, C. and Weller, P. (2006), *Review of the Evidence Base on Faith Communities* (London: ODPM).
Blackaby, B. (2004), *Community Cohesion and Housing: A good practice guide* (London: CIH).
Bonnett, A. (2002), 'The Metropolis and White Modernity', *Ethnicities* 2:3, 349–66.
Bowes, A., Dar, N. and Sim, D. (1998), *Too White, Too Rough and Too Many Problems: A study of Pakistani housing in Britain* (Stirling: University of Stirling).
Chahal K. and Julienne, L. (1999), *'We Can't All Be White': Racist victimisation in the UK* (York: Joseph Rowntree Foundation).
Commission for Racial Equality (1990), *Racial Discrimination in an Oldham Estate Agency* (London: CRE).

Community Cohesion Panel (2004), *The End of Parallel Lives? The Report of the Community Cohesion Panel* (London: Home Office).

Dorling, D. (2007), 'A Think Piece for the Commission on Integration and Cohesion', research report submitted to the Commission on Integration and Cohesion.

EUMC (2006), *Muslims in the European Union: Discrimination and Islamophobia* (Vienna: European Monitoring Centre on Racism and Xenophobia).

Flint, J. and Robinson, D. (2008), *Community Cohesion in Crisis* (Bristol: Policy Press).

FRA (2007), *Trends and Developments 1997–2005: Combating ethnic and racial discrimination and promoting equality in the European Union* (Vienna: EU Fundamental Rights Agency).

Gainer, B. (1972), *The Alien Invasion: The Origins of the Aliens Act of 1905* (London: Heinemann).

Goetz, E. (2003), *Clearing the Way: Deconcentrating the poor in Urban America* (Washington, DC: Urban Institute Press).

Goldberg, D.T. (1998), 'The New Segregation', *Race and Society* 1, 15–32.

Harrison, M., Phillips, D. Chahal, K. Hunt, L. and Perry J. (2005), *Housing, 'Race' and Community Cohesion* (Chartered Institute of Housing, London).

Henderson, J. and Karn, V. (1987), *Race, Class and State Housing: Inequality and the allocation of public Housing in Britain* (Aldershot: Gower).

Huttman, E. (1991), *Urban Housing Segregation of Minorities in Western Europe and the United States* (Durham, NC: Duke University Press).

Johnston, R., Forrest, J. and Poulsen, M. (2002), 'Are There Ethnic Enclaves/Ghettos in English Cities?', *Urban Studies* 29:4, 591–618.

Johnston, R., Poulsen, M. and Forrest J. (2005), 'On the Measurement and Meaning of Residential Segregation: A response to Simpson', *Urban Studies* 42:7, 1221–7.

Kaplan, D. and Holloway, S. (2001), 'Scaling Ethnic Segregation: Causal processes and contingent outcomes in Chinese residential patterns', *Geojournal* 53:1, 59–70.

Kempen, R. van and Ozuekren, S. (1998), 'Ethnic Segregation in Cities: New forms and explanations in a dynamic world', *Urban Studies* 35:10, 1631–56.

Kennett, P. and Forrest, R. (2006), 'The Neighbourhood in European Context', *Urban Studies* 43:4, 713–18.

Lewis, G. and Neal, S. (2005), 'Contemporary Political Contexts, Changing Terrains and Revisited Discourses', *Ethnic and Racial Studies* 28:3, 423–44.

McGhee, D. (2005), *Intolerant Britain? Hate, Citizenship and Difference* (Maidenhead: Open University Press).

Peach, C. (2006), 'Muslims in the 2001 Census of England and Wales: Gender and economic disadvantage', *Ethnic and Racial Studies* 29:4, 269–55.

Phillips, D. (1987), 'The Rhetoric of Anti-racism in Public Housing Allocation', in P. Jackson (ed.) *Race and Racism: Essays in Social Geography* (London: Allen and Unwin).

Phillips, D. (1998), 'Black Minority Ethnic Concentration, Segregation and Dispersal in Britain', *Urban Studies* 35:10, 1681–702.

Phillips, D. (2006), 'Parallel Lives? Challenging Discourses of British Muslim Self-segregation', *Environment and Planning D: Society and Space* 24:1, 25–40.

Phillips, D., Davis, C. and Ratcliffe, P. (2007), 'British Asian Narratives of Urban Space: Changing geographies of residence in multi-cultural Britain', *Transactions of the Institute of British Geographers* 32:2, 217–34.

Phillips, D., Simpson, L. and Ahmed, S. (2008), 'Remaking Communities', in J. Flint and D. Robinson (eds) *Community Cohesion in Crisis* (Bristol: Policy Press).

Phillips, T. (2005), 'After 7/7: Sleepwalking to segregation', speech to the Manchester Council for Community Relations, 22 September 2005 (available from http://www.cre.gov.uk).

Ratcliffe, P., Harrison, M, Hogg, R., Line, R., Phillips, D., Tomlins, R. and Power, A. (2001), *Breaking Down the Barriers: Improving Asian access to social rented housing* (Coventry: Chartered Institute of Housing).

Robinson, D. (2005), 'The Search for Community Cohesion: Key themes and dominant concepts of the public policy agenda', *Urban Studies* 42:8, 1411–28.

Sanjek, R. (1998), *The Future of Us All: Race and neighbourhood politics in New York City* (Ithaca, NY: Cornell University Press).

Simpson, L. (2004), 'Statistics of Racial Segregation: Measures, evidence and policy', *Urban Studies* 41:3, 661–81.

Simpson, L. (2006), 'The Numerical Liberation of Dark Areas', *Sage Race Relations Abstracts* 31:2, 5–25.

Simpson, L. (2007), 'Ghettos of the Mind: The empirical behaviour of indices of segregation and diversity', *Journal of the Royal Statistical Society A* 170:2, 405–24.

Simpson, L. and Gavalas, V. (2005), 'Population Dynamics within Rochdale and Oldham: Population, household and social change', CCSR Working Paper, Manchester: University of Manchester.

Simpson, L., Phillips, D. and Ahmed, S. (2007), 'Housing, Race and Community Cohesion in Oldham and Rochdale', report to Oldham and Rochdale Housing Market Research Programme.

Soysal Y. (1994), *The Limits of Citizenship: Migrants and post-national membership in Europe* (Chicago: University of Chicago Press, Chicago).

Vertovec, S. (2007), 'New Complexities of Cohesion in Britain. Superdiversity Transnationalism and Civil Integration', research report submitted to the Commission on Integration and Cohesion.

Won,g D. (2004), 'Comparing Traditional and Spatial Segregation Measures: A spatial scale perspective', *Urban Geography* 25:1, 66–82.

Worley, C. (2005), '"It's Not about Race. It's about Community": New Labour and "community cohesion"', *Critical Social Policy* 25:4, 483–96.

Chapter 14

After the Cosmopolitan? New Geographies of Race and Racism

Michael Keith

The Social Construction of Proximity and Distance

There is a conventional reading of the work of Robert Park that foregrounds the importance of the spatial through an equivalence made between social distance and spatial distance. Informed by strong empirical evidence that patterns of exogamy, education, identification and interaction are all conditioned by spatial configuration it is common to emphasise the spatial correlates of social form. In the British tradition of social geography, the influential work of Ceri Peach and his students in the 1970s and 1980s was important in this respect (e.g. Peach et al. 1981).[1] And there is a similar analysis of social interaction between peoples of different 'ethnicities' that can be found in contemporary mainstream readings of social psychology that figure space as the transparent territory on which interaction takes place over a transparent sense of the temporal (Hewstone 2006).

In such analysis the sense of the spatial and the temporal figure as causally significant but analytically transparent, the metricated measures of distance and history. Such a reading of spatial form might be said to be particularly influential in the positivist geographies of the 1970s (and its legacies) and a long tradition of geographical literatures has subsequently qualified such an understanding of the spatial through a reading of the plural spatialities *of produced cartography* (largely, though not entirely in the wake of Henri Lefebvre) and complex historicities of relativised temporality.

And yet. The public concern with patterns of racial segregation, social interaction and the measurement of proportionate dissimilarity in the residential configuration of racial or ethnic categories continues to be a subject of academic analysis, journalistic worry and social policy received wisdoms that promote the residential 'mixing' of peoples of different backgrounds (Peach 1996, 1999; Poulsen et al. 2001, 2002).

But it is worth remembering that Park was a student of the sociologist Georg Simmel in late nineteenth-century Germany and in Simmel's own work and its legacies we can find a slightly different, more narrative, more curatorial understanding of the role that space plays in mediating social life. In particular, in Simmel's seminal essay on 'that synthesis of nearness and remoteness which constitutes the formal position of the stranger' he addresses some of the core concepts that link the concerns of moral philosophy to the spatialisation of the urban that are central to this chapter. For Simmel the stranger organises the unity of remoteness and distance that defines all human relations so that 'distance means that he who is close by is far, and strangeness means that he who also is far is actually near' (Simmel 1923). In Simmel's work – and in literatures in architectural theory, sociology and cultural studies that have taken his lead – this interplay of perspectival movements between closeness and distance correlates directly with the privileging of different kinds of city view and different genres of urban knowledge.[2] And when we begin to think about the way perspective confounds metric readings of proximity and distance then we might think carefully also about the geographies of Simmels' paradox. And in particular we might consider the way in which the essay of the stranger and the paradoxical nature of what is close by and simultaneously distant, or removed and yet so proximate, influence the thinking of another former student of Simmel's; Walter Benjamin, the German chronicler of the rise of fascism in the early twentieth century.

The *convolutes* of Benjamin's great unfinished opus the Arcades Project are evidence of his refusal to privilege particular representational regimes in articulating the meaning of the city and attempting to come to terms with the new forms of metropolitan culture that so fascinated Simmel. Benjamin's take on the spatial questioned the modalities through which geography mediates knowledges, ethics and notions of the beautiful (Gilloch 1997; Keith, 2000). In one of the two essays on 'Paris, Capital of the nineteenth century' that was meant to introduce and explain the Arcades project he attempts to taxonomise culture by matching personalities with positionalities; a sixfold typology of the metropolitan that invokes in turn, Fourier, or the Arcades; Daguerre, or the Panorama; Grandville, or the World Exhibition; Louis Phillippe, or the interior; Baudelaire, or the streets of Paris and Hausmann, or the barricades (Benjamin, 1999a). The interplay and technological substitution of that which is close and that which is distant also become central to Benjamin's own famous and influential essays on photography and the work of art in the age of mechanical reproduction (Benjamin 1973, 1999b).

It is perhaps an exploration of this double interplay between closeness and distance, strange and familiar that might help us to understand not only Benjamin's unfinished work but also the particular incommensurabilities of the post-colonial present and its valorised urbanistic cognate the cosmopolitan. The cosmopolitan translates into the racialised geographies of the contemporary as both recognised difference and valorised hybridity. But how might it inflect the ways in which we think about the curatorial power of the spatial to describe and explain the 'new geographies of race and racism' that this volume takes as a central concern? Specifically, the ambivalence about the duplicity of relations of proximity and distance that Benjamin draws from Simmel's understanding of the stranger might illuminate our understanding of the moral geographies of segregation, social distance, and the obligations that inform an understanding of contemporary multiculture and might refigure the politics of the racial imaginary.

In this chapter I want to develop these thoughts through a consideration of both the ways in which the debate about racial geographies plays out in the public realm in the context of the British government's establishment of a Commission on Integration and Cohesion in the wake of the London bombings of July 2005 ('7/7') and also to contextualise such thinking through consideration of one of the more painful moments of the cosmopolitan by drawing on some work carried out by the author in the east London borough of Barking and Dagenham during 2006 and 2007.

Spatial Configuration: Racial Geography in the Public Realm and the Commission on Cohesion and Integration

The Commission on Integration and Cohesion was established by then Secretary of State Ruth Kelly in the spring of 2006 and was tasked with analysing some of the dynamics of multicultural social change and providing suggestions that might inform social policy intervention. For some, the genealogy of the Commission itself was questionable and for others the terminology invoked potentially displaced the impetus of generational struggles against injustice and racism (Keith 2007). But as a participant (and 'commissioner'), for this author the process also lent an interesting perspective on the manner in which the geographies of race and the geographical scales at which we considered the multicultural present informed the social policy imaginaries of politicians and bureaucrats alike.

The tone of public debate that preceded the Commission's work had been set partly by the rising controversy around new migrations to the United Kingdom, civil disturbances that had taken place in Birmingham and some rural areas immediately prior to the Commission's launch and ongoing continued security concerns in the wake of the London bombings. However, a related theme concerned the issues of proximity, distance and social interaction themselves and the suggestion that the spatial configuration of British multiculture was becoming increasingly polarised.

Following the 2001 disorders in the 'mill towns' of northern England the subsequent inquiry, chaired by Ted Cantle had suggested that:

> Whilst the physical segregation of housing estates and inner city areas came as no surprise, the team was particularly struck by the depth of polarisation of our towns and cities. The extent to which these physical divisions were compounded by so many other aspects of our daily lives, was very evident. Separate educational arrangements, community and voluntary bodies, employment, places of worship, language, social and cultural networks, means that many communities operate on the basis of a series of *parallel lives*. These lives often do not seem to touch at any point, let alone overlap and promote any meaningful interchanges. There is little wonder that the ignorance about each others' communities can easily grow into fear; especially where this is exploited by extremist groups determined to undermine community harmony and foster divisions. (Home Office 2001, 9, emphasis added)

The metaphor of parallel lives, closely identified with Cantle personally, became a reference point for journalistic discussion of British multiculture throughout the last seven years. The motif was amplified by a number of interventions in public debate by Trevor Phillips, the chair of the Commission for Racial Equality (and subsequently the chair of the Commission on Equality and Human Rights). Most notably in 2005 Phillips suggested that:

> We are a society which, almost without noticing it, is becoming more divided by race and religion. ... If we allow this to continue, we could end up ... living in a New Orleans-style Britain of passively co-existing ethnic and religious communities, eyeing each other uneasily over the fences of our differences ... we are *sleepwalking our way to segregation*. We are becoming strangers to each other, and we are leaving communities to be marooned outside the mainstream. (Phillips, CRE speech 22 September 2005, emphasis added)

Although the Commission on Integration and Cohesion preferred to caution against a sense of 'sleepwalking into simplicity' (DCLG 2007, 6), in an explicit rebuttal of Phillips's agenda, the balance between national and local senses of belonging and cohesion featured prominently in both the focus of work and the final report. In particular the objective understandings of proximity and distance echo the longstanding tensions between the very different heirs of Simmel's thought.

In mainstream social thought the default mode of thinking about the problems of integration and cohesion is at the level of the national. This chapter does not suggest that this is wrong but it does suggest that we might wish to inflect or nuance this default model. The nation, the state and the uncertain British articulations of both provide a grounding context for powerful imaginaries of identity and uncertain cartographies of citizenship. But we live locally, our identifications with our neighbourhood, with local government, with the public realm in our everyday lives mean that the scales through which we experience notions of citizenship and belonging are rarely singular. So we might wish to understand changes in processes of cohesion and integration alongside a sense of the dynamism and change that takes place simultaneously nationally and locally.

The nation state maybe weaker, and is certainly not disappearing, but can not ignore either the cultural or the economic logics of the global. In the British articulations of the national various attempts to provide a diagnosis of its ills, prognosis of its future or prescriptions for its rebirth in a new sense of Britishness (after prime minister Gordon Brown) or a 'new covenant' in British chief rabbi in Jonathan Sacks' (2007) *The Home We Build Together* have sparked a lively public debate about what it might mean to reinvent and promote a sense of Britishness that can survive the forces of devolution, globalisation and disillusionment with the political mainstream. But telling stories at the sentimental scale of the national is just one of the many registers through which communities are imagined, notions of belonging are mapped and loyalties plotted. In privileging the national, relatively little attention is paid to the sentimental others of the nation; the territories, homes and communities that revolve around alternative cartographies of identity and belonging.

In a straightforward sense it might be worth thinking the nation through not just its *national* cartography but simultaneously through other geographical scales.

How do we balance a sense of our neighbourhood, a sympathy with transnational social movements that want to 'make poverty history' and 'end global warming', diasporic obligations to past and present global networks along with a notion of national belonging that is 'fit for purpose' for Britain's twenty-first century?

In the final report of the Commission on Integration and Cohesion 'Our Shared Futures' the analysis suggests that the complexity of local realisations of contemporary multiculture confounds attempts to characterise a singular understanding of British multiculturalism at the level of the nation. The focus is consequently more weighted towards the local, both in terms of specific policy recommendations and also in defining an understanding of cohesion and integration that recognises a sense of moral principles that are not entirely commensurable as the basis for an understanding of contemporary multiculture.

The report argues that debate should flow from a set of four principles. The first principle of 'shared futures' valorises a sense of becoming that does not erase the imprint of history and memory over a sense of being. The second principle argues that we need to consider a framework of rights and responsibilities that recognises the incommensurabilities of the global, national and local senses of the citizen. The third principle argues for an 'ethics of hospitality' that acknowledges the moral place of the stranger in the rapidly changing landscapes of today's Britain. And the fourth argues that these forms of recognition need to be geared with a sense of visible social justice that stems from principles of equality and institutional transparency.

If these principles generate particular forms of incommensurability then it is through the temporal and the spatial configuration of the settlement of race and ethnicity that moral principles are transformed into political geographies and histories. This is best illustrated by working through both the abstract basis for this sense of a spatial fix of incommensurability and then working through a specific example in the east of contemporary London.

I have tried to argue elsewhere that the conventional ways of thinking about what is progressive and what reactionary in public debate and academic prose does not quite fit with the realities of everyday multiculture (Keith 2005, 2007). The spatial is important because it mediates regimes of rights and governmentality and curates our communitarian imaginaries. And these regimes of rights depend on a tradition of liberalism that is fundamentally in tension with the logics, the moralities and the epistemologies of the communitarian.

I have also tried to argue elsewhere (Keith 2005, chs 1 and 3) that there is a homology between three traditions of thinking in moral philosophy, urban theory and the study of race and ethnicity. Thinking about the tensions between the liberal and the communitarian in moral philosophy, the knowable and the unknowable city in figurations of urbanism, and the languages of identity ('ethnicity') and coercion ('racialisation', 'race') in multiculture can highlight similarities of pattern that can begin to substitute each for the other. The conventional ways of thinking about the city technocratically (to maximise efficiency and optimise social form) are closely identified with certain traditions of geographical intervention and should be properly distinguished from deconstructions of the fabricated nature of the metropolitan that expose the artifice of city form and the systemic injustice of its architecture. And if the latter is conventionally identified politically with 'the left' and the former with 'the right' then such a distinction is perhaps less useful than an analysis of what such thinking says about how we come to understand how the city works and how it might be 'otherwise'.

Social policy tends to conflate, displace or ignore this sense of the incommensurable but out there, 'on the ground', these incommensurabilities become the fault lines of popular sentiment and the logical drivers of 'ethnic competition' and the rational foundations of racist sentiment. To ignore what is incommensurable analytically is consequently to fail both our technocratic and our deconstructive readings of cosmopolitan city life.

In pragmatic terms, implicit in this set of policy prescriptions is a sense that we want to take the best of the communitarian sense of belonging, mutuality, links and bonds alongside the best of the honourable liberal promotion of regimes of human rights (and responsibilities) with a heavy dose of suspicion about the voices of authority that will tend to privilege themselves in the setting of rules.

Of course, in the abstract world of logical argument this wish list is 'incommensurably' utopian. Regimes of rights depend on regimes of power and these invariably subsume systems of injustice. What Les Back (1996) has described as 'the metropolitan paradox' is common to many sites of multicultural difference, where intense forms of intolerance flourish in the spaces that generate the most acute forms of intercultural dialogue. We get by in the rub, in the everyday world, because the abstract contradictions of the communitarian and the liberal are subsumed in the everyday. But understanding how these contradictions structure everyday life also might help us to understand how the spaces of the city 'curate' ethnic difference and generate particular forms of conflict. In the remainder of this chapter I shall try to illustrate this by drawing on recent experience in the London Borough of Barking and Dagenham, which attracted national attention in the spring of 2006 when the British National Party (BNP) won unprecedented success in the local authority elections.[3]

Spatial Curation; The Racial Geographies of Communitarian and Liberal Government in the London Borough of Barking and Dagenham

The London Borough of Barking and Dagenham is undergoing radical and rapid change. The population of the borough as a whole is growing fast and its ethnic composition is in flux. At the same time as the consequent pressures on public services of all kinds, especially social housing, have increased, the once impressive manufacturing economy of the area – which offered plentiful jobs to the less skilled – has been in severe decline. This combination of growth, decline and pressure is the back-drop to the high profile rise of the far right in the area. In 2006 local government elections the BNP became the official opposition party locally, winning all but one of the seats that they contested. It was commonly suggested that they might even have won control of the council, had they contested enough seats.

The borough sits at the heart of the major regeneration project of the Thames Gateway, the attempt to accommodate somewhere between 0.5 and 0.75 million people in a 'a city the size of Leeds' in the pie slice of London that radiates east from Tower Bridge. Barking and Dagenham contains one of the largest development sites within the Gateway project, the postindustrial and much polluted site of London Riverside, where proposals for several decades to build upwards of 10,000–20,000 new homes form a high profile regeneration backdrop to the patterns of change locally. The site is owned by English Partnerships, who are in a joint venture partnership with a major private sector developer to build out the site[4] that effectively forms one of three growth poles within the borough. The other two poles focus on the contrasting geographies of the old settlements of Barking and Dagenham respectively.

Barking and Dagenham was created in 1965 by the amalgamation of the two councils. To this day the historic divide between the two places and their quite different structures and morphologies to some extent guides destinies – reminding us of the challenge of building a third significant locus at Barking Riverside and of doing so well and for the long term, let alone without exacerbating the existing divides within the borough.

Barking Town centre is a historic town, originally a fishing port and market and subsequently a residential and service hinterland for London's Docklands. It is well connected by public transport to central London, the city and the new financial district of Canary Wharf. With a concentrated retail offer and other assets normally associated with a market town, Barking has a quite different urban structure from Dagenham. An upsurge in projects in and around Barking Town centre in the property boom of the 2000s is rooted in that structure and the transport connections which make Barking town centre a key new site in the build-out of the London Thames Gateway.

Dagenham's imaginary is much more in Essex. Though the village of Dagenham has been in existence since Saxon times, it never became in urban hierarchy terms, a town, and indeed all but a few remnants of the ancient village were obliterated by development by the 1960s. In fact, most of the population growth which gave 'Dagenham' its twentieth century profile, took place outside the historic village on land which became world-famous as the Becontree Estate. It was once reputed to be the biggest social housing estate outside the Soviet Union and given the absence of high rise flats in the model, the estate with the most low density family homes anywhere in the world. At its peak the estate had 27,000 homes and contained 90,000 of what the Council's website itself calls 'the better-off working class Londoner'. From the 1930s the destinies of the area were largely determined by the development of the Ford Motor Co. and its huge dispersed site on the Thames near Dagenham Dock. The two key forces acting on Dagenham for decades were thus established by the end of the 1930s: social housing and well-paid manufacturing employment as 'pull' factors though over time, the individual social mobility enabled by employment at Ford, the availability of private family housing further out in Essex and the impact of right to buy policies which had a unique impact on Dagenham with its large number of family-sized council homes, resulted in significant 'push' factors and flight from Dagenham of many aspirational families. The decline of decades is palpable and whilst both Barking and Dagenham exhibit characteristics of northern post industrial towns, Dagenham is a *locus classicus* of the ex-manufacturing mono-culture with more pockets of severe deprivation, ill health, unemployment, low skills and educational attainment and unemployment.

With a small and dispersed retail offer, lower residential density, no obvious town centre, an urban form disrupted by the development of Ford and its logistics requirements and more recently the proximity of the A13 and the M25 , and being less connected to opportunities in central London, the City or Canary Wharf, Dagenham is thus quite different from Barking. Barking Town centre is in marketing terms and reality 'beginning to happen' as developers discover its potential. Despite some recent innovative master-planning initiatives and the LDA focus on the area as a potentially significant modern manufacturing and logistics hub, little in development terms or in terms of 'making the place' is going on in Dagenham.

However, unlike northern towns which peaked economically and demo-graphically 100 years ago, both Barking and Dagenham are experiencing new population, economic and housing pressures, not least because they are both part of the Thames Gateway whose development is beginning to shape the destinies of these two places in related but differing fashions rooted in the differing labour and housing markets and market potentials of the two places. In this context, not only is there potentially a divergence between Barking and Dagenham that amplifies an old political division, there is also a significant difference in the drivers of development and change. The difference has also had different ethnic consequences – with an increasing Asian presence in the Barking part of the borough, as its development meets inner East End suburbanisation, melds into that of Newham, Tower Hamlets and Redbridge around it and an increasing, and aspirational, Black-African presence in Dagenham, drawn by the availability of former family-sized social housing at an affordable price. In the latter area in the last five years on the Council's school rolls, the proportion of children of school age in some of the borough's schools coming from a Black-African family has gone from 3 per cent to 15 per cent

Since the first mooting of the Thames Gateway, the local authority in Barking and Dagenham has embraced the regeneration agenda and the council is the accountable body for the Thames Gateway London Partnership. However, the success of the British National Party in the local authority elections has provided a challenge to this support. The BNP campaigned against the Gateway development, against development in, and migration to, the borough and against the development of Barking Riverside. They also campaigned for the local council to control both local housing stock and its allocation. In these circumstances it is clearly important to understand both the dynamics of change that inform the growth in BNP support and the policy levers that can be brought to bear in the face of the potential social division that results. It is of great interest that the overwhelming majority of their twelve councillors represent Dagenham – where there is no new residential 'Thames Gateway' development – which might suggest that where development has already taken place and where populations have grown and indeed changed, that the political impact is less or already factored in and where the fear of new development and new populations predominates and no new opportunities or development are actually on the ground, political reaction is at its most acute. However, this view is questioned by local politicians from the established parties who say 'if the BNP had stood a candidate in every ward in the borough in May they would have won them all'. And behind this, say the same sources, is the view that for many existing inhabitants in many places in the borough the Thames Gateway experience thus far has been 'all pain/no gain' with increased populations and pressures on public services and housing with insufficient new affordable housing in either the private or the public domain and new jobs which many locals cannot access.

This is the context which confronts both the local political establishment and mediates the narratives of major regeneration on the ground. Although there is some evidence that the threat of the BNP has actually acted as a galvanising force on the local leadership and borough management, the challenge is serious and the battle only just begun. It is a significant challenge in this context to combine market forces and state subsidy to build a new community at London Riverside whilst simultaneously managing the dynamics of change locally. These changes exemplify the tensions between the need to recognise existing community sentiments and the need to build a new sense of identity and place at Barking Riverside while simultaneously recognising the rights of existing and arriving citizens of the borough. Regeneration *curates* a spatialisation of change that sets the tension between the rights and responsibilities of welfare state rationing against the communitarian imperatives of place making, popular sentiment and imagined community.

In development control terms a site that was at the heart of the earliest plans for the Gateway has, after many false starts been given the green light for delivery at a time when concerns about the rate of housing delivery in London continue. It will possibly be one of the first major tests of the capacity of the Housing Corporation and English Partnerships, either acting in partnership or via their newly merged form, to deliver very large residential numbers at urban renaissance qualities. Equally, there remain difficulties around facilitating development on the site. Longstanding aspirations – and recent commitments – to have the Docklands Light Railway extended to Barking and the need for road improvements to the A13 both provide challenges and further demands on the public purse. But most importantly, the quality of the build out of Riverside, particularly in its earliest phases will attract significant public attention. It will be one of the largest single residential developments in the Thames Gateway and will attract significant public attention, akin to the focus on the new town at Thamesmead in the 1970s or the build out of Greenwich Peninsular more recently. The reputational implications can only be amplified by a democratically elected local authority BNP presence opposed to the build out of the site.

Development and Change

In the current Greater London Authority (GLA) London Plan proposals the borough of Barking and Dagenham is expected to see the development of over 11,900 new homes from 2007–2008 to 2016–2017[5] and (in varying estimates) well over 20,000 new homes over the next twenty years. Whilst a significant proportion of these new homes are intended to be developed on Barking Riverside it is becoming clear that a significant number of sites, with existing transport access are already coming to the market in Barking Town centre. Whilst Dagenham has capacity in several locations for housing capacity growth, development pressures reflect the poor transport links and are much less strong.

Housing Stock

Since the introduction of the 'right to buy' council owned homes by their tenants in the 1980s the profile of home ownership in the borough has changed significantly. A 1970s pattern of tenure saw approximately 60,000 households in the borough of which 40,000 were council tenants. By the turn of the century this pattern had changed to a situation where stock breaks down almost evenly with approximately 20,000 privately owned, 20,000 leaseholders and 20,000 council tenants. This is a very different market and context for the council to seek to understand let alone manage.

A strong municipalist political culture in the borough has led to a rejection of many of the institutional vehicles common to the last decade, including stock transfer and arms length management organisations (ALMO). In a context where remaining council tenants wanting to upgrade or up-size confront incoming homeless families, refugees or simply nominations from other council areas in the sub region in a competition for slimmed down housing assets – because 'right to buy' has denuded and reduced the diversity of the stock leaving council tenants, as the BNP puts it, 'on the fourteenth floor' when in the past they could move on to a family home in Becontree now lost to the stock – the common aspiration of the local political leadership to provide more social and affordable housing and more choice has been translated into a desire to either build or extend council management of housing stock. The BNP have indeed campaigned for the Council to control and build stock, in the wake of the May 2006 election result and claim responsibility (tendentiously) for the council's innovative emerging proposals to use council land for an equity share in new housing developments.[6]

Belonging and Becoming; The Realities of Ethnic Competition

The economic logics of housing led population growth need to be understood against the context of the multicultural dynamics of population change that has informed the growth of BNP support locally. Migration locally is multiracial and multicultural but not straightforwardly so. Objectively, there are at least three strands of competition between existing white communities and 'incomers' that are at the heart of BNP appeals for support.

The first strand of competition revolves around access to social goods such as education, GP surgeries, hospitals but particularly to social housing. The combination of right to buy reductions in stock availability, the advent of sub regional lettings and the suburban drift of inner London that has characterised migrations of the last century all increases the amount of competition for social housing in the borough. It remains the case that the borough has one of the lowest waiting lists for social housing in London and the growth agenda offers opportunity to grow the social housing stock for the first time in two decades. The competition for social housing occurs particularly in the west of borough and around Barking and will be most intense in sites where private development contributes social housing. The competition is 'racial' in as much as families from a diverse range of ethnicities compete for social housing in east London on the basis of social need. Different families have differing levels of 'social need' and have varying attachments, family histories and community links to the borough. Objectively, the numbers of inward migrants to the borough that stem from sub regional lettings have remained low. But the competition for the 'scarce good' of social housing is not an illusion and the balance between local 'belonging', communitarian sentiment and social needs is both ethnically nuanced and subject to easy and insidious racialisation in populist campaigns about who gets access to new homes as they become available for letting.

The second strand of competition revolves around access to private goods, and again particularly concentrates on the housing market but in this instance focuses around private housing for purchase. There is considerable evidence – particularly from school rolls in the east of the borough – of significant inward migration based on uptake in the cheapest housing market in London. The competition for housing to buy occurs particularly in the more desirable 'cottage estates', where low density, family homes (with gardens) have been particularly subject to right-to-buy growth of large numbers of leaseholders (such as on the Becontree estates and in the low density private ownership areas of Dagenham, in the east of the borough). The competition is 'racial' in as much as a significant number of buyers are from the first or second generation of successful British African communities, economically and socially mobile. Clearly, incomers are not all African and as the analytical work commissioned by the council demonstrates, as development continues and new unit supply increases this competition in the housing market will combine with the 'churn' within exiting stock, leaseholder properties and private new build; creating more complex market options locally. However, events such as the sale of the major Beam Reach site, close to Barking Riverside, by the London Development Agency to the Black Majority Church displaced from the Hackney Wick site through the Olympic development (the Kingsway International Christian Centre) will clearly enhance the public profile of the African and African Caribbean communities locally.

The third strand of competition revolves around the labour market. The socio-economic profile of the borough generally, and the leaseholders 'in particular is significantly concentrated on the skilled manual standard industrial classifications (SICs). A casual examination of estates such as Becontree reveals the concentration of residents in occupations that relate to plumbing, building roofers, glazers. These are precisely the areas of London's labour market that have been most susceptible to competition from inward migration from the new accession country migrations, particularly from Poland and Lithuania since 2004. So the labour market position of the local white working class, impacted on through the long term restructuring of Fords is subject to intense competition in building related trades and occupations

The point about this typology is to understand that objectively, the forms of competition between 'old white communities' in Barking and inward migrants are plural in nature. They involve established BME communities of two and three generations, African communities whose migration pattern frequently can be traced to the 1980s and an Eastern European migration of the last three years. There is also no necessary correspondence between the 'objective' forms of interethnic competition and either the level of BNP support or the forms of 'threat' that the BNP themselves have chosen to highlight in their campaigns at borough level as local research on BNP support demonstrates.[7] However, significantly the default model of change does indeed point towards increased segregation generated by the existing forms of competition for and access to public and private resources and opportunities in housing and the labour market. The default model points to increased BME access to social housing estates where there is most 'churn', development and regeneration – particularly around Barking Town centre – and increased BME private housing self segregation particularly in areas of low cost (largely African) purchase in Dagenham. This leads to tensions and poses challenges to community cohesion.

Conclusion

If we assume that BNP support is based on irrational populist sentiment alone – and not forms of rational self-interest as well – then we will not frame correctly policy interventions that structure priorities within the welfare state nationally or the shaping of place at the level of the local. Notions of integration and cohesion need to address the sometimes incommensurable and frequently uncomfortable ways in which the sentimental and powerful imaginaries of belonging and affiliation do not always sit easily with the rational organisation of rights and responsibilities. Geography is important here but not in the ways we might imagine from tabloid translations of indices of dissimilarity propagated by geographers of the more quantitative inclination. Separation and segregation are properly susceptible to certain kinds of measurement but we need to place such empirical information together with both a policy informed understanding of the complexities of tenure and market and a theoretical grasp of the sometimes paradoxical social construction of proximity and distance that Simmel and Benjamin teach us to address. This means taking seriously the ways in which cartographies and historical legacies of both rights and belonging produce spatialities that map the multicultural present and structure a sense of becoming that might lie at the heart of a more cosmopolitan future. Indeed it is precisely because we need to understand how both regimes of rights and our senses, imaginaries and everyday communitarianisms are structured by the local that a figuring of the geographies of racism is so important in making sense of the nature of contemporary multiculture.

Notes

1 This chapter in part develops a piece that was originally prepared for a one-day conference that celebrated the work of Ceri Peach, a scholar and individual to who the author owes many debts that remain unpaid.
2 The work is also useful in prefiguring an understanding of how the tension between the familiar and the strange links closely to a Freudian exploration of the uncanny and a psychoanalytic reading of the city. The focus of this chapter is on the parallelism between the spatialisation of the familiar and the strange and the changing configurations of city ethics and urban knowledges. This is not to devalue such psychoanalytic insights. For an exploration of such 'warped spaces' see Vidler 2001.
3 The British National Party represents the most successful element of far right political mobilisation in Britain in the last fifty years and are commonly described as a 'fascist' or 'nationalist' political party, although both of these terms are open to contest. Emerging from the old National Front they have been the most electorally successful 'far right' political party in the UK, first winning representative power with a by-election success on the Isle of Dogs in the London Borough of Tower Hamlets in 1993. Subsequent electoral successes had been focused in the north of England until, in the 2006 London local government elections the BNP made the major advances in the London Borough of Barking and Dagenham that are the focus of this section of the chapter.
4 This piece of the chapter draws on more extensive research carried out for English Partnerships by Tim Williams and Michael Keith that addressed the community cohesion dynamics of large-scale regeneration in Barking and Dagenham 'British People Live on the 14th Floor' (BNP): Building a new community in Barking and Dagenham following the May elections – The implications for English Partnerships of development at Barking Riverside', Michael Keith and Tim Williams (Consultants' Research Report).
5 Draft alterations to the London Plan: Housing Provision Targets, October 2005, GLA, London.
6 'BNP Scores First Victory in Dagenham', 23 May 2006, BNP press release.
7 *The Far Right in London: A challenge for local democracy?* (2005), is published by the Joseph Rowntree Reform Trust. The research was carried out by Professor Peter John (University of Manchester), Helen Margetts (Oxford University), David Rowland (University College London) and Stuart Weir (University of Essex). 'BNP's Barking Stance: The BNP's victory in Barking has led to calls for greater provision of affordable housing', *Property Week*, 26 May 2006.

References

Back, L. (1996), *New Ethnicites and Urban Culture* (London: UCL Press).
Benjamin, W. (1973), 'The Work of Art in the Age of its Technological Reproducibility', in H. Arendt (ed.), *Illuminations*, trans. H. Zohn (London: Fontana).

Benjamin, W. (1999a), *The Arcades Project*, trans H. Eiland and McClaughlin (Boston, MA: Harvard University Press, Belknap Press).

Benjamin, W. (1999b), 'A Little History of Photography', in W. Benjamin, *Selected Writings: Vol. 2 1927–34*, ed. M.W. Jennings, H. Eiland and G. Smith (Cambridge, MA: Belknap Press of Harvard University Press), 507–30.

Department of Communities and Local Government (2007), *Our Shared Future* (London: HMSO), <http://www.integrationandcohesion.org.uk/Our_final_report.aspx>.

Gilloch, G. (1997). *Myth and Metropolis. Walter Benjamin and the City* (Oxford: Polity Press).

Hewstone, M. (2006), 'When, How, and Why', Presentation to the Commission on Integration and Cohesion, 13 December.

Home Office (2001), *Building Cohesive Communities* (London: HMSO).

Keith, M. (2000), 'Walter Benjamin, Urban Studies and the Narratives of City Life', in G. Bridge and S. Watson (eds) *Blackwell Companion to Urban Studies* (Oxford: Blackwell).

Keith, M. (2005), *After the Cosmopolitan. Multicultural Cities and the Future of Racism* (London and New York: Routledge).

Keith, M. (2007), 'Don't Sleepwalk into Simplification: What the Commission on Integration and Cohesion really said', *Open Democracy* 12 August, <http://www.opendemocracy.net/ourkingdom/articles/commission_on_integration_and_cohesion>.

Peach, G., Ceri, K. Robinson, V. and Smith, S. (1981), *Ethnic Segregation in Cities* (Oxford: Oxford University Press).

Peach, G.C.K. (1996), 'Does Britain Have Ghettoes?', *Transactions*, Institute of British Geographers NS22, 216–35.

Peach, G.C.K. (1999), 'London and New York: Contrasts in British and American models of segregation', *International Journal of Population Geography* 5, 319–51.

Poulsen, M.F., Johnston, R.J. and Forrest, J. (2001), 'Intra-urban Ethnic Enclaves: Introducing a knowledge-based classification procedure', *Environment and Planning A* 33, 2071–82.

Poulsen, M.F., Johnston, R.J. and Forrest, J. (2002), 'Plural Cities and Ethnic Enclaves: Introducing a method for comparative study'. *International Journal of Urban and Regional Research* 27, 229–43.

Sacks, J. (2007), *The Home We Build Together: Recreating Society* (London: Continuum).

Simmel, G. (1923), 'Excurs uber den Fremden', in G. Simmel, *Soziologie* (Munich and Leipzig: Duncker und Humblot).

Vidler, A. (2001), 'Spaces of Passage: The architecture of estrangement: Simmel, Kracauer, Benjamin', in A. Vidler (ed.) *Warped Space: Art, architecture and anxiety in modern culture* (Cambridge, MA: MIT Press), 65–81.

PART 3
Race, Space and 'Everyday' Geographies

Chapter 15

The Precarious and Contradictory Moments of Existence for an Emergent British Asian Gay Culture

Camila Bassi

Preface

This chapter presents a series of vignettes that define critical moments in the early development of British Asian gay cultural identities, forms and practices on Birmingham's, predominantly white, commercial gay scene. In particular, it offers snapshots of the meanings denoted in, through and beyond Birmingham's first ever monthly British Asian gay club night, *Ultimate Karma*, and its second monthly night, *Saathi*, by a number of the regular participants.[1] Whereas the first set of vignettes exemplify the radical possibilities that are drawn from the use-values of the commodities of bhangra and Bollywood, the second reveal the constraints that derive from the threat of racism perceived to be heightened by certain British Asian gay representations. On the one hand then, the lyrics of bhangra and Bollywood music, the sights, sounds and tastes of the family wedding party hall, and the dance moves and aesthetics of Bollywood films, are collectively re-appropriated during the nights of *Ultimate Karma* and *Saathi* as integral parts of a conscious British Asian gay culture. On the other hand, amid this British Asian gay culture, specifically among those seen to be part of this culture, is a reaction to avoid racism on the gay scene through a subconscious internal-Otherisation; that is, some gay and bisexual British Asians attempt to differentiate themselves from the visible behaviour of other gay and bisexual British Asians who are judged to fuel a burdensome stereotyping of 'the Asian gays'. What this chapter thus unravels is a precarious, contradictory and impeding tension between radical possibility and constraint within an emergent British Asian gay culture. More than this, however, is a theoretical effort to explain this contradictory tension as indicative of, more generally, people's immediate reactions to unequal and oppressive conditions of existence.

Refashioning the Unfashionable: Introducing a Marxist Toolkit

In attending to a contradictory tension between radical possibility and constraint amongst an emergent British Asian gay culture, I draw primarily upon the theoretical inheritance of Antonio Gramsci (1971, 1985), specifically, his notion of *contradictory consciousness*. This notion offers an explanation of how people make sense of the world around them in ways that entail contradictory ideas and meanings (some of which are regressive and others more progressive), and it frames these ideas and meanings as part of a dialectical process.[2] Theoretical applications of Gramsci have largely disappeared from geography and the social sciences. Birmingham's Centre for Contemporary Cultural Studies (CCCS) popularised Gramsci, and the Centre's demise partly explains this disappearance. Any surviving applications of Gramsci have tended to be via a re-working into neo-Gramscian approaches by, for example, the post-structuralists Laclau and Mouffe (2001), and aspects of queer theory.[3] Deliberating the relationship present-day academia has to applications of Gramsci's work, Morton (1999, 2) concludes that both an 'austere historicism' (that relegates key ideas of Gramsci to their era of origination) and a neo-Gramscian perspective (that eliminates from analysis the political context of Gramsci's thoughts) are unsatisfactory. As an alternative method of contemporary application, he proposes Gramsci's dialectical understanding of absolute historicism (Gramsci 1971), that:

> the appropriateness of the position that past ideas, questions, or philosophies still have a bearing on the present, and may thus transcend social context and 'speak' to us, *can* be established but only by empirical investigation ... questions, ideas and problems will be determined by the particular culture or period under consideration. (Morton 1999, 4)

In this chapter, the explanatory usefulness or power (Skeggs 1998) of Gramsci's notion of contradictory consciousness is established by what it tells us about the ideas and meanings made of an emergent British Asian gay culture by those amid such a culture. I also work into my Gramscian toolkit both the Marxist Robert Miles' (1989; 2000) concepts of racism, racialisation and signification, and the Bourdieuian Beverley Skeggs' (1999; 2000) critique of commercial gay space as operating through appearance-based judgements of respectability, a politics of recognition and desires for legitimation. Miles departs from the main of 'race', ethnicity and racism literature in his original critique of the Race and Politics Group of the CCCS, under the guidance of Stuart Hall. Miles (2000, 137) argues that the failure of the CCCS to critically evaluate the idea of 'race' subsequently reifies 'race', i.e., effectively serves 'to freeze and legitimate that representation'. This critique remains important, I contend, since the legacy of enquiry left by the CCCS has foreclosed discussion on the constraints that are dialectically coupled with the radical possibilities of the urban cultural politics of 'racially' marked, ethnic minority groups. Miles' application in this chapter thus best explains the problematical tensions of 'racially' marked cultural identities, forms and practices that are bound up with counter-hegemonic cultural endeavours. Contemporary sexuality literature has tended to shift from theorisation that is open to Marxist ideas in favour of post-structuralist approaches. That said, Skeggs' (1999, 2000) work on commercial gay space adds an explanatory usefulness when worked into my Gramscian toolkit. Her work recognises the commercial gay scene as having developed through the organised desires of an elite few for visibility, recognition and territorialisation, thus as a space complicit with capitalist demands and a discourse of bourgeois individualism; and her analysis offers a nuanced understanding of the struggles, hierarchies and exclusions of marginal groups negotiating this space. This chapter follows up Skeggs' (1999, 229) assertion that '[i]t is only in the concrete articulations of living the everyday, in the struggles for future space, that one can begin to understand what is and what is not beyond incorporation'.

For Gramsci (1971, 324) all people have a 'spontaneous philosophy' (or contradictory consciousness) to understanding the world around them.[4] Furthermore, '[o]ne's conception of the world is a response to certain specific problems posed by reality, which are quite specific and 'original' in their immediate relevance' (ibid., 324). Gramsci points out that people's impulsive ideas about the world tend, in the first instance, to be 'not critical and coherent but disjointed and episodic' and so they will simultaneously belong 'to a multiplicity of mass human groups' (ibid., 324), i.e., to groups with opposing outlooks. With regard to their surroundings, people are active in 'modifying certain of its characteristics' while 'preserving others' (ibid., 265). Hence spontaneous philosophy is a 'contradictory consciousness' (ibid., 333) made up of what Gramsci refers to as both 'good sense' and 'common sense' (ibid., 323):

> one [good sense] which is implicit in his [*sic*] activity and which in reality unites ... and one [common sense], superficially explicit or verbal, which he [*sic*] has inherited from the past and uncritically absorbed. (Gramsci 1971, 333)

A key aspect of good sense is the 'struggle for a new culture' and 'a new intuition of life' to the point that it might become part of 'a new way of feeling and seeing reality' (Gramsci 1985, 98). Gramsci understands material reality as enabling and delimiting radical possibilities (Holub 1992), such that (paraphrasing Marx's famous dictum):

> people aspire to the adventure which is 'beautiful' and interesting because it is the result of their own free initiative, in the face of the adventure which is 'ugly' and revolting, because due to conditions which are not posed but imposed by others. (Gramsci 1975, in Holub 1992, 113–14)

Embryonic good sense then is the healthy nucleus of an anti-dialectical and dogmatic common sense – the latter of which consists of supposed truths and 'prejudices from ... past phases of history' that have 'left stratified deposits in popular philosophy' (Gramsci 1971, 324). '[V]ulgar common sense' is the part of contradictory consciousness that induces a state of submissiveness because of its eagerness 'for peremptory certainties' (ibid., 435). Its 'most fundamental characteristic' is its incoherence, 'in conformity with the social and cultural position of those masses whose philosophy it is' (ibid., 419). Common sense nonetheless appears to be 'a specific way of rationalising the world and real life' (ibid., 337) albeit 'a creation of concrete phantasy' (ibid., 126).[5] Gramsci's notion of contradictory consciousness provides an explanation of how people can 'accept' dominant ideas by not questioning them fully or not realising their own agency for change. Hegemonic conditions can thus be understood as reflecting a grudging and partial 'putting up with' the status quo.[6] Decisively, for Gramsci (1971, 350) contradictory consciousness is not a fixed state, rather it is dialectical since within any mass of people organic intellectuals can emerge to stimulate an 'active and reciprocal' dialogue aimed at bridging the gap from a starting point of embryonic good sense to the formation of a critical philosophy of inter-subjective engagement. Organic intellectuals are capable of feeling and responding to 'the elemental passions of the people' (ibid., 418), by working out and making coherent 'the principles and the problems raised by' a group 'in their practical activity' (ibid., 330) – see Bassi (2003).

Miles' concepts of racism, racialisation and signification are useful in more fully grasping the specific formation of common sense that is made by a number of gay and bisexual British Asians in response to particular representations of an emergent British Asian gay culture on Birmingham's commercial gay scene. Two different paths of enquiry to theorising race and racism are inherited from the work of the CCCS and Miles. The CCCS developed an approach to race which considers it as a construct with social and political relations in which its very meaning is struggled over (Solomos and Back 1996; 2000). Gilroy (1987), for instance, argues that the collective identities of Britain's urban, alternative public spheres, which speak by means of race, locality and community, are able to disperse and suspend the order of hegemonic culture. He goes so far as to assert that the term race should be retained as an analytical category because the collective identities to which it refers, 'in the forms of white racism and black resistance', have become 'the most volatile political forces in Britain at present' (ibid., 247). The metropolitan context promises an authentic, spontaneous and immediate politics of race that both differs from an older Marxist politics and is independent of class (Gilroy 1987). While reflecting a shift away from the view of a workable black politics to that of a politics grounded on new ethnicities (Hall 1992), later work by Back (1996) and Amin (2002) also directs attention to the political potential of the metropolitan, alternative public sphere (along with cautioning of possible closures) and retains the analytical category of race. Sharma, Hutnyk and Sharma (1996) seek to expand the black political project of Gilroy (1987) to include the politics of new Asian dance music (thus also retaining the analytical category of race), and they are more critical of work by, for example, Back (1996), which they see as abandoning serious politics. In contrast to this inheritance of the CCCS, Miles (1989, 2000) regards the very idea of 'race' – as a regulatory political construct concealing economic relations – of *no* analytical value which should instead be replaced by a more useful focus on racialisation. The concept of racialisation is considered to be part of a wider dialectical process of signification, that is, a process of depicting the world and making sense of how things are by ascribing characteristics to differentiate the Other from the Self (Miles 1989). In the case of racialisation meaning is attributed to certain biological features like skin colour. Racism is subsequently understood as 'a particular form of (evaluative) representation which is a specific instance of a wider (descriptive) process of racialisation' (ibid., 84). Miles (1989, 80) also points out that:

> [a]n emphasis upon racism solely as a 'false doctrine' fails to appreciate that one of the conditions of existence of ideologies (which by definition constitute in their totality a false explanation, but which may nevertheless also incorporate elements of truth) is that they can successfully 'make sense' of the world, at least for those who articulate and use them.

The instance of racism can be understood as a prime example of common sense. Moreover, it is also possible for common sense entanglements to occur – between evaluative racism, the descriptive process of racialisation and a descriptive process of signification – which can superficially function as an outlook. So, with the case of an emergent British Asian gay culture, an internal-Otherisation occurs as a signification in response to the wider process of racialisation and the perceived raised threat of racism. The internal-Otherisation that some gay and bisexual British Asians make of other gay and bisexual British Asians is also conditioned by the social relations of commercial gay space – in particular, by appearance-based judgements of respectability (Skeggs 1999, 2000). Research by Skeggs (1998, 1999, 2000) defines the concept of respectability as part of a core signifying process in how we interpret others and ourselves. The moral authority of respectability contains historically-grounded judgements about 'race', class, gender and sexuality (that have engendered certain ways of thinking about the world and its social order) and, crucially, there are those who struggle to possess respectability in and through social relations. Respectability hinges on appearance, that is, evaluative judgements regarding outward bodily markers that subsequently serve to naturalise social order as essential and fixed (Skeggs 2000). Acknowledging a shift in political demands from class-based redistribution towards a politics of recognition (Fraser 1995; Taylor 1994), Skeggs (1999) understands the territorialisation of commercial gay space as having developed through the organised desires of an elite few for recognition. More to the point, it is a space achieved by white, middle-class, gay men who have successfully converted cultural capital into symbolic and economic exchange (Bourdieu 1986; Skeggs 1999, 2000), and which is thoroughly embedded in a pervasive, judgement-fuelled, 'scopic economy' (Skeggs 2000, 129). It is within and through this scopic economy that common sense entanglements occur. Prior to discussing the common sense constraints of an emergent British Asian gay culture, I now proceed to present the radical, good sense possibilities that they are dialectically coupled with.

An Emergent British Asian Gay Culture and Moments of Embryonic Good Sense

> A new social group that enters history with a hegemonic attitude, with a self-confidence that it initially did not have, cannot but stir up from deep within itself personalities who would not previously have found sufficient strength to express themselves fully in a particular direction. (Gramsci 1985, 98)

For Gramsci good sense is a highly creative domain since oppressed people will seek moments of escape 'through fantasy and dreams' (Holub 1992, 113), which in turn stimulate the search for and cultivation of cultural arenas that permit the enactment of such fantasies and dreams. Furthermore, the 'elementary and primitive phase' of good sense is 'found in the sense of being 'different' and 'apart', in an instinctive feeling of independence' (Gramsci 1971, 333). In terms of urban cultural politics, Sharma, Hutnyk and Sharma (1996, 2) purport that British Asians have undertaken a shift to new Asian dance music that is suggestive of a future 'expressive culture unbound from the conventions and clichés of hegemonic order'. Kaur and Kalra (1996, 225–6) suggest that whilst bhangra and Bollywood music enabled second-generation British Asians to innovatively express their culture in the 1980s, this subsided by the end of the decade; as, for instance, '[m]any youth began to express a dissatisfaction with the Hindi film costuming – usually sequins, spangles and white trousers', in addition to 'complaints about [b]hangra's untrendiness … and lack of "socially meaningful" lyrics'. Paradoxical to this (predominantly heterosexual) British Asian youth trend, there is a desire amongst an emergent British Asian gay culture to return to and to re-consume the traditional sights and sounds of bhangra and Bollywood. But this too, I would suggest, is indicative of a future counter-hegemonic expressive culture, with (during the nights of *Ultimate Karma* and *Saathi*) use-values deriving from an active, imaginative and collectively re-appropriated take up of these more traditional commodities.

Exemplifying Gramscian good sense, bhangra and Bollywood music – as commodities with use-values evocative of memories and new emotions – are transported to sites that provide space for newly intuitive and self-confident, British Asian *gay* cultural identities, forms and practices. Rani, a deejay and co-organiser of *Ultimate Karma*, and Perdes, a deejay and co-organiser of *Saathi*, sum up aspects of this advance:

> The tunes, for many people it brings back a lot of memories, you can see it in people's faces, and they're like the ones who've been dancing really well to a particular song, I think 'I bet you've been practicing at home' or, 'I bet you've watched that film or that video over and over again', and then if you've watched it in a family setting right, and all those people at home have no idea why you're particularly focusing on a particular hand movement, cause they don't know that you're trying to re-create that at *Karma*, every song's got a memory for someone, and for some people it's about starting a new memory. (Rani)

> I was very anti-Asian music when I was growing up, because I found it very kind of sexist, I found it very homophobic, it was nothing that I could identify with, and it

was only when I identified as being Asian and gay, I needed some kind of connection and the music was that immediate connection for me, if you're in an Asian gay club … bhangra music is like gay, and there's that song by umm, Bally Jagpal, *Aja Sohneya*, and it's like, a man singing 'Come Here Beautiful' with another man, even though you know … the majority of the song is sung by a female except for the chorus, it's that kind of connection, you wouldn't feel uncomfortable singing it to another man … it's kind of that thing about having to self identify with the music and claiming it as yours, and you've got songs like from *Pakeezah*, things like 'Chalte Chalte',[7] now if you're a gay man, it's automatic. (Perdes)

The resourcefulness that people have 'to modifying the social environment' (Gramsci 1971, 265) is exemplified particularly well by the re-creation of the traditional family wedding party. Within the space of the commercial gay scene, the nights of *Ultimate Karma* and *Saathi* entail a conscious, dynamic and shared re-invention of the family wedding party hall. This, in part, is enabled by the commodities of bhangra and Bollywood music and aesthetics, together with food and drink, which are key ingredients of this wedding party culture (Werbner 1996). The following two quotes demonstrate the actualisation of this radical possibility (with Jazz, in his role as an early co-organiser of *Saathi*, referring to his attempt to re-create the sociability of the wedding party hall via some of its core commodities, and Tony referring to his first experience of *Ultimate Karma*):

What's in my head is thinking about Asian weddings, you know, you've got the shraab [the alcohol] and you've got the Bells whisky and you've got the bottle of coke and you've got the bottle of Barcardi and you've got your peanuts and your crisps and stuff, it's all kind of like very sociable stuff, and just recreating some of that, that kind of closeness, that kind of intimacy. (Jazz)

There was more of an offering there, I found listening to the music brought back memories of, it's really strange for me to be saying this, of weddings, the sort of congregations that you use to have within the culture, when a relative was getting married and you'd all congregate together, we'd dance and listen to music, I found it different and very, very refreshing, Camila, it was almost as though bringing my culture into an environment which is very, very taboo and which the culture sees as taboo, but again to, to extract that and bring it into another environment completely is invigorating, very liberating as well. (Tony)

The feelings of intimacy and invigoration all stem from the good sense capacity to culturally bring forwards what otherwise might be experienced as an oppressive environment. In the process self-assured personalities are cultivated that are able to experience and express themselves in novel ways.

Working with a definition of fashion which 'in a sense is change' (Wilson 1985, 3), the British Asian gay drag queen – which draws inspiration from Bollywood films (Leonard 1998), including the aesthetic and musical commodity forms – epitomizes the wider visible arrival of British Asian gay cultural expression onto Birmingham's commercial gay scene. The good sense aspiration for beautiful and passionate adventure, and the ability to carry this out by a new-found strength to express oneself in an innovative direction, is reflected in the subsequent account by Zaf on his first experiences of dragging up for *Ultimate Karma*:

> It was wicked, oh! It was such a laugh! Cause the first time I did it, I wore black kurta pyjama, thirty pounds it cost me to get it sewn, silver embroidery all over it, and it looked like really unique, the worst thing was getting the cleavage done when I done it in the gold lenga I wore, I had a proper bra on, I nicked it from my sister's bedroom! While we were getting dragged up the cleavage dropped, and I had to have sellotape across my tits, but it was wicked! Cause you know, Miss Binny, the drag queen of *Route Two*, she's been voted best drag queen of Birmingham, she saw me and she goes, 'wicked!', and I was so proud, because it's like, for like Amitabh Bachan to say, 'well done!' to an actor, it was like, Miss Binny to say that to me, I was like 'hi, no way!', the buzz of it was that, the laugh, the memories, and like now when we go out we can sit back and say, 'oh remember Zaf when your cleavage dropped?!', or 'Zaf, you remember when your lenga got caught onto the sink and it ripped!', it's the memories, like you can sit back and say, 'we did it, I did it, been there, did that'. (Zaf)

This account is evocative of 'a world intimately ingrained in "possible artists"' (Gramsci 1985, 98). Moreover, like Dwyer's (1999, 5) findings on the use of the veil by British Muslim women, the account demonstrates the capability of people to rework the 'meanings attached to different dress styles' in the production of 'alternative identities'. Interestingly, one significant new memory for Zaf is based on his moment of legitimation by a popular, white gay drag artist, who he compares to one of Bollywood's foremost and longstanding actors, Amitabh Bachan. Still, the 'scopic economy' of the commercial gay scene, which both fuels desire for legitimation and judges all visible representations of Otherness (Skeggs 2000, 129), also provokes more precarious and contradictory moments of existence. The attention of this chapter now turns from a focus on good sense that is 'displayed in effective action' (Gramsci 1971, 326) – and which engenders and is engendered by British Asian gay cultural identities, forms and practices – to a focus on common sense that is 'affirmed in words' (ibid., 326) – and which stems from an entanglement of signification, racialisation and the threat of racism, and effectively divides and impedes counter-hegemonic potential.

The Burden of Racialisation and Moments of Common Sense

> [T]he co-existence of two conceptions of the world ... is not simply a product of self-deception ... It signifies that the social group in question may indeed have its own conception of the world, even if only embryonic ... But this same group has, for reasons of submission and intellectual subordination, adopted a conception which is not its own but borrowed from another group .... (Gramsci 1971, 326–7)

The precarious and contradictory moments of existence for an emergent British Asian gay culture hinge on the broader descriptive processes of signification and racialisation, and the ongoing threat of evaluative racism (Miles 1989, 2000). On the predominantly white commercial gay scene, gay and bisexual British

Asians feel and carry the burden of racialisation via the visible marker of their skin colour. Moreover, *within* this racialised collective, there are two further visible groups – additionally marked by class and gender – which are felt by some to pose a raised threat of racism for all of those carrying this burden of racialisation. Consequently, these visible groups are internally-Othered (by means of a dialectical process of signification) in a subconscious reaction to avoid racism. The two groups comprise of those judged to be the most lacking in respectability: 'the butch lesbians' and 'the rent boys', i.e., those who are much less likely to have the ability to convert their cultural capital in symbolic and economic exchange (Skeggs 1998, 1999, 2000).

In the early stage of *Ultimate Karma*, recollections of an incident involving two working class, British Asian lesbians circulated amongst a number of (mostly male) gay and bisexual British Asians regularly accessing Birmingham's commercial gay scene. These recollections can be seen as a Gramscian common sense, modern-day folklore, which operates as a reactionary attempt to distance oneself from the perceived raised threat of racism. The following two quotes reflect typical recollections of the episode, which involved a disturbance at the end of an *Ultimate Karma* night that, because it occupied the street near the commercial gay scene's main taxi rank, was within the wider gaze of the predominantly white gay scene:

> You had all the bouncers from the main entrance and umm, a couple of lesbians had come out, they just started fighting, I think that it was just that they were drunk, they were feeling in the mood, that, you know, 'OK, we've gone to', you know, '*Karma*, and we've been really relaxed about ourselves', you know, 'about feeling gay and Asian', and they just started fighting, it was like everyone started getting involved, it was like they were just doing it to sort of show the white people that 'we're here', but my mate was saying that he was so embarrassed because the white people were just thinking, 'oh is this what Asian people are like?', you know, and 'if there was going to be an Asian night', and if they thought, 'there's always going to be trouble'. (Ranjit)

> It was a couple of short Asian lesbians, like really short hair and everything, and all that happened was, basically they just lashed out at each other, and then they started on the bouncers as well, saying that 'you're this, you're racist', and all this shit, and really, I mean, they shouldn't have brought their problems into *Karma*, it was really bad because everyone was looking down at Asians thinking, 'they come on the gay scene to fight', and you don't blame them because at the end of the day that's what we were doing. (Zaf)

These accounts (both by men) exemplify a mild disdain for the 'unrespectable' women, that is, by blaming the 'couple of lesbians' with 'really short hair' for being 'drunk' and 'in the mood' for a fight. What is more, the possibility that the fight was aggravated by racist comments made by the bouncers is mentioned but dismissed with dogmatic certainty as 'all this shit'. The emphasis of all of the recollections I encountered during this time was on the burdensome, embarrassing nature of the women's 'illegitimate' visible appearance and behaviour, and its implication for 'white' people's perception of 'Asian' people. An internal-Otherisation founded on judgements of respectability (in a reaction to distance oneself from the perceived raised threat of racism) thus excuses racialisation and the potential instance of racism therein by comments such as 'you don't blame them'.

The space of the commercial gay scene is thoroughly embedded in a pervasive, judgement-fuelled, scopic economy, which conditions an insecure existence for all of those who desire legitimation in and through its social relations but lack the right combination of capital to possess respectability (Skeggs 2000). A *visible* number of young, working class men in their late teens and early twenties working as rent boys in Birmingham are British Asian. The burden of racialisation felt by their visibility on the commercial gay scene, and the subsequent internal-Otherisation that occurs within an already racialised 'Asian gay' collectivity, are laid bare in the next quotes by Dan and Anup (the latter of whom, at the time of the quote, was the bar manager of *The Village Inn* – a prominent venue on Birmingham's gay scene):

> It gives a bad name to everyone innit, if they [the British Asian rent boys] wanna behave like that that's really entirely up to them, if they wanna be drunk and do whatever, if people wanna stereotype them they can do, but it's also true to say people will stereotype them, and I'm not saying you should go out and be on your best behaviour but, I mean, I've been drunk in the past but some of these young Asian people, they're really, like act outrageously. (Dan)

> When I came out ten years ago there was me and a couple of other Asian guys, and that was it, and now there's so many, they're all camp, that's just the annoying thing, it's been in the past couple of years when this big explosion happened, this new crowd that come in *The Village*, there's still more coming out, they see me there and they think they're safe there, and they won't get hassle, but then they create their own problems, they're without a drink half the time, some of them have been barred from like half the bars, causing trouble, because people come down for like a quiet drink sometimes, or sit down the bar, and like they're sitting there clogging up all the tables, people can't sit down and just like chill out, half of them are dressing up and stuff like that, which is fine, each to their own, but some of them just go over the top ... there's this stereotype now that all of the gay Asians are camp, that's the one thing I'd like to change because we're not all like that, there's more and more of them that I'm seeing now dressing up and it's just like, why do they need to dress up for? I don't mind it, but it's giving the rest of us a bad name, so, they've even asked me, saying 'why don't you dress up?', and it's like, 'sorry but I'm not that way inclined, so don't try and fit me into your category' ... it's the rent boys mainly. (Anup)

The underlying references here are to groups of 'unrespectable' British Asians – 'the rent boys mainly' – either 'act[ing] outrageously' or 'clogging up' the space of gar bars. Further still Anup's comments (denoting an outlook that derives from more than one social group) subconsciously absorb a metaphor of the 'racial' Other taking up too much room.

Conclusion

The precarious and contradictory moments of existence that define an emergent British Asian gay culture on Birmingham's commercial gay scene reflect, in a Gramscian sense, the nucleus of embryonic good sense to modify one's social surroundings that subsists within a common sense state of submissiveness to preserve the existing hegemonic environment. What this chapter has sought to do is, on the one hand, emphasize the radical possibilities that flow from ethnic and sexual minority cultural identities, forms and practices, which are suggestive of a more promising counter-hegemonic future; and, on the other hand, not take for granted the 'racial' minority status of this emergent ethnic and sexual minority culture, but to expose its confinements. The predicament in this case is that some individuals in a racialised collectivity are especially marked and vulnerable to racism, while other individuals in this collectivity, in turn, internally-Otherise the former in a reaction to avoid racism. The future challenge for an emergent British Asian gay culture is perhaps then part of a more general problematic facing all existing cultural groupings: 'to make the people join in a criticism of themselves and their own weaknesses without making them lose faith in their own strength and their own future' (Gramsci 1985, 251).

Notes

1   The empirical material analysed in this chapter was generated as part of a wider ESRC-funded doctoral study (award number R00429934028) which investigated the lived and material phenomenon of a British Asian gay club scene and its development on Birmingham's commercial gay territory (Bassi 2004). In total, twenty-seven regular participants of Birmingham's British Asian gay club nights were formally interviewed, twenty-three of which were British Asian and four British white. Twenty-two of the participants defined themselves as gay men, two as bisexual men and three as lesbian women. A further eight interviews were held with organisers of the British Asian gay club nights and venue owners of Birmingham's commercial gay scene. I conducted participant observations of the first monthly British Asian gay club night, *Ultimate Karma*, from February 2000 to April 2001, and its successor monthly club night, *Saathi*, from July 2001 to November 2002. During this time, Ultimate Karma occupied the attic room of *The Nightingale* nightclub and Saathi occupied the conservatory room of *The Village Inn* pub (both of which are prominent, longstanding, commercial gay venues). Throughout this duration, I also frequented and recorded participant observations of Birmingham's key commercial gay scene venues. For a trajectory of Birmingham's gay political economy, and therein the British Asian gay club scene, see Bassi (2006).

2   It is worth briefly outlining the version of dialectics upon which Gramsci's contradictory consciousness is premised and advances. Hegelian dialectics is perhaps best defined as a set of tools for thinking about movement. For Hegel (1830), one can only make sense of the world through an ongoing (or in-motion) discussion of ideas, which inevitably involves techniques of argument (or reasoning tools) based on the opposition between appearance and essence in a range of forms. Whilst Marx rejected the idealism of Hegel, by insisting that ideas are nothing but the material world reflected in human minds and translated into forms of thought, he upheld Hegel's dialectics by seeing all

historically developed phenomena as in opposing-yet-interrelated and fluid states (see, for example, Marx 1951). Gramsci's notion of contradictory consciousness is dialectical in that it deals with the human activity of argument which, in turn, is seen to reflect conditions of existence. It summates how people reason about the world around them by drawing upon both regressive *and* more progressive ideas and meanings. What is more, Gramsci pushes further by stressing the role of 'organic intellectuals', that is, an rising layer of intellectuals who act as an organ of the body of people (and also emerge from this body) to both challenge the reactionary elements and encourage the development of the progressive ones through dialogue and debate.

3   For instance, the queer theorist Cooper (1995) discusses Gramsci's notion of counter-hegemony alongside an additional notion of anti-hegemony (which is defined as deconstructing the status quo without putting an alternative political project in its place). She proposes a compromise between counter-hegemony and anti-hegemony, so while the former encourages aspiration for progressive societal transformation, the latter places this in check by a caution that such change can only be a strategic narrative.

4   For Gramsci (1971, 324) it is imperative to banish 'the widespread prejudice that philosophy is a strange and difficult thing' since everyone possesses a 'spontaneous philosophy'. When referring to Gramsci's writings on philosophy, Donald and Hall (1986, xii) insert the word philosophy in inverted commas and argue that 'ideology cannot be confined to its "philosophical" moment'. In contrast, I suggest that Gramsci considers philosophy and ideology as inseparable concepts, with ideology signifying a more immediate and emotive form of philosophy (see Gramsci 1971).

5   Hall (1986) implies that common sense is a permanently existing sphere of conflicting views, so does not give credence to the notion of *two* competing philosophies. Instead, he calls attention to 'the level of casual, everyday, contradictory, common-sense' as an 'ideology [that] cannot be reduced to its philosophical essence' (ibid., 36). Further still from Gramsci's distinction between common sense and good sense, Joll (1977, 101) argues that common sense is what Gramsci saw as the highest aspiration for the philosophy of praxis – suggesting that revolutionary parties should be 'subjected to the test of common sense'.

6   Hegemony tends to have been interpreted as referring to how political power in society is maintained via social consent (Anderson 1976–77; Holub 1992; Joll 1977). Simon (1985, 21), for instance, states that '[h]egemony is a relation, not of domination by means of force, but of consent by means of political and ideological leadership'. In some instances, this has been interpreted as consent achieved by means of ideological social control (reminiscent of Althusser), for instance, Joll (1977, 68) states that 'hegemony [is] the way in which a minority can impose its leadership and its values on a majority'.

7   In the film *Pakeezah*, there is a sympathetic focus on the main character of a courtesan and her experience of isolation and pain, as well as the theme of romance and sexual desire (Joshi 2001). Gupta (1993) suggests that whilst there are no openly gay characters in the film *Pakeezah*, there is a unspoken gay meaning. Using the example of the song 'Chalte, Chalte', he explores its theme of looking and cruising with the song's literal translation of 'Walking, I met someone'.

References

Amin, A. (2002), 'Ethnicity and the Multicultural City: Living with diversity', *Environment and Planning A* 34, 959–80.

Anderson, P. (1976–77), 'The Antinomies of Antonio Gramsci', *New Left Review* 100, 5–80.

Back, L. (1996), *New Ethnicities and Urban Culture: Racisms and multiculture in young lives* (London: UCL Press).

Bassi, C. (2006), 'Riding the Dialectical Waves of Gay Political Economy: A story from Birmingham's commercial gay scene', *Antipode: A Radical Journal of Geography* 38:2, 213–35.

Bassi, C. (2003), *Asian Gay Counter-Hegemonic Negotiation of Birmingham's Pink Pound Territory* (unpublished thesis, University of Sheffield).

Bourdieu, P. (1986), *Distinction: A social critique of the judgment of taste* (London: Routledge).

Cooper, D. (1995), *Power in Struggle: Feminism, sexuality and the state* (Buckingham: Open University Press).

Donald, J. and Hall, S. (1986), 'Introduction', in J. Donald and S. Hall (eds) *Politics and Ideology: A reader* (Milton Keynes: Open University Press).

Dwyer, C. (1999), 'Veiled Meanings: Young British Muslim women and the negotiation of differences', *Gender, Place and Culture* 6:1, 5–26.

Fraser, N. (1995), 'From Redistribution to Recognition? Dilemmas of Justice in a "Post-socialist" Age', *New Left Review* 212, 68–94.

Gilroy, P. (1987), *'There Ain't No Black in the Union Jack': The cultural politics of race and nation* (London: Routledge).

Gramsci, A. (1971), *Selections from Prison Notebooks* (London: Lawrence and Wishart).

Gramsci, A. (1985), *Selections from Cultural Writings* (Boston, MA: Harvard University Press).

Gupta, S. (1993), 'India Postcard, or Why I Make Work in a Racist, Homophobic Society', in M. Gever, J. Greyson and P. Parmar (eds) *Queer Looks: Perspectives on lesbian and gay film and video* (London: Routledge).

Hall, S. (1986), 'Variants of Liberalism', in J. Donald and S. Hall (eds) *Politics and Ideology: A reader* (Milton Keynes: Open University Press).

Hall, S. (1992), 'New Ethnicities', in J. Donald and A. Rattansi (eds) *'Race', Culture and Difference* (London: Sage).

Hegel, G. (1830), 'Encyclopaedia of the Philosophical Sciences', *Marxists Internet Archive*, <http://www.marxists.org/reference/archive/hegel/works/01/encycind.htm>, accessed 21 August 2007.

Holub, R. (1992), *Antonio Gramsci: Beyond Marxism and postmodernism* (London: Routledge).

Joll, J. (1977), *Gramsci* (Glasgow: Fontana).

Joshi, L. (ed.) (2001), *Bollywood: Popular Indian cinema* (London: Dakini Ltd).

Kaur, R. and Kalra, V. (1996), 'New Paths for South Asian Identity and Musical Creativity', in S. Sharma, J. Hutnyk and A. Sharma (eds) *Dis-Orienting Rhythms: The politics of the new Asian dance music* (London: Zed Books).

Laclau, E. and Mouffe, C. (2001), *Hegemony and Socialist Strategy: Towards a radical democratic politics* (London: Verso).

Leonard, T. (1998), 'One Nation: Under a groove', *Gay Times*, September, 28–30.

Marx, K. (1951), *Capital: Vol. 1* (London: Everyman's Library).

Miles, R. (1989), *Racism* (London: Routledge).

Miles, R. (2000), 'Apropos the Idea of "Race" ... Again', in L. Back and J. Solomos (eds) *Theories of Race and Racism: A reader* (London: Routledge).

Morton, A. (1999), 'On Gramsci', *Politics* 19:1, 1–8.

Simon, R. (1985), *Gramsci's Political Thought: An introduction* (London: Lawrence and Wishart).

Sharma, S., Hutnyk, J. and Sharma, A. (1996), 'Introduction', in S. Sharma, J. Hutnyk and A. Sharma (eds) *Dis-Orienting Rhythms: The politics of the new Asian dance music* (London: Zed Books).

Skeggs, B. (1998), *Formations of Class and Gender* (London: Sage).

Skeggs, B. (1999), 'Matter Out of Place: Visibility, Violence and movement in the city', *Leisure Studies: The Journal of the Leisure Studies Association* 18:3, 213–32.

Skeggs, B. (2000), 'The Appearance of Class: Challenges in Gay space', in S. Munt (ed.) *Cultural Studies and the Working Class* (London: Cassell).

Solomos, J. and Back, L. (1996), *Racism and Society* (Basingstoke: Macmillan).

Solomos, J. and Back, L. (eds) (2000), *Theories of Race and Racism: A reader* (London: Routledge).

Taylor, C. (1994), 'The Politics of Recognition', in D. Goldberg (ed.) *Multiculturalism: A critical reader* (Oxford: Blackwell).

Werbner, P. (1996), 'Fun Spaces: On identity and social empowerment among British Pakistanis', *Theory, Culture and Society* 13:4, 53–79.

Wilson, E. (1985), *Adorned in Dreams: Fashion and modernity* (London: Virago).

Chapter 16

Encountering South Asian Masculinity through the Event

Jason Lim

Introduction

Hindsight is a fine thing, or so they say. With the benefit of wisdom and reflection after the event, we are supposed to be able to make sense of exactly what happened in the event itself. When we reflect upon events sometime after they have happened, we tend to place them into certain historical narratives in order to make sense of them. These narratives are often offered up to 'explain' the event and what happened within it. In this way, historicising can often suggest a specific set of supposedly determining factors that led to the event turning out as it did.

What I want to do in this chapter is to problematise how we come to understand events through historicising them. This questioning of the event and its relationship to history arose in my considerations of a particular event – an encounter between a young man of South Asian ancestry and a young white woman in a pub in the early 2000s, an encounter I witnessed while conducting my doctoral research. Now returning some years later to contemplate this encounter in order to think about South Asian masculinity in a British context, I find that this task has been complicated by the ways in which public discourses surrounding South Asian masculinity have changed since the encounter took place. The most obvious of these changes has been a whole series of contested events and debates surrounding the changing threat of terrorism, in particular since the London bombings of July 2005, and the impact this has had on the perceived relationship between British South Asian Muslims and 'Britishness'. In some ways, these subsequent events are a red herring because they already invoke a narrative within which to understand the encounter in the pub, a narrative that closes down other ways of understanding that encounter. On the other hand, the changing politics of the relationship between British South Asians and Britishness is a very good reason to try to think about how difference is encountered in the event and about how events become related to race discourses and memories.

Indeed, in order to be a worthwhile exercise, thinking about the relationship between history and events in order to intervene in the racialised relationship between British South Asians and Britishness must have some impact on how we go about encountering difference in future events. In turn, this task requires us to think about the question of, to paraphrase Deleuze (1988), what anybody might be able to do in a given event and, consequently, the question of how such action is constrained. If, despite the lines of expectation and constraint that we and others might have in a given situation, we are able to contemplate how our interactions are not spelt out for us word by word, action by action, then we might open up the question of how to act in the event itself. This is a question that allows us the potential to interrupt the rememorialising – the bringing back to life – of histories of race, gender and class.

In this chapter, then, I explore how even though a specific event resonates with various racialised histories, we cannot say for certain what relations were actualised between those involved in the encounter. I then discuss some implications for how we might approach future events – future encounters with difference – in an ethical manner. Finally, I conclude with some brief thoughts about what thinking through the relation between the event and history might mean for thinking about cosmopolitanism at the interstices of what Paul Gilroy (2004) calls post-colonial melancholia and convivial culture. But, first, I must tell you the story about an encounter in a pub ...

The Encounter

I was at the bar, buying a round of drinks. It was a busy Friday evening in suburban London, and the crowd in this O'Neill's pub – one amongst many in a chain of ersatz Irish pubs – was already a mass of pushing and swaying bodies. Mostly these bodies, shoehorned into the sweaty space, were white and young. The man who, rather drunkenly, decided to start talking to me at the bar, however, immediately struck me as being of a South Asian appearance. He was gregarious, rather loud, rather demonstrative. Something I noticed straight away was that, unlike most young South Asian men on a night out in this part of London, he had neither chosen to sport the smart-casual shirt and trouser combination so popular among many of his contemporaries, nor chosen to adopt and adapt the styles of clothing or mannerism that mark blackness through an embodied citation of US hip hop culture. Rather, he wore a somewhat battered and old-fashioned brown leather jacket and his mock cockney accent was betrayed when it occasionally slipped into something less extravagantly performed.

We talked for quite some time, and after a while I learnt that his name was Minesh and that he was of 'Indian' ancestry – his family were Hindus from Gujarat – and that this mattered to him. Throughout our conversation, Minesh would make a show of knowing lots of other people in the pub, bantering with the barman and, every now and then, introducing me to other young men he knew who happened to be passing by the bar or coming to buy some drinks. He seemed to be good friends with these men, chatting and joking with them, laughing loudly and apparently very relaxed.

A little later a young woman, who appeared to me to be white, pushed her way past us to join the throng at the bar. It seemed that Minesh thought he knew her. He tried to start talking to the woman, but she ignored him. As he continued to try to start up a conversation, even going so far as to tap her on the shoulder, she remained poker faced, avoiding eye contact, trying not to respond. Turning her body away from him, her closed stance seemed to erect some kind of barrier to any potential contact. It struck me that perhaps she thought that Minesh was attempting to chat her up and that perhaps she also felt somewhat harassed by this. Even when Minesh guessed at her name – 'Sarah?' – she barely reacted. Minesh gave it one last shot: 'Aren't you a friend of Ryan Gardner's? Weren't you at his party last weekend?' Yes, she replied, but she still didn't recognise Minesh, nor have any intention of talking to him. The conversation ended abruptly, but Minesh remained unfazed.

Memory and History

It would seem that Sarah was somewhat discomfited by Minesh's attentions to her. With her body, in the simple act of turning away from him, the simple act of remaining inattentive to him, she performs a social distance and reduces the potential for further affective entanglement between them. Beyond this, though, what might we say or think about this encounter? Is this encounter amenable to interpretation?

If we were to attempt to interpret this encounter, we might ask whether Sarah thinks that Minesh is not worth talking to. Another interpretation might be that she considers him to be a sexual threat. After all, in pubs and bars young women are often faced with men making sexual advances towards them, and these women might encounter such advances as a hazard, an opportunity, a threat, or an amusement. Regularly, in such situations any hint of openness by women, in their embodiment as much as in what they might say, might be seized upon by men as a sign of sexual interest. This might be thought of in terms of a kind of memory that is both social and embodied. It is the minutiae of bodily actions that establish the nature of social relations within a situation, for example whether they are sexualised, desexualised, friendly, indifferent, authoritative or deferring. So, we might think of a shared 'recognition' that in such spaces – a pub, a bar, a nightclub – certain gestures, postures and expressions tend to signal a receptiveness to a sexual advance. Similarly, Sarah might have apprehended Minesh's actions as an attempt to chat her up because his actions 'recalled' a whole history of styles of men trying to chat women up in such spaces. It should be noted, though, that I use the terms 'recognition' and 'recalled' rather loosely here. As many other thinkers have noted, in the midst of everyday interactions we don't stop to think about the kinds of actions we and others are performing or about what kinds of social relations these actions are establishing. Instead, as with Bourdieu's (1977; 1986) conception of the *habitus*, these memories seem to be unthought by the mind and, rather, learnt and remembered by the body. And if these memories are not recalled by individual minds, neither are they remembered by *individual* bodies because they have to be shared in order to be effective, to establish social relations. They have to be social memories, memories that cite a collective history. Given that this citation is implicit rather than explicit – we don't go around citing the precedents for our actions all the time – such a memory might be conceived along the lines of Judith Butler's rendering of the performative (1990, 1993, 1997, 2004).

Sarah's turning away from Minesh and remaining impassive to his attempts to start conversation, then, might be read as not only preventing any potential contact between them from developing, but also as specifically intended to stop the encounter from developing as a sexual advance. Moreover, that Sarah ignored Minesh is suggestive of how interaction in such spaces relies on an apprehension of surfaces. All Sarah can apprehend is a young man, slightly drunk, trying to talk to her, and she precludes this contact in case what is desired is sexual. By taking Minesh as a surface, though, might she also reduce what is taken to his skin colour?

In addition to thinking about it in terms of sexuality and gender, then, another way we might interpret this encounter is in terms of race and ethnicity. Indeed, when I came to analyse this event in the course of my doctoral research (Lim, 2003), I placed it in the context of other findings I had made during my research. Combined with some of the texts I had been reading (Anderson 1987; Hutnyk 2000; Willis 1977), this seemed to suggest something problematic in the way that Asian masculinities have become constructed in the West generally and in Britain in particular. Another research respondent, Darren, who was of Indian and Anglo-Indian ancestry, talked about how he found it difficult to find a way to perform an appropriate gendered and sexualised masculinity when he went out to pubs or bars with his predominantly white friends. He remarked upon having found it difficult to become attractive, to embody himself in a way so that he would be able to 'pull' a female sexual partner on a night out at a pub, bar or nightclub. There seemed, for him, to be something ineffable about how these ethnicised distinctions in performing masculinity were enacted. Darren's friend Henry, who was of Chinese ancestry, was a little more explicit when remarking upon how Asian men in general – South and East Asian – were often considered to be quiet and effeminate and, by implication, somehow less than fully masculine.

In the context of Darren and Henry's comments, one possible interpretation of Minesh's encounter with Sarah might place the encounter within two parallel sets of discourse, one a (distinctively British) public discourse and the other an academic discourse reflecting upon the histories of the public discourse. So, on the one side, the public discourse reproduces understandings of young Asian men as passive, soft and weak (Nayak 1999; Hopkins 2006), and as effeminate, well-behaved over-achievers in school and in the workplace (Archer 2001), especially if they are considered middle class. In a conflation of a certain version of masculinity with what it is to be sexually attractive, young Asian men are often considered to be less 'hard' and less stylish and consequently less attractive than young white men and, in particular, young black (African-Caribbean) men (Nayak 1999, 89–91; Hopkins 2006, 338). Such a conflation might arise from an understanding of Asian cultures and religions as requiring a certain asexuality among unmarried men as well as women (see Alexander 2000). On the other side, then, are a set of academic discourses that historicise and problematise these public discourses. These might include historical analyses of how hierarchies of representations of differently racialised sexualities have developed (Gilman 1985;

Kabbani 1994), and how such developments might have related to colonial and post-colonial discourses, policies and practices that have sought to control the production of gendered and sexualised subjects (Sinha 1999) and to legitimise the development of ethnicised communities as political constituencies and polities (see Das Gupta 1997). They might also interrogate the historical co-development of white (British) masculinities and heterosexualities (Willis 1977; McDowell 2000, 2002, 2003).

Given that such discourses might suggest that South Asian men find it difficult to embody an appropriately masculine position within British public cultures, I think it is interesting that Minesh, before the encounter with Sarah, seemed to be acting out an almost parodic version of white working-class masculinity. Despite his having attended a prestigious British university, he made a great deal of his disdain for academic knowledge, instead playing up the virtues of 'real', practical work. More than this, he embodied a particular style of boisterousness, accompanied by the appropriate performances of accent, gestures and bonhomie, all of which were redolent of London's East End. An abdication of class privilege in some ways, but perhaps an attempt to appeal to a somewhat unruly, physical, virile and culturally British heterosexual desire (Nayak 1999; McDowell 2003; Willis 1977)?

Of course, Sarah never talked to him for long enough to gauge the full impact of his performance. Perhaps, however, she did not want to talk to Minesh, not because she perceived him to be a soft and effeminate Asian, but because she perceived him to be overbearing. In this respect, it is important to recognise the academic discourses that note the separation of two attributes – effeminacy and patriarchal power – within public discourses about South Asian men (Alexander 2000, 2004; Hopkins 2006). Alexander (2000; 2004) notes how Asian men, these days visible in both public and academic discourse primarily as young Muslim men, have been constructed as expressing an aggression born out of a failure to negotiate the twin demands of their expected patriarchal role in their ethnic communities and their place within contemporary British society. Commendably, Alexander (2004) challenges academic discourses that propound this view, pointing out how they advance a static and absolutist view of culture as ethnicity, eliding other dimensions of life, and deflecting away from the racism faced by young South Asian men. Moreover, Alexander (2000) also draws our attention to the divergence between constructions of the patriarchal power wielded by South Asian men and the real limits of these men's control over South Asian women's sexual practices, especially those of older female relatives, including older sisters. She also points up the divergence between the fictions of asexuality on the part of unmarried South Asian men and women – fictions that circulate both inside and outside South Asian communities – and the more fraught realities of having to negotiate specific lines of permissibility, for example having relationships with women from different communities and maintaining certain silences about sexual relationships.

As Alexander (2000, 2004) suggests, the shift in public and academic discourses surrounding South Asian masculinities that has taken place since the early 1990s – a shift from viewing South Asian men as quiet, weak victims of racism to seeing them as aggressors with a problematic relationship to patriarchal power – has depended upon a separation of 'Indian' (non-Muslim) masculinities from Muslim Pakistani and Bengali/Bangladeshi masculinities. While the former have remained stereotyped as quiet, perhaps effeminate achievers, it is the latter who have been seen as a threat. Nonetheless, this is not to say that Indian men are not perceived by many whites as invested in patriarchal forms of power within their communities.

The Violence of Interpretation

In several ways, attempting to place Sarah's possible perception of Minesh within the broader context of my doctoral research findings and within the context of both public and academic discourses regarding South Asian masculinities amounts to an act of attribution that is impossible to substantiate. Not only would we be attributing intentions to Sarah that we cannot validate, but we would also be attributing a set of potential senses of Asian masculinities to Minesh's body, none of which we can be sure came into play. Given that racism works by effacing difference – by lumping together different bodies under the sign of an ethnicity or a racial appearance – we might ask how sophisticated Sarah's reading of male South Asian bodies would have been. And yet we can have no answer. There are also many other potential ways in which we might account for why she brushed Minesh off. She might have been tired. She might just have had a bad day at work. Perhaps she had been arguing with a friend. I would not have known if she had met Minesh here before (or, indeed, at Ryan Gardner's party) and found him to be a boor or a bore.

Recalling the first of the problematics I started this chapter with – the problematisation of how we come to understand events by historicising them – we can see that already our process of reflection after the event has placed the event into various historical narratives. Of course, this happens all the time, but it is often done for particular reasons, and always has particular effects. The event becomes enrolled into stories that can be put to use politically. Stories about the apprehension of a young South Asian man might be mobilised to justify certain actions (measures to help 'integrate' different ethnicised communities, perhaps), close down other potential lines of action, demarcate belonging ('Are South Asians an alien culture within Britain?'), or allocate blame for enmities or distributions of resources ('Why are Indians getting ahead?'). In some senses, all interpretation involves violence (Foucault 1991), appropriating systems of relation and subjecting them to secondary rules, and appropriating bodies and ideas and making them stand in for other ideal meanings, collectives and histories. Interpretation asks that bodies and their relations obey a system of proper, explainable relations. By placing Minesh's encounter with Sarah into a set of discourses surrounding South Asian masculinities, we ask the event to conform to the relations and rules, the causes and effects, and the origins and destinations put forward by each respective discourse as an explanatory framework.

We can think about both the implications and the limitations of this kind of process of historicisation by turning to the second problematic I set out at the beginning of the chapter, namely how to act ethically in the (future) event. In any given situation, it might be said that action and interaction is constrained by any number of institutional forces, custom and etiquette, and other ways in which social relations are regulated through policing and self-policing of what is normally or properly done – 'the done thing'. All of these institutional and regulatory forces might be thought of as both depending upon and reproducing histories of how social relations have been organised, how bodies and their interactions have been organised. However, as we have already seen, the question of how to act in any given situation can also be thought about in terms of social and embodied memories. And if remembering acts out a social and embodied memory of normal, proper modes of interaction, it might be said that remembering re-enacts these very institutional forces, customs and etiquette, and policing and self-policing of everyday actions. The preceding section's discussion of the different potential histories that might help us to understand the encounter between Minesh and Sarah can also be understood as an attempt to think about the different social and embodied memories that may have been brought into play in the encounter: what memories do both Minesh and Sarah bring to the encounter regarding what kinds of bodies they find attractive? What desires and antipathies are carried within the body such that the body attempts a certain style of engaging another in conversation or attempts a certain style of rebuff?

However, when thinking about how such memories work to re-enact normal practice and historical distributions of power and difference, it is one thing to say that in the event other bodies are encountered as their surface and that, for example, fear of an other's skin remembers a fantasy of the fearfulness of the other's body, thus constituting the other (as fearsome) and the self's relation to the other (afraid) (Ahmed 2004). It is much more problematic, though, to say what histories play out over the skin (*why* this fearful relation?). We can, of course, attempt to trace these histories through eliciting more explicit discourse about the event: recordings of talk in the event itself; interviews or statements after it; or documents, archives, or commentaries. All of these sources can help. Yet, no matter how much discourse is collected, there would still be the problem that discourse is often not a reliable account of practice (Frith and Kitzinger 1998, 2001; Malson et al. 2002) because it is often blind to the unconscious and is often deployed, within specific social contexts, to re-enact the politics of the normal and the proper. It is important, therefore to think further about the relationships among the event, memory and history, especially to see how these relationships take shape through our practice, both in everyday life and in the academy.

One way of thinking about the ethics of encountering difference – whether or not, in the event, we repeat memories that place bodies into relations of unequal power, antipathy or conflict – might be to consider how both *perception* and *affect* can themselves be thought of as events. Here, the conceptions of perception and affect developed by Gilles Deleuze (1988, 1992, 1993, 1994, 1995; Deleuze and Guattari 1988) can be of great help in thinking about what happens in racialised encounters. *Perception* can be understood as the capacity of bodies to apprehend other bodies, properties or ideas. If one is thought to be 'prejudiced', it could be said that one is all too ready to apprehend a differently racialised body, for example, as disgusting or frightening, and, in doing so, one's body repeats a kind of perception, a memory of a perception, or rather a perception as memory. Doing so, one forecloses other ways of perceiving that differently racialised body, the potential to take the encounter with that body on its own merits, seeing whether one might perceive it differently. *Affect* can also be a helpful term here. Very simply, it can be thought of as the capacity or power of bodies to affect other bodies and be affected by other bodies (Deleuze 1988). It is a slightly broader term than perception because there are ways that a body might be affected by another body that might not involve perception, or of which perception might only be a part. Dancing with somebody, for example, involves both perception (of their movement and their rhythm) and something more than that – *being moved* by their movement or by your shared movement with them. If we think about the encounter between Minesh and Sarah, we might say that in very broad terms we are simply thinking about how they affected one another and how they were affected by one another.

According to Deleuze, in every situation, we can distinguish between, firstly, the many different things we might perceive, the many different ways we might affect other bodies or be affected by them, and the many different things that we might do – what he calls the *virtual* field – and, secondly, the one or small number of these things that become perceived, affected or acted out – the *actual*. As bodies are always entering into new situations and new interactions – always doing something different from moment to moment – so the virtual field is always changing. The percepts available for experience and the affects available to be acted out between bodies are thus constantly changing. In the sense that such affects and percepts arise from the interaction *between* bodies, we might say that percepts and affects do not belong to individual bodies themselves (Deleuze 1995, 127).

One implication of thinking of perception and affect in this way is to think of them as *events*. Events, for Deleuze (1990), are virtual and are thus distinct from actual 'states of affairs'. Events might be conceived of as the senses, attributes, affects and percepts that pertain to any particular state of affairs. If we think of the encounter between Minesh and Sarah in terms of the event, we might come up with several lists suggesting some of the virtual percepts, affects and senses that might have been available for becoming actual. Some potential percepts (which are already affects) might have been: of Minesh's forcefulness; of Minesh's boorishness; of his gregariousness; of his attractiveness; of his effeminacy; various percepts of Minesh's movement, proximity and intonation of voice; percepts relating to Minesh's skin colour or dress; percepts of Sarah's disinterest; of her facing the other way; of her closing her stance; of her impassive expression; of the brevity and intonation of her responses. Some other affects might have involved but complicated any number of these percepts. To take a small number of examples: a becoming fearful of a forcefulness; a becoming annoyed by a forcefulness; a becoming threatened by Minesh's proximity; a becoming closed to his skin colour, a surface of his face. And, some of the senses that might have been available in the encounter might have partaken in some of these affects: a sense that Minesh was chatting up Sarah; a sense that Minesh was being drunkenly overbearing; a sense that he was boring and unattractive; a sense that Sarah did not recognise him; a sense that Sarah was fearful of him; a sense that she would talk to him if he could provide a point of common reference; a sense that she felt repulsed by him. All of these percepts, affects and senses may have been available for actualisation, although only a small number of them would actually have been. Given the multitude of potential percepts, affects and senses available, Minesh and Sarah would not only have potentially experienced what was happening quite differently from one another, but would also each have experienced different percepts, affects and senses simultaneously and from moment to moment. The virtuality or eventfulness of these percepts, affects and senses means that they are not reducible to the actions of bodies, although they do *pertain* to them and they do arise because of what bodies are doing in relation to other bodies. In turn, what actually happens – actual actions and interactions – can be thought of as a process of selection by which only some of these virtual percepts, affects and senses are acted out or experienced (Massumi 2002). The actual enacts some of these virtual events, repeating them but with a difference. So, while we are able to easily say what *actual* bodily interactions took place between Minesh and Sarah, it is much more difficult to discuss with any certainty what may have been important amongst the multitude of *virtual* perceptions, affects and senses that were available to them.

The Ethics of Encountering Difference: The Event, Memory and History

If we consider perception and affect as events, their virtuality has certain ethical implications for how we approach encounters with difference. Despite the lines of expectation and institutional forces that constrain us, the multiplicity of different things that might nonetheless happen might be taken as demanding of us an ethical stance whereby we attempt to become more open to all the different things that might be done, perceived, and experienced in each event (Massumi 2002; McCormack 2003). Such an openness to the contingencies of the event might be understood as warning us not to let memories, narratives and expectations of patriarchal, racist or sexist injunctions against 'improper' connections with certain other bodies interrupt out potential friendships, desires and relationships. Yet, it might also be taken as an implicit critique of those who are always aware of the effects of racism, sexism, patriarchy, homophobia, religious intolerance, class oppression etc. – especially those who are victims of intolerance. Such bodies might be considered as all too ready to be affected in particular ways – a defensiveness against violence, hatred, discrimination that starts to see a pattern of systematic intolerance in all situations (Sedgwick 2003) even when the motivations, affects or processes immanent to any individual event might not be known to them. However, making such a critique fails to offer a vision of how to think ethically to those who need to be defensive in order to get by or indeed to survive. We might ask how one is to approach encounters ethically if one is always finding that one is taken as undesirable or, moreover, if one is always fearful with good cause. In privileged society, such perennial fear is thought of as paranoia (Sedgwick 2003). Yet, for many people, perennial fear has a real object or, more pertinently, a real subject.

There is a need to think about how to act and think ethically whilst living in a world in which affect is already caught up in the necessity of lines of expectation where these lines are not yet consciously realised or reflected upon. Such a vision might also learn from how those who have lived through the effects of suffering or oppression are already practised in intertwining an openness to the event with ethical modes of thought that move beyond (and not just in opposition to) the kinds of representational thought that dictate an economy of proper relations and desires. Many people are already practised at interrupting the affordances and lines of expectation of their own bodies, for example, with respect to who they find attractive, or in the face of religious injunctions about what cultural practices are proscribed, or in needing to negotiate going to work safely and without censure or guilt. Many academics have already written suggestively about how to think of such an ethics (Gilroy 2004; Grosz 2005a, b; Hemmings 2005; Sedgwick 2003).

In the remainder of this section, I want to sketch out a way of approaching encounters with difference ethically that simultaneously allows for the role of memory, history and thought as we have discussed it so far. This vision of an ethical approach to the encounter must address the two sets of questions raised in the preceding sections. The first set of questions asks how to act ethically in the encounter so that we do not necessarily repeat memories that offer us lines of expectation – templates for how to interact properly and thus politically with other bodies. The second set of questions asks how to think about the process of understanding the event through historicisation in a way that does not posit apparently exhaustive historical explanations nor posit a proper relationality regarding how differently marked bodies are supposed to relate to one another. Crucially, I want to address these questions without disqualifying the kinds of transformative historical understandings brought to bear by many of those who have been subject to fear, marginalisation and oppression.

In order to sketch out such a vision, I take as inspiration Deleuze's (1993) conceptions of *folds* and *monads*. Amongst other things, we might think of the fold as a way to think about the relationship between the inside and the outside. There are all sorts of interiorities, including perceptions and affects – which, after all, are supposed to be 'inside' our heads or bodies – but also social encounters, architectural spaces, textural spaces etc. The idea of the fold suggests that each of these interior spaces is an infolding of the outside – the world. Together with the idea of the monad, by which we can think of the envelopment of the world from a particular point of view (Deleuze 1993, 25–7), this is one way to think about how multiplicities of histories and memories – all that has happened and all that might happen – are immanent to any specific event of thought, perception or social action.

It might seem that this idea of the *infolding* of the outside allows us to reconcile the idea that a multiplicity of potential percepts, affects and senses becomes infolded into the event (hence affording us a multiplicity of things that might be potentially perceived, experienced or done) with the idea that we bring memories of how to act, perceive, affect and be affected to each encounter. If we think of memory itself as virtual – a multiple collection of percepts, affects, actions and thoughts subsisting outside, in-between bodies, ready to be remembered in the body's interaction with others (Colwell 1997; Connolly 2002) – then we could say that it is only upon memories becoming repeated (actualised, remembered) that they can become historically narrated and politically structured. At this point, it could be argued that there are always institutional, customary and disciplinary constraints upon our actions, percepts and affects, and that these constraints are why we tend to *actually* remember how to act within certain norms. Yet, this would leave us with an ethics that amounts to an imperative to act *in defiance* of all these constraints in order to become more open to all the different potentials and memories folded into and offered up by the event.

Deleuze (1993), however, insists that there are not just folds, but rather a continual production of folds – folding, unfolding, refolding – bringing series of perception, affect, memory and thought into contact with one another. Given this, acting ethically might involve folding actual memories of how to act properly and conventionally into a relation with new perceptions and affects so that an unfolding might take place – something (an action, a perception, a thought) that is aware of the memory of propriety and power and yet is already different. Moreover, such processes of folding, unfolding and refolding might also be suggestive of a productive *confoundedness*. Sarah, for instance, might have perceived Minesh as being confident, gregarious, or boorish and overbearing. If, however, she had also remembered an expectation of a South Asian man as effeminate, quiet and unattractive, she might have had a sense of confoundedness. This is not only a question of the event confounding a system of recognition. It is also a question of a tension between a memory of perception (a memory of recognising a South Asian man as quiet or effeminate) and the multiplicity of perceptions and affects that resonate with and yet contradict, and thus *confound*, this memory of quietness or effeminacy. Thinking through folding allows us to apprehend the multiplicity of memories and histories that *might* politicise the event. With such an awareness, the moment of confoundedness might allow for something else to be done.

Indeed, above all, thinking through folds and monads opens up a space for those who bring to the event histories of fear, oppression or undesirability to fold these histories together with new affects (for example, of persistence or friendliness rather than shame) and thus to unfold new modes of relating in the world. Such a practice of thought is not solely about forging an ethical approach to acting in the (future) event, but also involves how we make sense of the event, in retrospect, through our historicising practice. As we have seen, in the process of placing events within historical narratives, explanations become offered that posit causes and factors, and that suggest origins and destinations. A more ethical historicising practice would be wary of offering something that posed as an exhaustive explanation. Rather, it would engage with the multiple histories and memories folded into any event.

Moreover, we might also be wary of the way that historical narratives can impose a range of structures that provide a framework within which the relations under consideration must take place. For Deleuze, structures are generated from an external point of view and always involve a set of negative differences (Colebrook 2004). It is important to think outside of negation and structure because if we take perception, for instance, to proceed through negation, then we already invoke a memory of how to perceive that places, say, Asian and white in opposition. Race discourses are implicit within such ways of understanding perception. The danger is that what the process of narrating perception does is, in effect, to say that this form of negative difference is the one that matters most, and that we must therefore organise our bodies and their relations – and our understandings of our bodies and their relations – around it.

Deleuze (1993) refuses negation by borrowing Leibniz's conception of the monad, which allows him to affirm how perceptions, affects, thoughts and bodies of all kind develop along series. Taking perception as an example: rather than thinking that we are able to perceive something *because it is not* something else (i.e. through negation), monads invite us to think of differences among perceptions as simply the movement from one perception to the next along a series of perceptions (for example, your perception of your body's temperature as it continually changes). Perceptions in a series succeed one another in process of change or development. Just as with perception, the expression or actualisation of affect, thought, action and bodies takes place through a continual process of folding and unfolding that develops along series and from one series to another. Deleuze's account of monads, then, suggests that the bodies (and their relations) that become subject to structuring are already unfolded by a process of differentiation as series of difference without negation (Colebrook 2004). Rather than impose the negative differences of explanatory structures upon these already differentiating bodies and relations, we might understand the history of a body, thing or perception in the positive terms of the co-existence and succession of virtual forces and memories that struggle for possession of it (Deleuze 1983, 3, cited in Baugh 1997, 135).

We might, then, envisage an ethical historicising practice that eschews proceeding through negation or pre-given structures and any of the lines of expectation that such structures would entail. Such a practice of ethical historical thought might bring together different series of memory and history within practices of narration that continually produce new folds and that unfold difference. A historicising practice along these lines would be aware of what it does in its own event (of historicising) (see Foucault 1991): it would be aware of the potential violence of reciting historical interpretations but would also be aware of the potential to unfold histories that undo and go beyond prevailing modes of interpretation. Importantly, this vision does not disregard those whose experiences of fear, oppression or undesirability afford them knowledge of the violence of prevailing histories: this knowledge is precisely that of the potential violence of reciting these interpretations. Rather, this vision suggests a historicising practice that folds histories of fear, oppression or undesirability together with memories of other affects in order to unfold the development of positive but confounded differences.

Conclusions: Thought, Conviviality and the Event

Several writers have suggested that what we might call 'conviviality' offers the best hope for dealing with and superseding racial antagonism within modern Western societies such as Britain (Amin 2002; Gilroy 2004; see also Thrift 2005). In this sense, conviviality is thought of as a way of both encountering difference and superseding it. Conviviality, it is suggested, arises in the forms of practical social activities in which people are just getting on with things together or even just getting by in the same spaces. This shared activity or sharing of space takes place irrespective of the invocation of memories and histories of racial difference and antagonism, and so might be said to involve a kind of suspension of these histories and memories.

Yet, in this chapter, I hope to have shown that thinking about histories and memories of encountering racial (as well as other kinds of) difference can augment an understanding of how to act ethically in future events. This might suggest that taking forward conceptions of conviviality such as Gilroy's (2004) might involve attending to exactly the relationship between the event of encountering difference and the processes of historicising, creating genealogies and multiple counter-histories. In other words, the task of exploring how we inhabit multicultural or cosmopolitan spaces cannot stop at thinking about how we are disciplined into certain kinds of affective relationships with racialised others; or thinking about how biopolitics stakes out a division of interests with respect to life; or thinking about how we interiorise narratives of racial difference and relation. Nor can it stop at thinking about the everyday forms of contact and getting along that might offer moments in which antagonism is suspended but not entirely forgotten (Valentine 2007). Rather, the task of conducting ethical and political practices that stake out less antagonistic ways of living might instead involve rethinking our historicising and remembering practices. Not only might we do well to broaden the scope of our attention towards the multiplicity of potential memories that the event affords us, but we might also *think* through historicising and remembering practices that fold antagonisms and expectations into different affects, percepts and thoughts, actively confounding racialised sense-making and unfolding something different in the process.

Acknowledgements

Many thanks to Steve Legg, Heidi Nast, Kezia Barker, Kanchana Ruwanpura, Divya Tolia-Kelly, Leo Minuchin and, of course, Claire and Caroline. I am also indebted to 'Minesh', 'Sarah', 'Darren', 'Henry' and my other PhD research participants. Finally, as always, many thanks to Ruth.

References

Ahmed, S. (2004), *The Cultural Politics of Emotion* (Edinburgh: Edinburgh University Press).

Alexander, C. (2000), *The Asian Gang: Ethnicity, identity, masculinity* (Oxford and New York: Berg).

Alexander, C. (2004), 'Imagining the Asian Gang: Ethnicity, masculinity and youth after "the riots"', *Critical Social Policy* 24:4, 526–49.

Amin, A. (2002), 'Ethnicity and the Multicultural City: Living with diversity', *Environment and Planning A*, 34: 6, 959–80.

Anderson, K. (1987), 'The Idea of Chinatown: The power of place and institutional practice in the making of a racial category', *Annals of the Association of American Geographers* 77:4, 580–98.

Archer, L. (2001), '"Muslim Brothers, Black Lads, Traditional Asians": British Muslim young men's constructions of race, religion and masculinity' *Feminism and Psychology* 11, 79–105.

Baugh, B. (1997), 'Making the Difference: Deleuze's difference and Derrida's différance' *Social Semiotics* 7:2, 127–47.

Bourdieu, P. (1977), *Outline of a Theory of Practice* (Cambridge: Cambridge University Press).

Bourdieu, P. (1986), *Distinction: A social critique of the judgement of taste* (London: Routledge).

Butler, J. (1990), *Gender Trouble: Feminism and the subversion of identity* (London: Routledge).

Butler, J. (1993), *Bodies That Matter: On the discursive limits of 'sex'* (London: Routledge).

Butler, J. (1997), *Excitable Speech: A politics of the performative* (London: Routledge).

Butler, J. (2004), *Undoing Gender* (London: Routledge).

Colebrook, C. (2004), 'The Sense of Space: On the specificity of affect in Deleuze and Guattari', *Postmodern Culture* 15:1, <http://muse.jhu.edu/journals/postmodern_culture/v015/15.1colebrook.html>, accessed 7 September 2007.

Colwell, C. (1997), 'Deleuze and Foucault: Series, event, genealogy', *Theory and Event* 1:2, <http://muse.jhu.edu/journals/theory_and_event/v001/1.2colwell.html>, accessed 29 July 2006.

Connolly, W. (2002), *Neuropolitics: Thinking, culture, speed* (Minneapolis: University of Minnesota Press).

Das Gupta, M. (1997), '"What is Indian About You?" A Gendered, Transnational Approach to Ethnicity', *Gender and Society* 11:5, 572–96.

Deleuze, G. (1988), *Spinoza: Practical philosophy* (San Francisco: City Light Books).

Deleuze, G. (1990), *The Logic of Sense* (London: Continuum).

Deleuze, G. (1992), *Expressionism in Philosophy: Spinoza* (New York: Zone).

Deleuze, G. (1993), *The Fold: Leibniz and the Baroque* (London: Athlone).

Deleuze, G. (1994), *Difference and Repetition* (New York: Columbia University Press).

Deleuze, G. (1995), *Negotiations: 1972–1990* (New York: Columbia University Press).

Deleuze, G. and Guattari, F. (1988), *A Thousand Plateaus: Capitalism and schizophrenia* (London: Athlone).

Foucault, M. (1991), 'Nietzsche, Genealogy, History', in P. Rabinow (ed.) *The Foucault Reader* (London: Penguin), 76–100.

Frith, H. and Kitzinger, C. (1998), '"Emotion Work" as a Participant Resource: A Feminist analysis of young women's talk-in-interaction', *Sociology*, 32:2, 299–320.

Frith, H. and Kitzinger, C. (2001), 'Reformulating Sexual Script Theory: Developing a discursive psychology of sexual negotiation', *Theory and Psychology* 11:2, 209–32.

Gilman, S.L. (1985), *Difference and Pathology: Stereotypes of sexuality, race, and madness* (Ithaca and London: Cornell University Press).

Gilroy, P. (2004), *After Empire: Melancholia or convivial culture?* (London: Routledge).

Grosz, E. (2005a), *Space, Time, and Perversion* (London: Routledge).

Grosz, E. (2005b), *Time Travels: Feminism, nature, power* (Durham, NC and London: Duke University Press).

Hemmings, C. (2005), 'Invoking Affect: Cultural theory and the ontological turn', *Cultural Studies* 19:5, 548–67.

Hopkins, P. (2006), 'Youthful Muslim Masculinities: Gender and generational relations', *Transactions of the Institute of British Geographers* NS31, 337–52.

Hutnyk, J. (2000), *Critique of Exotica: Music, politics and the culture industry* (London: Pluto Press).

Kabbani, R. (1994), *Imperial Fictions* (London: Pandora Press).

Lim, J. (2003), 'Ethics and Embodiment in Racialised, Ethnicised and Sexualised Practice' (unpublished PhD thesis, University of London).

Malson, H., Marshall, H. and Woollett, A. (2002), 'Talking of Taste: A discourse analytic exploration of young women's gendered and racialized subjectivities in British Urban, multicultural contexts', *Feminism and Psychology* 12:4, 469–90.

Massumi, B. (2002), *Parables for the Virtual: Movement, affect, sensation* (Durham, NC: Duke University Press).

McCormack, D. (2003), 'An Event of Geographical Ethics in Spaces of Affect', *Transactions of the Institute of British Geographers* NS 28, 488–507.

McDowell, L. (2000), 'Learning to Serve? Employment Aspirations and Attitudes of Young Working-class Men in an Era of Labour Market Restructuring', *Gender, Place and Culture* 7:4, 389–416.

McDowell, L. (2002), 'Transitions to Work: Masculine identities, youth inequality and labour market change', *Gender, Place and Culture* 9:1, 39–59.

McDowell, L. (2003), *Redundant Masculinities? Employment Change and White Working Class Youth* (Oxford: Blackwell).

Nayak, A. (1999), '"Pale Warriors": Skinhead culture and the embodiment of white masculinities', in A. Brah, M.J. Hickman and M. Mac an Ghaill (eds) *Thinking Identities: Ethnicity, racism and culture* (Basingstoke: Macmillan), 71–99.

Sedgwick, E.K. (2003), *Touching Feeling: Affect, pedagogy, performativity* (Durham, NC and London: Duke University Press).

Sinha, M. (1999), 'Giving Masculinity a History: Some contributions from the historiography of colonial India', *Gender and History* 11:3, 445–60.

Thrift, N. (2005), 'But Malice Aforethought: Cities and the natural history of hatred', *Transactions of the Institute of British Geographers* NS30, 133–50.

Valentine, G. (2007), '"Living with Difference": Reflections on geographies of encounter', *Progress in Human Geography* lecture at the RGS-IBG Annual International Conference 2007, 30 August 2007.

Willis, P. (1977), *Learning to Labour: How working class kids get working class jobs* (Farnborough: Saxon House).

Chapter 17

Everyday Multiculture and the Emergence of Race

Dan Swanton

The Force of Race[1]

Climbing Oakworth Road you turn left at a small cluster of shops that include a butchers, a laundrette, and a newsagents. Walking into a neighbourhood of somewhat tired Victorian terraces there is a palpable shift in the mood of the place. Salwar kemeez dry on washing lines strung across back streets; aromas of toasting spices fill the air; dormer windows betray extensions of living space inviting – seemingly automatically – Orientalist imaginaries of the 'Asian family'; the rhythmic thuds of bass beats escape a Honda Civic pulled up by the side of the road; children play in the street; park gates; a shop displaying posters for a Lollywood film; small groups of men leaning on a wall chatting, greeting passers-by they recognise with 'Salaam' and handshakes ... These elements all contribute to the distinct resonances of this space. Lund Park. A no-go area? After a series of attacks on white people raced memories and affects seem to hang heavily over the park. An intimidating space. The territory of violent Asian gangs. A place to be avoided after nightfall. But under the still intense sunlight of a summer evening the park hardly corresponds to the terrorising imaginative geographies conjured through urban myth, gossip and newspaper reports. But race still emerged immanently through the arrangements of bodies, things and surroundings in the park. Raced differences surfaced through the relative slowness of Asian bodies clustering by the park gates, hanging out and chatting; through the brisk walk, pale skin and clothing of a lone woman cutting through the park, her gaze fixed in front of her; or through the vigilance of bowling club members, hurriedly finishing games and escaping the park in their cars before dusk falls.

A purposeful drift. An imaginative reconstruction stitching together many fragments. My narrative of passing through and feeling contact zones and the fleeting emergence of race uses participant observation, newspaper archives, go-along interviews, gossip, research diaries, walking, interview material, local politics, and urban myth to evoke the turbulent socialities of multiculture in northern mill towns in England.[2] It begins to animate a particular encounter with Keighley, a mill town a few miles from Bradford, that in recent years has gained a reputation for uneasy intercultural relations through the activism of British National Party, widespread talk of segregation and 'no-go areas', terror alerts, and stories of gang violence, racist attacks or the grooming of white girls by Asian men. This encounter is necessarily partial, constructed through the intensely embodied practices of ethnographic fieldwork (and complexly marked by my body's whiteness; its maleness; my clothes; my satchel, notebook and camera; my accent; my inquisitiveness; my speed; the often awkward ways in which my body related to other bodies) and the techniques I employed to encounter this space with others (often white women).[3] The momentum of raced differences coursing through these encounters with/in Keighley was compelling, and this brief sketch seeks to elicit both the force and plasticity of race as it temporarily fixes bodies, things and spaces. For me, at least, the force of race on the ground in Keighley began to disturb the routine framing of race as a problem of epistemology across the social sciences (Saldanha 2006). Here, I argue that both presenting race as a social construct (Miles and Brown 1989; Jackson and Penrose 1993; Kobayashi and Peake 2000) and recent abolitionist arguments for the transcendence of race (Gilroy 1998, 2000; Ware and Back 2002) have had a 'deadening effect' on our academic talk about race by narrowing our empirical focus to discourses, narratives and representations, and constraining analysis to the categorical politics of identity and the extraction of textual meaning (Saldanha 2006; Thrift and Dewsbury 2000). But I want to suspend the moment of deconstruction in a move away from questions of interpretation (how do we know race?) as I focus on more practical questions of what Deleuze and Guattari (1987) would call experimentation (that is, what does race do? How does race function?).[4]

By repeatedly returning to practical questions of what race does and how it functions, I explore the possibilities of developing an assertively non-essentialist and non-determinist concept of race that takes seriously the materialities through which race takes form in moments of intercultural encounter. Tracking the force of race in moments of encounter in Keighley, I am arguing for an empiricism that traces the heterogeneous and immanent processes through which race takes form, rather than fixed identities, representations, genes or oppositions. And so I chart the temporary – but often recursive – fixings of race through souped-up cars, veils, street names, skin, graffiti, calls for prayer, Lollywood posters, pubs, kebab skewers, salwar kemeez, newspaper headlines and so much more. This empiricism is absolutely not a return to deterministic regimes of differentiation and classification. However, by thinking about the promiscuity with which drug dealing, Islam, segregation, sexual predation, terrorism, dirtiness, cultural difference, etc. sticks to some bodies (both human and non-human) in places like Keighley we can begin to confront a new form of racism (Saldanha 2006; Amin 2007) by tracking how loose racial summaries ground heterogeneous processes of differentiation and race comes to mediate encounters, align bodies, infuse dispositions, induce tendencies to attraction of repulsion, stir antagonisms or inspire engagement and exchange.[5] In short, this is an argument for a materialist engagement of race that recognises what Jane Bennett (2004) calls the 'force of things' and how raced memories and affects congeal around particular bodies, things and spaces and how these things then exert a kind of agency in the event of an encounter.

Mapping the intense materialisations of race and approaching urban multiculture as an accumulation of 'billions of happy and unhappy moments of encounter' (Thrift 1999, 302) that continuously take form and pass by in prosaic contact zones has important implications for thinking about race and the multicultural city. It shifts our analytical gaze to the affectivity and turbulence of intercultural interaction and foregrounds perspectives on multiculture from below. And from below urban multiculture looks quite different from recent accounts of multicultural cities – and particularly northern mill towns after urban unrest in Burnley, Oldham and Bradford in the spring and summer of 2001 and the 7 July 2005 terror attacks in London – that have tended to dwell less on the realities of how urban multicultures are lived out (for notable exceptions see Back 1996; Alexander 2000; Amin 2002a) and more on questions of how state sponsored policies of multiculturalism have played out in and through urban spaces. This, then, is not an account of segregation and 'parallel lives' (Bradford Race Review Team 2001; Simpson 2005; Peach 2006), building 'cohesive communities' (Community Cohesion Review Team 2001; Cantle 2005), paranoid and panicked urbanisms worrying about 'unruly strangers' (Amin 2003) and home-grown suicide bombers (Virilio 2005; Graham 2004; Flusty 1997), identity politics (Modood 2005), agonistic publics and counterpublics (Amin 2002a), or the racist activities of the far right (Keith 1995). My argument is not that we should overlook or dismiss these accounts. However, I do think that they often fail to fully appreciate the messy, challenging underside of interethnic relations and conflict, and as such are less well placed to generate innovative perspectives to disrupt the current force of race as it surfaces, for example, in emotive and muddled political discourses that routinely couple urban unrest with segregation and unruly strangers, and, after 7 July 2005, failure to integrate with terrorism and home-grown suicide bombers. In Keighley daily talk (by journalists, politicians, people in the street, at work, in cafés, etc.) of segregation and 'parallel lives', irreconcilable cultural difference, institutionalised racisms, moral panics about 'Asian gangs' and worries about unruly strangers attacking whites, dealing drugs, or grooming white girls, and the activities of the British National Party are all testimony to interethnic suspicion and resentment, and entrenched local cultures of exclusion and exclusivity. Here, however, I want to animate some fragments of the fleshy, visceral and practised realities of living with difference in Keighley by grasping the promiscuous and sticky operations of race as it is variously enrolled in continuous processes of differentiation.

Race Matters

Social constructionist and abolitionist critiques of race have – with notable success – undermined biological essentialisms, punctured taken-for-granted categories, and challenged racial oppression and multiple racisms (Miles and Brown 1989; Jackson and Penrose 1993; Gilroy 1998, 2000; Ware and Back 2002). However, both constructionist and abolitionist arguments tend to restrict social scientific engagements with race to questions of epistemology and interpretation (Saldanha 2006). Such representational approaches have a 'deadening effect' on our understandings of race, tending to a 'conservative, categorical politics of identity and textual meaning' (Lorimer 2005) that often disavows the entangled materialities through which race operates.[6] Here, I trace three lines of flight that challenge our academic talk about race, as I work towards a non-essentialist, emergent concept of what race is, what race does, and what race might be.

First, social constructionist accounts of race can be seen to vibrate between two poles. At one extreme race is either 'natural' and therefore unconstructed, while at the other race is a fabrication inscribed on raw and passive matter, a 'prediscurive field' that is itself constituted by the concept of race (Cheah 1996; Butler 1993). And in the effort to deconstruct the naturalness of race, social constructionist accounts sidestep the materiality of race, while simultaneously replacing one essentialism with another, as they reify the nature of matter, nature and the biological sciences.[7] Here, I argue that tendencies to privilege the discursive and the social and erase the real, material differences between bodies mean that social constructionist accounts of race are ultimately insufficient for a truly anti-racist politics. This insufficiency rests in the attempt to wipe out material differences 'which are not eradicable without disfiguring the body' (Grosz 1994, 18), and it surfaces most acutely in what Gilroy (1998) identifies as the 'pious ritual' in which the illusory nature of race is agreed upon before then deferring to the social embedded ness of race, such that it might be afforded existence-as-such. However, by theorising race as both a cultural interweaving and a production of nature it is possible to think about real differences between bodies without descending into essentialism (Grosz 1994). Appreciating the complex, immanent mixing of culture and nature as race takes form involves a concept of race which is absolutely not about imposing grids that divide human bodies into groups, categories or hierarchies. Rather race is plastic, dynamic and immanent to the processes it expresses (Massumi 1992). And so any ontology of race needs to labour the point that race is simultaneously fluid and fixing (Saldanha 2007), taking different forms immanently through synergies of bodies, things and surroundings in moments of encounter, but always in ways constrained by micropolitical technique, geologies of memory, and sedimentations of routine.

Building on this ontology of race, I ask what recent challenges posed by non-representational theory that seek to engage more fully with our embodied and practised worlds (Thrift 1996, 2004; Thrift and Dewsbury 2000) might do to academic engagements with race. Nigel Thrift (1997, 126) describes non-representational theory as a constellation of ideas that is:

> meant to provide a guide to a good part of the world that is currently all but invisible to workers in the social sciences and humanities, with their intellectualist bent, that part which is practical rather than cognitive.

For me, non-representational theories encourage a reconsideration of what counts as knowledge and explanation. In terms of thinking about race, non-representational theories move away from social constructionist arguments that start with subtractions of movement – its explanatory point of departure involves 'catching the body in cultural freeze frame' (Massumi 2002, 3) – by posing the challenge of thinking in terms of formation, or the becoming, of race in a field of emergence.

Third, framing race through its materiality and becoming requires a rethinking of what constitutes racism. While there is widespread agreement across the social sciences that there is not one racism, but multiple racisms (Jackson 1987; Balibar 1991; Blaut 1992; Pred 2000), racisms continue to be framed as the enactments of racist beliefs. Accordingly racisms – as enactments of fallacious investments in the existence of races – become a problem of epistemology, and remain wedded to representational modes of thought that fail to engage satisfactorily with the tenacity of racisms. However, if we take seriously non-representational theory's argument that we should address the practical as well as the cognitive, racisms begin to look quite different. By asking what race does we might, for example, think more explicitly about the force of race – where race is always conceived as a multiplicity, always becoming – as it bleeds into practices in moments of intercultural encounter. Here our focus shifts from whether ideologies, narratives or representations – however pernicious – are 'good' or bad' sociological explanations (Hage 1998) to more practical questions of how race surfaces immanently in indifference, verbal abuse, incivilities, physical violence, antagonisms, rejection, avoidance, abjection, discrimination, misanthropy, exploitation, desire, fear …

These lines of critique force us to think differently about race. Instead of asking how we know race, or how we might transcend race, I argue for an ontology of race that grasps how race is simultaneously fluid and fixing, momentarily taking form through arrangements of bodies, things and spaces in moments of intercultural encounter.[8] In this way we can think of race as transversal and rhizomatic.[9] Race is a multiplicity that takes form immanent along lines of movement through a relational field (DeLanda 1997). Race, then, is creative and capricious. It is – as Arun Saldanha (2006, 20) has argued – 'constantly morphing', surfacing immanently as Islamic terrorism; as clothing styles; as sexual desire; as refusals to integrate; as 'stranger danger'; as denuded civil liberties; dietary requirements; as male chauvinism, as standards of hygiene … as it sticks to particular bodies, things and spaces'.[10] Allying this promiscuous concept of race with recent theorisations of thinking as a layered, practical and distributed activity (Connolly 2002; Thrift 1996), we start to comprehend the force of race in the everyday, habitual production, arrangement and interpretation of human difference. Race thinking is a form of thought-in-action, a way of acting into the world that displaces any notion of thinking as an intellectual activity confined to the brain, nervous system or body. Race thinking is definitely not another form of knowing race, but relates to the particular doing of race at each moment of intercultural encounter. It is the outcome of, and is spaced across, an entanglement with the world that envelopes intensities of skin pigmentation, accents, salwar kemeez, designer stubble, pimped rides and taxis, street signs, pubs, mosque domes, parks gates at dusk and so much more. This conception of race thinking opens up the half-second delay not only as a 'space of bodily *anticipation*' (Thrift 2004, 67, emphasis in original) but also as a potential space of prejudice, during which the force of race sorts bodies, things and spaces, and coordinates thinking, judgment and action. The push of race punctures this half-second delay through the rapid work of virtual memories laid down through the formatting of perception by micro-political technique and past experiences and events.[11] Politically, this conceptual manoeuvre opens the analysis of race to new political registers and intensities (Thrift 2004), where by harnessing micro-political technique a new anti-racist politics might engineer an ethics of generosity and mutuality that form part of an agonistic politics of getting along (Amin 2002a; Amin and Thrift 2002). Practically, these conceptual arguments require an empiricism that begins with what DeLanda (2002) calls 'intensive difference', and the immanent processes through which race takes form.

This empiricism seeks to grasp the wildness of race, and the turbulence of urban multicultures by directing our attention to the countless temporary fixings of race in the ill-tempered rub and indifferent contact of the street, over newsagent counters, in rush hour traffic, at the supermarket, at pedestrian crossings; in the coming together of bodies in cafes, taxis, student unions and other sites of prosaic intercultural contact; through the distributions of bodies in pubs, libraries, gyms, playgrounds, evening classes, allotments and shopping centres; through encounters with street signs, calls for prayer, pimped rides, rucksacks, dormer windows all of which stir raced memories and affects; and through distinct moods and atmospheres of particular neighbourhoods, parks and alleyways. But this empiricism does not simply chart the immanent emergence of racial formations (i.e. becoming-brown, becoming-white, becoming-'Paki', becoming-terrorist, becoming-separate ...), it also asks how bodies, things and spaces are sorted through these temporary fixings so that we might apprehend the different intensities and modalities of intercultural contact, the specific distributions of differently raced bodies in particular social settings, or tendencies to aggregation and repulsion (cf. Saldanha 2007). And it is through this sorting that bodies of all kinds come to scatter specific affective energies (fear, desire, hatred, resentment, suspicion, indifference) that then infect judgments (about people, threats, civility, schools, desirability, neighbourhood safety, etc.) and practices (including the avoidance of certain parks, streets or schools; assuming indifferent, fearful or suspicious dispositions in interaction; venting petty prejudices and resentments; etc.). By way of exemplification, I briefly trace two encounters, and how the fixings of race impinge on the capacities of bodies to affect and be affected. Firstly, I sketch two tales of becoming-dealer that ask how moral panics around 'Asian gangs', drugs and bad boys stick to some bodies, things and spaces. This empiricism asks why, for example, skin + shiny Lexus + mobile phone x handshake = becoming-dealer. Moreover, in grasping how moral panic and suspicions coagulate around bodies, things and spaces, this empiricism encourages us to ask how in becoming-sticky the capacities of these human and non-human bodies to affect and be affected are transformed. My second story returns to Lund Park, arguing that the becoming-no-go of this park lies less in representations of this setting as a scary space, a 'Muslim ghetto' or the site of racist attacks on whites, and more in the ephemeral and varied ways in which race comes to matter through the resonances radiated by particular arrangements bodies, things, architecture, sunlight and surroundings.

Encounter 1: Brown Skin + Flash Car X Handshake = Becoming-Dealer

> Election 2005: Where candidates vie for race supremacy
> One day is much like any other at the Eclipse hairdressers' salon at Oakworth, an affluent, stone-built village in West Yorkshire's Bronte country, so the regulars were drawn from their blow-dryers when a new sports car parked outside in the street.
>
> The salon's young proprietor Sharon Wiseman was the first to clock that the driver and his passengers were young Asian men, and 'up to no good', she assumed. 'The first thing I thought was "drug dealers"', she admitted later.
>
> Her assistant added that she will not let her daughter into nearby Keighley because of stories about 'Asians grooming teenage white girls'. She also believes it is time to 'do something about' Keighley's immigration problem. From beneath a mass of pink rollers, Joan, aged 67, nodded her head vigorously. 'Up to no good,' she said. (*Independent*, 13 April 2005)

This tale of suspicion playing out at the hairdressers exposes the immanent sorting of bodies on the ground, and how connections between brown skin, flash cars and location stir suspicions and prejudices, judging these bodies 'up to no good': *They're probably dealers.* Might this rapidly slide into: *They might be paedophiles.* Here we have moral panic in action, filtering perceptions and directing conduct. A back catalogue of sensationalist headlines about 'racist attacks on whites', 'gangland executions', gun and knife crime, mug-shots of 'rioters'; hearsay about drugs; gossip about attacks; documentaries about grooming, far-right politicking; and urban myth is immanent to the event of this encounter, moving thought and perception in some directions and not others, in ways that bleed into gesture, disposition and action. And we see tales of suspicion playing out all over Keighley:

> Strange plants growing in the cracks between the paving stones in a suburban street. Cannabis? Jan's attention is grabbed by a car horn. A Subaru Impreza. Gold alloys. Whale tail spoiler. Twitching curtains. 'Dial-a-drug?' Two teenage lads. Well dressed. Brown skin. 'What are they up to?' 'How can they afford a car like that?' 'I know where they get the money from!' These lads are just sitting in the car with the windows down. Occasionally a local white kid approaches the car. There's an abrupt verbal exchange. A handshake. And then they walk off. 'A deal?' Vegetation + flash car + skin + music x handshake + surroundings = becoming-dealer.[12] 'Why are they shaking hands?' 'They must be exchanging something.' 'Drug dealers!' 'So blatant!' Suspicions turn to fear and rage. 'Why don't the police do something about it?'[13]

In these encounters we begin to see how race – surfacing immanently as gang violence, as male chauvinism, as sexual deviance, as styles of conspicuous consumption, as unruly strangers, as drugs trafficking – sticks to some bodies and things in particular settings. Racial formations temporarily, but recurrently, cohere through connections between brown skin, jewellery, clothing, flash cars, vegetation, handshakes, swaggers, Hip Hop, expressions, loitering, fireworks, styles of interaction, language etc. But these stories of becoming-dealer are about more than temporary fixings of race that chain bodies (Saldanha 2007), they emphasize the sorting of bodies that shapes the capacities of bodies to affect and be affected.

Here, I am interested in how raced memories and affects coagulate and condense around these conjunctions of skin, cars, drugs and settings, and how in becoming-dealer bodies scatter specific affective intensities that infect judgment, disposition and action. For example, suspicion appears to be distributed across heterogeneous elements and machinic connections (cannabis plants, brown skin, an expensive car, etc.), but some connections seem to amplify suspicions (for example Jan's suspicions are intensified by witnessing a handshake between the occupants of the car and white kids). And intensifications of suspicion initiate a contagion of other intensities – most often rage, abjection and fear. Pupils dilate. Adrenaline surges. Paces quicken. Curtains twitch. Eye contact is avoided. Affective intensities shape judgments that then feed into a host of tactics that include avoiding particular areas, worrying about children, calling the police, exuding corporeal confidence through gesture and disposition, sticking to well-lit areas, carrying mobile phones, retelling stories of what you witnessed to friends and colleagues, and so on. And so tracking the temporary fixings of race and the visceral intensities that the immanent sorting of bodies stirs, we begin to apprehend how possibilities for intercultural encounter and exchange are closed down as what Saldanha (2007) calls the viscosity of racial aggregates thickens, as the work of suspicion in these accumulated moments of becoming-dealer strengthens tendencies for differently raced bodies to repel.[14]

Encounter 2: Park Gates + Loitering X Dusk = Becoming-No-Go

In this second encounter I examine how the force of race operates through the atmosphere and moods of certain spaces. My aim is to supplement current thinking around race and urban space that tend to stop with a notion of space as an 'enabling technology' through which race is produced (Delaney 2002, 7; Anderson 1991; Dwyer and Jones 2000). These arguments about the spatial production of race proceed through a series of exemplary geographies – the ghetto, the camp, the inner city, the no-go area, the suburb – that fold difference into distance, and think about race territorially through distance, boundaries and partitions that fabricate distinctions between 'inside' and 'outside', 'familiar' and 'strange', 'same' and 'Other', 'white' and 'brown', 'us' and 'them' (Said 1979; Dwyer and Jones 2000; Gregory 2004). Here, I introduce another dimension to understanding multiculture that explores how race takes form through the sensuous and affective relief of urban spaces, the structures of feeling and abrupt shifts in mood and ambiance. As so returning to Lund Park – a park in Keighley flanked by Victorian terraces many of which are now home to Kashmiri families – we can begin to assemble an alternative account of how race surfaces through the resonances of this space. Lund Park is often talked of as an 'Asian area', and, after a number of well-documented and much-discussed attacks in the park, even a 'no-go area' for whites. We could settle on an argument that deconstructs the imaginative geographies – fuelled perhaps by moral panics about 'Asian gangs' – through which race is at once inscribed and produced through this locale. But stopping here would fail to engage substantively with the ephemeral and varied ways in which race might come to matter in and around Lund Park. Alternatively, considering Lund Park topographically – and not territorially – starts to grasp how provisional processes of racial differentiation operate through arrangements of things, bodies and light in this space.

At dusk, as the light fades and stillness descends on the park, interrupted only by a small gathering of joshing Asian lads at the eastern entrance and the occasional thumping bass of a passing car, the park might become a threatening space. Resonances emitted by the declining intensity of sunlight and lengthening shadows collude with the relative slowness of the bodies clustered by the wall, chillin' and joking, producing a distinct mood. Fragmentary raced affects and memories of racist attacks on whites might hang heavily in this atmosphere. An intimidating, scary space? A 'no-go area'. And under these specific conditions of light and viscosity, this space might reverberate with raced affects. Vigilance? Fear and loathing? For a lone body passing through the park – marked out, perhaps, by pale skin and clothing, but certainly by its relative speed – race might come to matter in distinct ways, through senses of being out of place, of being outnumbered, of not belonging. But under different conditions of light race might emerge in and through this space in very different ways. The park continues to vibrate with raced affects, but at each moment of encounter with this space the substance and intensities of these affects might be quite different. For example, in the midday sun processes of racial differentiation proceeding through the arrangements and speeds of bodies, things, and distributions of sunlight in Lund Park might inspire melancholia as faces, smells and salwar kemeez drying on washing lines disrupt childhood memories of a homely white working class neighbourhood, or reassurance and comfort at the recognition of familiar faces, sights, and sounds.[15] Then again these arrangements might provoke indifference in a self-absorbed dog-walker, stir anxieties (shaded, perhaps, with excitement) in some 'white' bodies anticipating intimidation, or agitate exhilaration at opportunities for multicultural touch on a bench or at the playground. In this way, we cultivate a more profound appreciation for the abrupt shifts in mood and ambiance that constitute our experiences and sensings of everyday multicultures, and the ephemeral processes through which race comes to matter as it bleeds into perception, judgement and action, disturbing the confident partitioning and ordering of urban spaces through exemplary geographies.

Conclusion

In an effort to come to terms with the apparent momentum of raced differences on the ground in Keighley, this chapter has opened the conceptual terrain for thinking differently about race. Arguments for a non-determinist, non-essentialist ontology of race that momentarily congeals in arrangements of bodies, things and spaces in moments of intercultural encounter, has provided ground from which we can confront a new form of racism where loose racial summaries (which extend well beyond the skin to include souped-up cars, swaggers, salwar kemeez, loitering, toasting spices, street signs, calls for prayer ...) become the basis for affectively imbued processes of differentiation that shape judgment, disposition and action. This conceptual work then translates into an empiricism that explores the intense, continuous sorting of bodies, things and spaces on the ground. My stories of becoming-dealer and becoming-no-go begin to exemplify the force of race in moments of intercultural encounter, and how temporary fixings of race stir specific affective intensities that help apprehend distributions of bodies in certain social settings, the tendencies of bodies to aggregation or repulsion and the many modalities of intercultural contact. These stories also point to innovative ways in which we can animate urban multicultures to better appreciate the visceral, fleshy and practised realities of living with difference.

Notes

1   My thanks to Caroline Bressey, Claire Dwyer and Susan Smith for their helpful readings of an earlier draft. This research was conducted between December 2004 and October 2005 as part of a PhD funded by the Economic and Social Research Council, and the chapter was written during an Economic and Social Research Council Postdoctoral Fellowship.

2   The presentational styles I adopt in this chapter need to be situated in relation to both the force of post-structural, post-colonial and feminist critique in anthropology (and its reverberations through the social sciences) that have shaken epistemological foundations and undermined authorial claims to truth and objectivity (Clifford and Marcus 1986; Haraway 1991; Hutnyk 2002), and more recent charges that the research practices of cultural geographers have failed to keep pace with our theoretical talk (Pratt 2000). In this opening narrative I weave together fragments from conversations, interviews, participant observation, and archival research to complicate the practices through which 'data' is brought back from the field and then re-presented, and to emphasise that writing is as much about the creation of effects and affects as the work of representation. And so this purposeful drift through Lund Park – that draws inspiration from psychogeography and its attempts to register and experience the impact of an environment on human emotions (Sinclair 1997; Pile 2005) – is a fiction, but a fiction based in ethnographic engagement and one that seeks to evoke the affective, sensuous, embodied and occasionally visceral contours of urban multiculture. The work of this presentational style resides in the attempt to provoke different ways of engaging with and understanding multiculture in places like Keighley. My style seeks to problematize seemingly automatic readings of race and space through lenses of segregation or deconstruction, and begins to ask how the lived, embodied and affective dimensions of everyday multiculture might be researched, understood and animated.

3   The knowledge generated through this research is necessarily produced from certain places and particular embodied practices. Here I track how this research was done through a specific body (mine) to emphasise the partiality and situated nature of the knowledge produced. First, fieldwork placed me awkwardly in complex, heterogeneous and fluid processes of differentiation on the ground that I was studying. This awkwardness relates to how I felt I stood out through the material connections of my skin, clothes, camera, accent etc. or the practices in which I was engaged (walking through particular neighbourhoods, hanging out at cafés, parks or benches, often paying a little too much attention). In addition to this awkwardness, I found it difficult to forget that I was 'an ethnographer with a mission' (Saldanha 2007), and it is impossible to tell how this inflected on my experiences in Keighley, what I saw and

heard, what I retrieved and recorded. Second, this research is necessarily the outcome of various performances (playing on whiteness, maleness, my status as an outsider and researcher, etc.) as I negotiated access to particular field sites and tried to engage potential respondents. But just as it is important to recognise that this research is shaped by the ways I played up academic credentials, downplayed local knowledge or cultivated approachability, I was also never fully in control of these performances. Third, the ethnographic encounters that frame this research need to be situated in the twists and turns of swirling power relations. This is not to make claims for 'transparent reflexivity' (Rose 1997) or to reify research practices as inescapably objectifying and hierarchical (Thrift 2003), but simply to register that gendered, raced and classed power relations mediated the research process. The insights and accounts presented in this chapter are then necessarily the outcomes of negotiations that are never fixed or transparently knowable, but the conditions of their co-production nevertheless must be acknowledged in ways that engender modesty for the claims made on or by the research.

4   Deleuze and Guattari's (1987) call to replace questions of interpretation with questions of experimentation requires a shift in 'ontological priority' as we recognise that 'passage precedes construction' (Massumi 2002, 8) and focus, therefore, on processes of formation. In terms of thinking about race, we need to rely less on concepts of race tied to categorical identities, oppositions or genes and think more in terms of the conjunctions, relations, connections, intensities and affects that race enables.

5   Here, I work towards a materialist perspective on urban multiculture that recognises what Bennett (2004) calls the 'force of things' and how raced memories and affects accumulate around particular things (like cars, salwar kemeez, calls for prayer, and so on) and how these things then exert a kind of agency.

6   Social constructionist accounts do engage with questions of materiality, but these engagements tend to be restricted to the material effects of the idea of race and how race is materially expressed through segregation, differential employment opportunities, or unequal access to health care and education (Jackson 1987; Jackson and Penrose 1993; Alexander and Knowles 2005).

7   DeLanda (2002), and Prigogine and Stengers (1984) introduce dynamic, nondeterministic conceptions of nature into the human sciences, such that 'a new formulation of nature is now possible, a more acceptable description in which there is room for both the laws of nature and novelty and creativity' (Prigogine and Stengers 1984 citing Connolly 2002, 56).

8   Precisely how race is fluid and fixing is complexly marked by the power geometries of colonialism, globalisation, terrorism, migration, familial relations, and so on. Recent deconstructions of whiteness have tended, for example, to emphasise it is precisely through its fluidity – its invisibility, transcendence, even emptiness – that whiteness maintains its power (Dyer 1997; Saldanha 2007). Conversely, Fanon's (1967) raw account of being fixed in the 'fact of blackness' by the child on the train ruthlessly demonstrates how race fixes brown and black bodies. Here, without overlooking the power geometries of how race is simultaneously fluid and fixing, I want to avoid the suggestion that whiteness equals fluidity whereas blackness or brownness equals fixity, by emphasising the creative and capricious nature of materializations of race that temporarily fix bodies, things and spaces in heterogeneous practices of differentiation.

9   For Deleuze and Guattari (1987) rhizomes disrupt the arborescent model of thought that characterizes State philosophy. 'State philosophy,' Massumi (2002, 4) writes, 'is another name for the representational thinking that has dominated Western metaphysics

since Plato ... State philosophy is grounded in a double identity: of the thinking subject, and of the concepts it creates and to which it lends its own presumed attributes of sameness and constancy'. In distinction to hierarchies of arborescent thought, rhizomes foreground principles of connection and heterogeneity – 'any point of he rhizome can be connected to anything other, and must be. This is very different from the tree or root, which plots a point and fixes an order' (Deleuze and Guattari 1987, 7).

10 Saldanha (2006, 20) writes: '... race is devious in inventing new ways of chaining bodies. Race is creative, constantly morphing, now disguised as sexual desire, now as *la mission civilatrice*, all the while weaving new elements in its wake. Deleuze and Guattari might say that what defines race is not rigidity or inevitability, but its "lines of flight".' Race can be as stark as apartheid, but mostly it's fuzzy and operates through something else.'

11 Raced memories and affects might be laid down as virtual memories in two ways. Firstly, perceptions are formatted by micropolitics. Raced memories and affects do not accumulate haphazardly; rather nonconscious thinking is susceptible to 'intensive institutional disciplining' as institutions (the family, the nation, genetics, friendship, consumption, neighbourliness, policing, etc.) infuse dispositions, perceptions, practices and resistances (Connolly 2002; Thrift 2004). Secondly, past events and memories colour perceptions. Here, we acknowledge how events can take a toll on those who witness and experience them, such that they shade perception and judgement in subsequent encounters. And so we can conceive of a distinctive bank of virtual memories weighing down on each and every one of us at every moment of encounter.

12 Here we see how suspicions stick to, and are distributed across the heterogeneous elements and machinic connections that constitute some bodies. Moreover, by including addition and multiplications I suggest that suspicions are not simply the outcome of a linear or symmetrical accumulations but are amplified through some connections leading to 'phase changes' in registers of intensity (DeLanda 2002).

13 This narrative reworks interview material and many conversations with Jan to draw out, and dramatise, the circulation, modulation and contagion of affective intensities as she witnessed an encounter from behind her living room curtains. Recognising that all writing is about the creation of effects, this narrative uses experiments with style and rhythm to throw into relief the affectivity of this encounter and the movement of thinking, but at the same time remaining committed to the ethnographic method. And so, this narrative is a kind of fiction grounded in ethnographic engagement that works to cultivate a particular way of apprehend an encounter.

14 Saldanha (2007) develops the concept of viscosity to come to terms with how race comes to matter through the turbulence of ongoing interactions by questioning how some bodies tend to hold together in immanent groupings, while other bodies fail to stick. Located somewhere between fluidity and fixity, viscosities enables us o account for the 'surface tensions' that envelop immanent racial clusters without letting go of the possibilities that boundaries can be, and often are, transgressed.

15 This diffuse sense of melancholia resonates with writing on white working class neighbourhoods, most notably in South East London. For example, parallels might be drawn with Michael Collin's (2004) defence of the white working class in Southwark in *The Likes of Us: A Biography of the White Working Class* or Avtar Brah's (1999) moving meditation on a particular narration of Southall in Tim Lott's account of his mother's suicide, *The Scent of Dried Roses*.

References

Alexander, C. (2000), *The Asian Gang: Ethnicity, identity and masculinity* (Oxford: Berg).

Alexander, C. (2004), 'Imagining the Asian Gang: Ethnicity, masculinity and youth after "the Riots"', *Critical Social Policy* 24:2, 526–49.

Alexander, C. and Knowles, C. (2005), *Making Race Matter: Bodies, space and identity* (Basingstoke: Palgrave Macmillan).

Amin, A. (2002a), 'Ethnicity and the Multicultural City: Living with diversity', *Environment and Planning A* 34, 959–80.

Amin, A. (2002b), *Ethnicity and the Multicultural City: Living with diversity*, report for the Department of Transport, Local Government and the Regions.

Amin, A. (2003), 'Unruly Strangers? The 2001 Urban Riots in Britain', *International Journal of Urban and Regional Research* 27:2, 460–63.

Amin, A. (2007), *The Racialization of Everything* (mimeo.).

Amin, A. and Thrift, N. (2002), *Cities: Reimagining the urban* (Cambridge: Polity Press).

Anderson, K. (1991), *Vancouver's Chinatown: Racial discourse in Canada, 1875–1980* (Montreal: McGill-Queens University Press).

Back, L. (1996), *New Ethnicities and Urban Culture: Racisms and multiculture in young lives* (London: UCL Press).

Back, L. and Nayak, A. (1999), 'Signs of the Times?: Violence, Graffiti, and Racism in the English Suburbs', in T. Allen and J. Eade (eds) *Divided Europeans: Understanding ethnicities in conflict* (London: Kluwer Law International), 243–84.

Back, L., Keith, M., Khan, A., Shukra, K. and Solomos, J. (2002), 'The Return of Assimilationism: Race, Multiculturalism and New Labour', *Sociological Research Online* 7:2.

Balibar, E. (1991), *Race, Nation, Class: Ambiguous identities* (London: Verso).

Bennett, J. (2001), *The Enchantment of Modern Life: Attachments, crossings and ethics* (Princeton, NJ: Princeton University Press).

Bennett, J. (2004), 'The Force of Things: Steps toward an ecology of matter', *Political Theory* 32:3, 347–72.

Blaut, J. (1992), ''The Theory of Cultural Racism', *Antipode* 24:4, 289–99.

Butler, J. (1993), *Bodies that Matter: On the discursive limits of sex* (London: Routledge).

Bradford Race Review Team (2001), *Community Pride not Prejudice: Making Diversity work in Bradford* (Bradford: Bradford Vision).

Brah, A. (1999), 'The Scent of Memory: Strangers, our own, and others', *Feminist Review* 61:1, 4–26.

Cantle, T. (2005), *Community Cohesion: A New framework for race and diversity* (London; Palgrave).

Cheah, P. (1996), 'Mattering', *Diacritics* 26:1, 108–39.

Clifford, J. and Marcus, G. (1986), *Writing Culture: The poetics and politics of ethnography* (Los Angeles: University of California Press).

Cohen, S. (1972), *Folk Devils and Moral Panics: The creation of the Mods and Rockers* (London: MacGibbon and Kee).

Collins, M. (2004), *The Likes of Us: A biography of the white working class* (London: Granta).

Community Cohesion Review Team (2001), *Community Cohesion: A report of the Independent Review Team* (London: HMSO).

Connolly, W. (2002), *Neuropolitics: Thinking, culture, speed* (Minneapolis: University of Minnesota Press).

DeLanda, M. (1997), *A Thousand Years of Nonlinear History* (New York: Zone Books).

DeLanda, M. (2002), *Intensive Science and Virtual Philosophy* (London: Continuum).

Delaney, D. (2002), 'The Space that Race Makes', *The Professional Geographer* 54:1, 6–14.

Deleuze, G. and Guattari, F. (1987), *A Thousand Plateaus: Capitalism and schizophrenia* (London: Continuum).

Dwyer, O. and Jones, J.P. III (2000), 'White Socio-Spatial Epistemology', *Social and Cultural Geography* 1:2, 209–22.

Dyer, R. (1997), *White* (London: Routledge)

Fanon, F. (1967), *A Dying Colonialism* (Algeria: Grove Press).

Flusty, S. (1997), 'Building Paranoia', in N. Ellin (ed.) *Architecture of Fear* (New York: Princeton Architectural Press).

Gilroy, P. (1998), 'Race Ends Here', *Ethnic and Racial Studies* 21:5, 838–47.

Gilroy, P. (2000), *Against Race: Imagining political culture beyond the colour line* (Cambridge, MA: Harvard University Press).

Graham, S. (2004), *Cities, War and Terrorism: Towards an urban geopolitics* (Oxford: Blackwell).

Gregory, D. (2004), *The Colonial Present* (Oxford: Blackwell).

Grosz, E. (1994), *Volatile Bodies: Toward a corporeal feminism* (Bloomington, IN: Indiana University Press).

Hage, G. (1998), *White Nation: Fantasies of white supremacy in a multicultural society* (Annandale, NSW: Pluto Press).

Hall, S., Critcher, C., Jefferson, T., Clarke, J. and Robert, B. (1978), *Policing the Crisis: Mugging, the state, and law and order* (Basingstoke: Palgrave Macmillan).

Haraway, D. (1991), *Simians, Cyborgs and Women* (London: Routledge).

Home Office (2001), *Building Cohesive Communities: A report of the Ministerial Group on Public Order and Community Cohesion* (London: Home Office/Her Majesty's Government).

Hutnyk, J. (2002), 'Jungle Studies: The state of anthropology', *Futures* 34:1, 15–31.

Jackson, P. (ed.) (1987), *Race and Racism: Essays in social geography* (London: Allen and Unwin).

Jackson, P. and Penrose, J. (eds) (1993), *Constructions of Race, Place and Nation* (Minneapolis: University of Minnesota Press).

Keith, M. (1995), 'Making the Street Visible: Placing racial violence in context', *New Community* 21:4, 551–66.

Keith, M. (2005), *After the Cosmopolitan? Multicultural Cities and the Future of Racism* (London: Routledge).

Kobayashi, A. and Peake, L. (2000), 'Racism out of Place: Thoughts on whiteness and an antiracist geography in the new millennium', *Annals of the Association of American Geographers* 90:1, 392–403.

Lorimer, H. (2005), 'Cultural Geography: The busyness of being "more-than-representational"', *Progress in Human Geography* 29:1, 83–94.

Massumi, B. (1992), *A User's Guide to Capitalism and Schizophrenia: Deviations from Deleuze and Guattari* (Cambridge, MA: MIT Press).

Massumi, B. (1993), *The Politics of Everyday Fear* (Minneapolis: University of Minnesota Press).

Massumi, B. (2002), *Parables for the Virtual* (Durham, NC: Duke University Press).

McCormack. D. (2003), 'An Event of Geographical Ethics in Spaces of Affect', *Transactions of the Institute of British Geographers* 28, 488–507.

Miles, R. and Brown, M. (1989), *Racism* (London: Routledge).

Modood, T. (2005), *Multicultural Politics: Racism, ethnicity and Muslims in Britain* (Edinburgh: Edinburgh University Press).

Peach, C. (2006), 'Muslims in the 2001 Census of England and Wales: Gender and economic disadvantage', *Ethnic and Racial Studies* 29:4, 629–55.

Pile, S. (2005), *Real Cities* (London: Sage).

Pratt, G. (2000), 'Research Performances', *Environment and Planning D: Society and Space* 18, 639–51.

Pred, A. (2000), *Even in Sweden: Racisms, racialized spaces and the popular geographical imagination* (Berkeley/Los Angeles/London: University of California Press).

Prigogine, I. and Stengers, I. (1984), *Order out of Chaos: Man's New dialogue with nature* (London: Heinemann).

Rose, G. (1997), 'Situating Knowledges: Positionality, reflexivities and other tactics', *Progress in Human Geography* 21:3, 305–20.

Said, E.W. (1979), *Orientalism* (London and New York: Vintage Books).

Saldanha, A. (2005), 'Vision and Viscosity in Goa's Psychedelic Trance Scene', *ACME* 4:2, 172–93.

Saldanha, A. (2006), 'Reontologising Race: The machinic geography of phenotype', *Environment and Planning D: Society and Space* 24:1, 9–24.

Saldanha, A. (2007), *Psychedelic White: Goa trance and the viscosity of race* (Minneapolis: University of Minnesota Press).

Simpson, L. (2005), 'On the Measurement and Meaning of Residential Segregation', *Urban Studies* 42:7, 1229–30.

Sinclair, I. (1997), *Lights Out for the Territory* (London: Granta).

Thrift, N. (1996), *Spatial Formations* (London: Sage).

Thrift, N. (1997), 'The Still Point', in S. Pile and M. Keith (eds) *Geographies of Resistance* (London: Routledge).

Thrift, N. (2003), 'Practising Ethics', in M. Pryke, G. Rose and S. Whatmore (eds) *Using Social Theory: Thinking through research* (London: Sage).

Thrift, N. (2004), 'Intensities of Feeling: Towards a spatial politics of affect', *Geografiska Annaler B* 86:1, 57–78.

Thrift, N. and Dewsbury, J.D. (2000), 'Dead Geographies – and How to Make Them Live', *Environment and Planning D: Society and Space* 18, 411–32.

Virilio, P. (2005), *City of Panic* (London: Berg).

Ware, V. and Back, L. (2002), *Out of Whiteness: Color, politics, and culture* (Chicago: University of Chicago Press).

Chapter 18

Everyday Geographies of Marginality and Encounter in the Multicultural City

John Clayton

Introduction

In the context of widespread public debate over the future of multiculturalism in the UK, this chapter tackles the issue of inter-ethnic relations as found in a specific urban area of the UK; the city of Leicester in the East Midlands of England.[1] With debate increasingly focussed upon the ills of residential segregation and the separate development of communities alongside a need for greater inter-ethnic contact and shared values,[2] I show how the geographies of everyday multiculturalism work to condition forms of racism and acceptance through both cultures of marginality and practices of inter-ethnic encounter.

As part of a growing body of work which challenges the normalised status of whiteness (Frakenberg 1993; McGuiness 2000), but which also draws attention to the intersection of class and race (Ruddick 1996; Byrne 2006), the empirical focus here is upon contestations of multiculturalism and limited forms of accommodation experienced and expressed by 'white' research participants. The chapter examines the ways in which racialised thinking and action may be reinforced or disrupted through everyday spatial trajectories in a place recognised as a beacon of multicultural success. I concentrate on the interplay of two related positionings: marginality (in classed and geographical terms) and inter-ethnic encounter (within as well as beyond marginal neighbourhoods).[3] The relations between these everday geographies are recognised as critical aspects of an approach which takes seriously the productive capacity of multiculturalism as lived through 'the everyday'.

Everyday Geographies of Multiculturalism

As Goldberg (1994) notes multiculturalism should not only be considered as a political philosophy of community recognition, representation and provision, but also as an everyday social phenomenon, what Amin (2002) refers to as the condition of 'living with difference'. In the context of the juxtaposition of an array of differences, or what Massey (2005) calls the 'throwntogetherness' of cities, ethnicity is seen as emergent and negotiated, but also a frequently 'stubborn' aspect of identity (Valins 2003). A focus upon the everyday geographies of multiculturalism looks to unpack the active negotiation between the engrained and the possible, illustrating what Secor (2004, 357) calls '… the variously fluid and fortified boundaries of urban space'. It is a way of appreciating that ethnic difference and similarity is constituted through the rhythms and spaces of the routine, which are embedded in wider social, political and economic processes.[4] As Schein (2002, 4) notes:

> Ideas such as race and racism do not emerge unprompted from individual minds, but are thoroughly embedded in our collective everyday lives and in the very structures of our social, political and economic activities.

This is an understanding of space and spatial practices as productive of differences, not just a reflection of them. It is through the symbolic and experiential geographies of the city, that difference and similarity is given shape, reinforced and also re-configured.

In what follows I identify a sense of unease in relation to the public culture of place in Leicester by young 'white' research participants. These individuals express a sense of distance between themselves and a promoted image of Leicester as a beacon of multicultural harmony. With reference to two clearly identifiable (yet inter-related) everyday positionings of marginality and encounter, I show how the spatial trajectories of individuals impact on inter-ethnic outlooks and practices. Not only do everyday encounters reinforce ideas of 'us' and 'them', established through cultures of marginality, they also have the capacity to upset established ideas of social incompatibility.

The Case of Leicester: Multiculturalism for All?

Negotiations of belonging, do not take place in a spatial void. Understandably this gap has often been filled by an examination of inter-ethnic relations as a national phenomenon. However, there is also a need to consider the ways in which such negotiations take their form not only in, but also through various places. Whilst a fear of immigrants and ethnic 'outsiders', racially inflected conflict and reactions to the spectre of terrorism are certainly conceived of as national issues, they are all interpreted, experienced and practiced differently through different places.

One place in particular, the provincial city of Leicester in the East Midlands of England, has been framed as a success story by local politicians, other influential figures in the city, the national government and academics alike (Vertovec 1996; Winstone 1996; Singh 2003). Leicester's success in this regard hinges upon the absence of overt inter-ethnic conflict in spite of the fact that the city is home to a relatively large 'non-white' population.[5] This situation has been sharply contrasted with other areas of the country, particularly the north west of England, blighted by images of the 'race riots' in 2001 (Cantle 2001). As a consequence, the media as well as the British government have looked to this city as an example of how positive inter-ethnic relations might operate and might be translated elsewhere.[6]

The apparent peaceful co-existence of communities has been attributed to a number of prevailing social, political and economic trends in the city. These include: specific immigration patterns, particularly the movement of a middle class self-supporting African-Asian population from East Africa in the early 1970s, a relatively buoyant local economy, the absence of the formal far right, a local political scene which has incorporated ethnic minority groups and delivered on an equalities agenda, as well as the pro-active work of non-political agencies, organisations and networks. Paradoxically, as Mai (2007) points out the much lauded achievements in this place have been seen as the result of the successful management of separate, distinct and identifiable ethnic communities and not necessarily through the development of a cohesive, inter-cultural consensus which has brought these communities any closer.

While this success story has a large degree of credibility, agreement with this image of racial harmony is certainly not unanimous. I do not want to focus here on forms of opposition expressed by individuals and groups who rightly stress the continual existence of racism, discrimination and racially inflected disadvantage in the city[7] (Westwood 1991; Benyon et al. 1996; Chessum 2000; Pakistani Youth and Community Association 2002), rather I will show how the notion of multicultural success might represent a threat to those who see themselves as being outside of this success story on the basis of their everyday positionings.

There has been no overt and widespread 'white' backlash to a project of multiculturalism in Leicester. Many of the more vocal and extreme opponents to a growing ethnic minority presence have relocated to the country[8] and there has been a limited basis for dissent, given the prevailing economic conditions in the city, the class profile of many immigrants and as Singh (2003) puts it, the limited claims made on public resources by migrant communities. However, this does not mean that opposition has disappeared or that new forms of opposition cannot be (re)produced, particularly given wider discourses of exclusive national identity circulated in the British media and contemporary processes of inequality and alienation.

It is not that such outlooks do not exist in this place, rather they are often not given the space to be openly performed. In many ways this can be seen as beneficial in terms of the intolerance of racism, however it also masks the tensions within which given specific circumstances emerge. As Emily, who grew up in a predominantly 'white' working class estate explains, resentment towards the diversity of the city and racist stances are not uncommon amongst her fellow 'white' students at her college of further education. However, given the context of a multicultural college which has an ethnically mixed intake and places a strong emphasis on anti-racism and racial equality, there is little opportunity for these sentiments to be openly expressed.

> I think a lot of like white people that come to the college, I think they are quite racist, but you can't really say anything if like you've got a mixed group ... I suppose because

you are told that if you say anything then you can be done for racism and you have to watch what you say or watch how you say it.

Again, the following excerpt highlights a sense of disagreement and opposition.

John: Some people see Leicester as quite a successful place.
Warren: I don't, I don't agree, from what you hear around here.

As with Emily's comments, Warren is talking from a specific geographic standpoint when he speaks of 'around here', that is, his own neighbourhood, a predominantly 'white' working class estate on the western edge of Leicester. He suggests that positive inter-ethnic relations are not part of the everyday reality for those living on the margins in raced and classed terms.[9] Indeed the manner in which multiculturalism is promoted in this place as a form of middle class cosmopolitanism where diversity is deemed attractive on certain economic and cultural terms (Skeggs 2003), may compound this. For those who feel they have no place in this endeavour, the idea and experience of multiculturalism represents more of a cultural threat than a cultural opportunity. Support for and involvement with widely accepted forms multiculturalism is then geographically uneven.

Geographies of Marginality and Territoriality

I now want to look in greater detail at the geographies through which such ideas might take hold, in particular through the experience of marginal 'white' working-class neighbourhoods, parts of the city (but not the only parts) which in different ways have been 'left behind' (Shields 1990). Though very much economically and culturally disconnected from the city, marginality and the defence of territory is not just about the geographic isolation of these communities. The active reconstitution of the identities of these neighbourhoods and their residents, relies upon reference to, connection with, and experience of 'other' spaces.

Ideas and practices which racialise space and fix the identity of ethnic others work to '… secure us "over here" and them "over there"' (Hall 1992, 16), forming part of the taken for granted fabric of urban life. One of the main ways in which this operates in Leicester is through the racial coding of the city's neighbourhoods in the local popular imagination. These geographical imaginations are tied into a comprehension of the built environment, assumed levels of poverty and 'roughness', the racial, ethnic and religious make-up of an area, as well as stories of danger and memories of 'racialised' experiences. As Craig, who lives in a 'white' working class neighbourhood in the north of the city notes, the coding of areas of the city on racial terms is a way of making sense of Leicester's complex forms of diversity but also works to fix the identity of 'others', and himself.

Cos you've got the big estates. Like you've got Melton Road is the whole Asian estate, you've got St Matthews' is the black estate, you've got Highfields: black estate … St

Matthews', Rushey Mead, and Hinckley, they've all got different cultures behind them, for example Hinckley whites, St Matthews' blacks, Melton Road slash Rushey Mead Asians, my area's white.

This racial coding of the city is more strongly associated with some neighbourhoods than others. Those areas which feature most heavily are those, firstly, which are racially marked as 'not-white', areas found mostly in central Leicester, seen to be home to substantial ethnic minority groups, caught up in discourses of decline and criminality. Secondly, those areas towards the outskirts of the city, identified as unintegrated and dominated by 'white' working-class populations hostile to the incursions of those who are not seen to belong on a number of social and cultural criteria. Taking into account the classed dimension of multicultural inclusion it is no surprise that both 'white' working class areas and 'non-white' working-class neighbourhoods figure so heavily in a local racial vernacular and a fear of difference. As Clive, a 'white' student who spent some years living in one of these 'white' neighbourhoods puts it; the story of Leicester's harmonious relations does not take into account the experiences of these marginal neighbourhoods: '... so they've given it a reputation judged on the areas that look good.'

Whilst overt exclusionary language and behaviour is often hidden given the history of official multicultural accommodation in the city, in situations deemed to be 'safe' it can be seen to be an important aspect of everyday outlooks and practice. On some occasions this is expressed by participants commenting in an objective manner, (as Emily does above) often related to a fear of expressing controversial views, but in other situations it is more personalised. In discussions with young people on one estate in the west of the city, one female participant made her feelings clear about ethnic diversity in the city and the possibility of this upsetting the ethnic balance in her own area. Her neighbourhood was one with which she felt a strong sense of affiliation based upon its 'whiteness'. In her eyes this is an identity and an area to be defended from ethnically threatening incursions. As she explains, 'Braunstone's ours! It belongs to white people.'

In discussions with another group of young people living in the same neighbourhood, the existence yet muted performance of racist views was made clear. The body language of two young 'white' females in particular indicated a level of frustration at the inability to express how they 'really' felt about the idea of Leicester as a successful multicultural place. Given the presence of someone whom they identified as racially different in the group, the social dynamics of this situation meant that overt and verbal expressions of racism were deemed inappropriate in this specific context (Back and Nayak 1999; Hyams 2004). I was only aware of their feelings once the session had ended when I asked one of these females if there was anything she had wanted to say during the course of the discussion, but felt she couldn't. She clearly felt she could confide in me as a 'white' researcher that: 'I'm racist! I couldn't say anything because I was sitting right next to that Indian lad. I'm racist and I admit it!'

This tallies with the experiences of most 'non-white' young people involved in the research, whereby certain areas including this neighbourhood, epitomise geographies of exclusion and acceptance, where they felt 'out of place' (Cresswell 1996). However, it is far too simplistic to label such neighbourhoods as 'racist'. Whilst it is certainly the case that those living through materially disadvantaged residential contexts sometimes appear antagonistic towards the presence of racially marked 'others', there is also a suspicion and resistance to incursions from outsiders regardless or race or ethnicity. Indeed, this has been part of the reason for the relative absence of the British National Party in such places. This tells us something more about the negotiated state of insiders and outsiders and definitions of 'us' and 'them' in this city, something I will return to in the following section. As Emily explains, those living in more disadvantaged neighbourhoods feel a need to protect their interests and each other from possible threats embodied in the figure of the outsider.

> When I lived up New Parks, if like any coloured people walked by, I suppose like estates kind of stick together, don't they? And they look out for one another. And it's like, I don't know really, if anybody came up the area that was, not just not coloured but like you knew that weren't from like around there, they'd probably get some stick or something.

Warren also discusses how for people like himself who see themselves in positions of marginality, lacking adequate access to work opportunities; they are adversely affected by the presence of racially marked new arrivals in the city (in the guise of 'asylum seekers' and 'Somalians'). Multiculturalism, in this way, is seen as something which does not contribute to a quality of life, as in cosmopolitan definitions of the 'good multicultural city' but are seen to take away.

> Where people like ourselves that are poor, haven't got much money, have got to scrimp and save, look for jobs, get jobs. People who I know have been driven out of work because asylum seekers, Somalians are saying 'ah we'll do it for 60 quid a week', when mates of mine getting a good income off £150 they've been laid back and they've been taking on these other people, because they want to work for cheaper, which is not fair on other people.

The defence of identity, of the city and of 'white' neighbourhoods is clearly woven through the material and economic realities of the lives of some of these participants. Here Craig also offers details of his own fragile economic position and routine, working long hours to look after himself and his family. His own circumstances are again blamed upon the presence of ethnic 'others'. However, this stance is not *just* a product of his marginal economic positioning, it both feeds from and into his encounters with ethnic 'others'. For instance, in this example he makes reference to his encounters with these ethnic 'others' in the space of his college of further education, set within the context of his own socio-economic experience. He identifies the attitude of some ethnic minority students at the college as his justification. If anything these forms of inter-ethnic contact seem to exacerbate his negative feelings towards those not seen to rightfully belong.

But my dad's ill, we're struggling to make by here, I'm struggling to make by, all the part-time money I'm earning is going towards my family. I can't afford to eat myself, because every time I get home, I'm out again ... my usual scenario is I ain't gotta work tonight, but go to work straight off, then I'm working to 10, 11, don't get in till midnight and what about 14 hour day, 13, 14 hour day, and they wonder why people are getting stressed. And then again it all boils down to people who come here for free, abuse the system and they just think: 'Oh! it's a place to doss'. I've seen some of the people's attitudes here and there are absolutely disgraceful and they shouldn't be here.

Cultures of exclusion clearly emerge from opposition to and anxiety over the idea of a multicultural city which is supportive of the needs of ethnic minorities above the needs of the 'white' working class population. However they are also related to the everyday reality of living with difference, shared space and the possibility of inter-ethnic encounter. Inter-ethnic encounters play off and into experiences of multiculturalism. As Ahmed and Stacey (2001, 7) explain, through their use of 'inter-embodiment', this '... is a way of thinking through the nearness of other others, but a nearness that involves distancation and difference', a way of rethinking how the spatial proximities and social relations might operate in diverse urban areas.

Encountering Difference and the Negotiation of Identities

From what we have seen, 'difference' is understood through the cognitive, practical and situated geographies of the city in relation to a public culture of place. For these 'white' research participants the ability and desire to manoeuvre across racialised boundaries remains problematic and limited. However, it is also the case that individuals come to understand themselves and others through limited but critical forms of inter-ethnic knowledge and embodied exposure. In the following excerpt Emily continues to discuss her childhood and illuminate the importance of both her upbringing in the neighbourhood of New Parks, but also her limited experiences of 'other' parts of the city.

One of the areas Emily identified as visiting most often beyond her current neighbourhood was that of Thurmaston on the northern edge of the city to visit relatives. The car journey which takes her there passes along Melton Road, an area which retains a number of racialised associations based upon past experiences. Her memory of travelling along this same stretch of road with her parents as a youngster, playing the 'game' of 'spot the white man', appears as a symbolic and practical resource which she still draws upon.

I remember when we was younger, my mum and dad like, we'd go out on a Sunday or something and we'd drive up there and my mum and dad would say: 'Right, spot the white person time!' And if like, as many white people as we'd see my mum and dad would give us a pound, but we'd never get more than two.

While she now recognises this practice as 'awful', it has had an enduring impression upon her outlooks into adulthood and continues to organise her impressions of the city and her interactions with those identified as ethnically different within her current neighbourhood in the city centre.

Strict ideas of self, the ethnic 'other' and the gap between these two positions are clearly reinforced through the very process of encountering difference. This seems to go against the idea that the forced propinquity of urban life and the various forms of inter-cultural contact enabled by the city reaps progressive rewards (Zukin 1998). This is because there is a need to take into account the sorts of experiences and cultural baggage brought into such encounters and the power relations which operate through such meetings. In the example above Emily and her family appear as explorers in some foreign and exotic land, yet firmly aware that this was their city, with the power to gaze and name whilst still remaining very much removed from the realities of the area within the safety of their car.

Encounters are riven with power relations and personal positionings which may not be conducive to progressive relations, but even in the spatial contexts of marginality, territoriality and 'white' defensiveness, forms of identity negotiation are possible. This has often been noted in terms of the shifting cultural styles of 'white' young people through dress, language and musical tastes (Nayak 2003). But it is also clear that such negotiations take other forms, evident through the everyday trajectories of individuals. As with Emily above, Calvin (also identifying as white) indicates that in his experience, the territoriality of the city did not always operate purely on racial terms, with reference to practices of surveillance between marginal 'white' neighbourhoods seen in terms of their working class-ness and their 'whiteness'. Fear of these 'other' white estates and encounters with their residents, with distinct identities seemed to be more of an issue than those seen in terms of their ethnic difference.

Calvin: There are certain areas that I wouldn't go on my own
John: Yeah? OK
Calvin: Like deep into New Parks and Beaumont Leys because you've always got
           that, you're from Braunstone, you're an outsider.
John: How do they know that you're from Braunstone?
Calvin: Sometimes if they don't know your face, they're like where's he from? And
           if you say: 'oh I live down there' they'll say: 'no you don't!'

For some 'white' young people, the importance of and identification with their neighbourhood trumps ethnic differentiation, what Back (1996) refer to as 'neighbourhood nationalism'. Some of the participants living in these parts of the city stressed how it was important to become known to others on their estate for the very reason that they would be identified as an insider, as Clive noted: 'You have to build up a reputation'. Becoming known and used to each other is clearly something which takes time, and acceptance is something which has to be worked at, particularly when racial, ethnic and religious differences block the path. Whilst it may be slow going, the terms of membership and belonging even in the most marginal and ethnically segregated neighbourhoods of the city are open to some forms of inter-ethnic negotiation, even if that means that other forms of racialised tensions remain and the targets of racism shift. As David, identifying as of dual heritage ('African-Caribbean' and 'white'), noted from his experiences:

> You do see though, particularly in the rough areas that a lot of the time, it's not always necessarily each cultures on it's own, it could be a group of cultures against another group of cultures. Like a lot of the time you see a lot of black and white people and then they might be like troubling some Asians or whatever, might have trouble with some Indian people or whatever if you know what I mean. It's not always just each culture for their own, it could just be a group of people that don't like another culture.

This same idea of conflict between 'mixed' groups from rival neighbourhoods coded in terms of their 'whiteness', was also discussed by two young people (Nic, identifying as of mixed Sikh-Hindu heritage and Rebecca as 'white') as part of a group discussion in one of these neighbourhoods in the west of the city. When asked to identify what in their experience divided young people, they were keen to stress that although people in these neighbourhoods could often be seen to display racist attitudes, 'trouble' was more often based upon territorial divides.

> Nic: Yeah it's like area against area, it's not like religion versus religion.
> Rebecca: It's, yeah, you're right, you're right, it's areas! Because like you'll get a gang of youths from Beaumont Leys say and its the gang will be totally mixed and then you'll get a gang from Braunstone and that gang will be totally mixed and they're not having a problem with each other through the cultures that they're from, it's just by the area ... it's territory innit!

A focus on these practices of territoriality and encounters with difference through the neighbourhood illustrates that the boundaries of the city are subject to ongoing struggle (Silk 1999), whereby acceptance into a collective but fragile 'we' is often, but not always based upon a rigid marker of racial difference. However, with time such a critical factor in these negotiations, it is no surprise that those seen as newer arrivals in the city are amongst the most stigmatised, marginalised and differentiated in local racist discourse (Back 1996).

It is also clear that encounters beyond the boundaries of the neighbourhood have the capacity to unsettle ideas of difference established elsewhere, particularly when they are of an intensity and duration which enables other forms of solidarity and identification to be established. While I have emphasised the importance of the neighbourhood in the everyday lives of these individuals, the spatial trajectories of young people should not be seen as solely limited to places of residence and face to face 'community' socialisation. Indeed as has been shown above, the role of other everyday spaces including 'other' neighbourhoods, but also the city centre, leisure spaces, educational institutions and workplaces are key to the shifting character of identification and differentiation. Craig's account illustrates this point, as well as the fact that forms of racism and more inclusive sentiment very often exist alongside each other, and in specific spatial contexts are more openly performed or restricted. As Craig seems very aware:

> Views tend to change you know to suit their friends. One person could have been
> incredibly racist with one group of people and then [not] to a different [group], and
> they could be hanging around with black people.

On this occasion, his own views and outlooks of the city are particularly
influenced it seems by his experiences of his workplace, mentioned above in his
description of the long hours he worked. On reflection he explained how the
forms of interactions he was almost forced into in this environment, with those
previously seen as threateningly different, had a discernible impact on the changing
character of his own outlooks.

> If they are friendly yeah, I'll talk to them, I don't mind, but that state of mind has
> come strictly from my job that I have to do. I have to talk to people and when you
> deal with a lot of people on the basis that I do, then you become broader minded, you
> think, wait a minute why are you thinking these thoughts in the first place? It makes
> me think deeper if anything, yeah I was racist, I do admit that, that's before I started
> work, then I started to deal with Asians, Black people and then you think, you think
> to yourself they're not so bad.

These limited forms of accommodation however must be seen in the context
of the more exclusionary views expressed above in relation to other encounters
with ethnic difference in the space of the college, where very different power
relations and social dynamics were at play.[10] What is clear is the unstable, often
contradictory, but spatially negotiated basis of inter-ethnic relations as well as the
ongoing personal struggles of identification and differentiation enacted through
and viewed through the geographies of marginality and encounter.

Summary and Concluding Remarks

While the case of Leicester illustrates the importance of considering inter-ethnic
relations as a placed phenomena and offers a sense of hope in terms of the
historical development of relatively progressive inter-ethnic relations, a closer
examination reveals that those who consider themselves on the margins, in
this case working class 'white' youths, may not be part of, or desire to be part
of, an image of multicultural harmony. These experiences have been examined
through the relationship between the 'geographical positionings' of marginality
(in geographical, socio-economic and cultural terms), and everyday forms of
inter-ethnic contact, illustrating the ways in which social and spatial im/mobility
are intimately related and influence the terms upon which the negotiation of
identity take place.

The current UK policy agenda of community cohesion, with its emphasis
upon bringing parallel communities together, must not disregard the divisive
inequalities of the urban experience, nor treat neighbourhood segregation as
a problem of distance and isolation removed from its connection to other city
spaces and its active re-production as a defensive resource. It must also consider
that knowledge of and physical co-presence with those seen as 'different' is no

guarantee of progressive relations, particularly for those in fragile economic and social positions who have not accrued the social and cultural capital (Bourdieu 1986) required to deal with such encounters. There is then a need to address the terms on which such encounters take place, the power relations involved and the experiences brought into these situations. From this position we may be better placed to seize the potential of opportunities which engage individuals, promote respect for difference, generate more inclusive ideas of identification and belonging and address the everyday needs of those on the margins.[11]

Notes

1   This work is taken from ESRC/ODPM sponsored doctoral research conducted between 2002–2006, which examined inter-ethnic relations and everyday geographies amongst young people in the city of Leicester. Leicester was selected for this research on the basis of its established reputation as a model for successful race relations. The project involved working with and speaking with young people from a variety of ethnic backgrounds in school and youth centre contexts in order to gather their perspectives on the idea of Leicester as a multicultural success and to document their everyday lives and experiences within this specific multicultural urban environment. The project was approached through a theoretical framework which stressed the productive and relational character of space and everyday spatial practices.

2   I do not have room in this chapter to expand upon my arguments in relation to the idea of imposed British values and narrow conceptions of national identity as a solution to the problems of inter-ethnic tension. However, I wish to make it clear that I emphasise here the contextually negotiated character of identities which cannot be legislated for.

3   The focus here is 'white' working class subjectivities; however this is not to deny the very real existence and practice of a variety of middle class racisms.

4   Back and Nayak (1999: 257) argue that for the white working classes the performance of racism, can be seen as a 'ritual expression', but one which is deeply embedded in wider processes of 'individualisation, anomic consumerism and geographical dislocation'.

5   36 per cent according to the 2001 census. Office of National Statistics.

6   For instance Leicester City Council was awarded Beacon status for Promoting Racial Equality in 2002, and Beacon status for Community Cohesion in 2003. In February 2003 the city was also selected as one of 15 national Pathfinder Areas for community cohesion on the basis of the development of strong projects and networks seen to deliver crucial facets of the governments' cohesion agenda.

7   Recent research shows 'racist' incidents are common place for young people in Leicester (Rupra, 2004). According to this report, between April 2003 and March 2004, 1089 incidents of racial harassment were reported, a sign both of the willingness of individuals to highlight these incidents but also of the extensive practice and force of racism.

8   As one key informant explained: 'A lot of NF (National Front) supporters moved physically out of the city to the county. They didn't want to be in an Asian city as they saw it and whole families, racist families, moved out physically to get away from Asians into Eastern Leicestershire.'

9   Indeed, whilst not the focus of this chapter it is worth stressing that 'non-white' research participants also expressed a sense of marginalisation and exclusion from

positive forms of multiculturalism. This especially emerged through the accounts of those belonging to newly arrived communities in the city, such as those identifying as Somali. For these groups racial and economic exclusion combine to push them to the peripheries.

10 For instance Craig was part of an 'all white' Electrical Installations class, which meant that his interactions and encounters with those seen as ethnically different in that space, did not involve the same level of contact but also intensity and mutual respect required in the workplace context.

11 I would like to thank Claire Dwyer and Caroline Bressey for their helpful and constructive comments in the preparation of this chapter. I would also like to thank all those participants, who made this research possible. In particular I would like to give special thanks to the young people involved for their time but also for allowing me to gain an insight into their worlds.

References

Ahmed, S. and Stacey, J. (2001), *Thinking Through the Skin* (London: Routledge).
Amin, A. (2002), 'Ethnicity and the Multicultural City: Living with diversity', *Environment and Planning A* 34:6, 959–80.
Back, L. (1996), *New Ethnicities and Urban Culture: Racisms and multiculture in young lives* (London: UCL Press).
Back, L. and Nayak, A. (1999), 'Signs of the Times? Violence, Graffiti and Racism in the English Suburbs', in T. Allen and J. Eade (eds) *Divided Europeans: Understanding ethnicities in conflict* (Amsterdam: Kluwer Law International).
Benyon, J., Dauda, B., Garland, J. and Lyle, S. (1996), *African Caribbean People in Leicestershire – Summary of the Final Report* (Leicester: Scarman Centre for the Study of Public Order).
Bourdieu, P. (1986), *Distinction* (London: Routledge).
Byrne, B. (2006), *White Lives: The interplay of 'race', class and gender in everyday life* (London: Routledge).
Cantle, T. (2001), *Community Cohesion: A report of the Independent Review Team* (London: Home Office)
Chessum, L. (2000), *From Immigrants to Ethnic Minority: Making Black community in Britain* (Aldershot: Ashgate).
Cresswell, T. (1996), *In Place/Out of Place: Geography, ideology and transgression* (London: University of Minnesota Press).
Frankenberg, R. (1993), *White Women, Race Matters: The social construction of Whiteness* (Minneapolis: University of Minnesota Press).
Goldberg, D. (ed.) (1994), *Multiculturalism: A critical reader* (Oxford: Blackwell).
Hall, S. (1992), 'Race Culture and Communications: Looking backward and forward at cultural studies', *Rethinking Marxism* 5, 10–18.
Hyams, M. (2004), 'Hearing Girls' Silences: Thoughts on the politics and practices of a feminist method of group discussion', *Gender, Place and Culture* 11:1, 105–19.
McGuinness, M. (2000), 'Geography Matters? Whiteness and Contemporary Geography', *Area* 32:2, 225–30.
Mai, N. (2007), *Between 'Parallel Lives' and 'Community Cohesion*, The Centre on Migration, Policy and Society Annual Conference, University of Oxford, 5–6 July.
Massey, D. (2005), *For Space* (London: Sage).

Nayak, A. (2003), *Race, Place and Globalization: Youth culture in a changing world* (Oxford: Berg).

Pakistani Youth and Community Association (2002), *Research into the Needs of the Pakistani Community in Leicester* (Leicester: Chief Executive's Department of Leicester City Council).

Ruddick, S. (1996), 'Constructing Difference in Public Spaces: Race, class and gender as interlocking systems', *Urban Geography* 17, 132–51.

Rupra, M. (2004), *I Ain't Racist But …* (Leicester: Leicester Racial Equality Council).

Schein, R.H. (2002), 'Race, Racism and Geography', *The Professional Geographer* 54:1, 1–5.

Secor, A. (2004), 'There is an Istanbul that Belongs to Me: Citizenship, space and identity in the city', *Annals of the Association of American Geographers* 94:2, 352–68.

Sheilds, R. (1990), *Places on the Margin: Alternative geographies of modernity* (London: Routledge).

Silk, J. (1999), 'The Dynamics of Community, Place and Identity', *Environment and Planning A* 31, 5–17.

Singh, G. (2003), 'Multiculturalism in Contemporary Britain: Reflections on the "Leicester Model"', *International Journal on Multicultural Societies* 5:1, 40–54.

Skeggs, B. (2003), *Class, Self, Culture* (London: Routledge).

Valins, O. (2003), 'Stubborn Identities and the Construction of Socio-Spatial Boundaries: Ultra-Orthodox Jews living in contemporary Britain', *Transactions of the Institute of British Geographers* 28, 158–75.

Vertovec, S. (1996), 'Multiculturalism, Culturalism and Public Incorporation', *Ethnic and Racial Studies* 19:1, 49–69.

Westwood, S. (1991), 'Red Star over Leicester: Racism, the politics of identity and Black youth in Britain', in P. Werbner and M. Anwar (eds) *Black and Ethnic Leaderships: The cultural dimension of political action* (London: Routledge).

Winstone, P. (1996), 'Managing a Multi-Ethnic and Multicultural City in Europe: Leicester', *International Social Science Journal* 147, 33–41.

Zukin, S. (1998), 'Urban Lifestyles: Diversity and standardisation in spaces of consumption', *Urban Studies* 35:8, 825–39.

Chapter 19

Young People's Geographies of Racism and Anti-racism: The Case of North East England

Anoop Nayak

Introduction

British social geographers have produced a series of vivid accounts of race and racism in urban spaces. This includes research on race and residential housing (Smith 1987), urban social segregation (Peach 1975), post-colonial carnival (Jackson 1988), multi-ethnic living (Amin 2002; Keith 2005), Muslim schooling (Dwyer 1993) and neighbourhood discrimination (Phillips 2006; Hopkins 2007), as well as inner-city race and class conflict (Burgess 1985; Keith 1991, 1993). Such studies have provided a strategic intervention in debates on race and migration in the discipline.

However, it is noticeable that the bulk of geographical research on race has concentrated on urban areas, and in particular the spaces and neighbourhoods where ethnic minorities reside. A constraint with this approach is that it can lead to a presumption that race is a 'problem' and one that is located in multi-ethnic areas (Gilroy 1987). To this effect, 'a myth is being perpetuated of *The Inner City* as an alien place, separate and isolated, located outside white, middle-class values and environments' (Burgess 1985, 193) and requiring stringent policing (Keith 1993). The focus on cities may even give rise to the false premise that rural and mainly-white areas are immune to racism. Reports on racism in the countryside emphatically challenge these preconceptions. Empirical studies undertaken in Devon, (Bonnett 1993a), Norfolk (Derbyshire 1994) and Shropshire (Nizhar 1995) challenge the 'no problem here' (Gaine 1987, 1995) denials of racism in remote white areas and demonstrate that white children and youth can hold deeply prejudice ideas on race (Jay 1992; Troyna and Hatcher 1992a). Exploring racism and anti-racism in mainly-white areas counters 'the hegemonic status of the inner city discourse in relation to race and space' (Watt 1998, 688) and its tendency to dismiss 'out of the way' places with their unique race histories (Nayak 2003).

This chapter aims to investigate young people's perceptions of race and ethnicity in the mainly-white region of North East England. I begin by documenting the changing face of the region in the light of recent migrations and new patterns of global settlement, before going on to briefly discuss the ethnographic methods deployed. The research explores how white youth perceive race and ethnicity living in a largely white, post-industrial setting. I then go on to consider their attitudes to anti-racism and identify the barriers preventing a more cosmopolitan and multicultural outlook befitting of global times. Finally, I seek to demonstrate how young working-class people living in mainly-white areas can begin to connect with anti-racism through an informed cultural geography, or 'pedagogy of place'. The attempt to make anti-racism meaningful in a mainly-white region and inspire new understandings of place and identity is a response to McGuinness's (2000) concern that in their unveiling of a global sense of place geographers all too easily lapse back upon the familiar, multi-ethnic, hybrid metropolis. Instead, he enquires 'might such predominantly white places as the British New Towns and the seemingly endless suburbs of Middle Britain also be sites of an equally dynamic and (perhaps) exciting reinvention of post-colonial ethnicities?' (228). This chapter seeks to address this possibility through an engagement with race, place and globalisation in a predominantly white locality.

The Changing Face of the North East: Race, Migration and Asylum

The North East has a history of shipbuilding, chemical engineering, coal-mining and heavy industry (Robinson 1988). Since this time it has undergone major post-industrial transition that has seen the regeneration of old industrial areas, development in creative arts and cultural industries, the growth of a knowledge economy and in particular the rise of a service sector economy designed around call centres, consumption, tourism and hospitality. Although the North East is historically one of the whitest regions in the country there is some evidence to suggest that this portrait is gradually changing. According to the last 2001 Census approximately 2.4 per cent of the North East population are from black and minority ethnic (BME) backgrounds compared with 9.1 per cent of England as a whole, reflecting the sparse nature of non-white diasporic settlement. The largest preponderance of BME settlement is to be found in Newcastle (6.9 per cent) and Middlesborough (6.3 per cent). South Asian communities constitute the largest BME group in the region (1.3 per cent) with the majority heralding from Pakistan (0.6 per cent) and being of Muslim faith. These communities are made up of people with secular and multi-faith attachments including those with Christian, Muslim, Sikh, Hindu, Jewish, Buddhist and other religious identifications.

On the surface the ostensibly white composition of the area may appear to suggest that questions of racism, ethnicity, migration and settlement are peripheral issues to a region that contains the lowest proportion of minority residents in the country. However, the number of migrants who identify as other than white British is increasing more rapidly than elsewhere in the country with an annual growth of just under 10 per cent between 2001–2003. The globalisation of migration is also having a significant impact upon the locality. In recent times the North East has become one of the allocated destination zones for the dispersal of asylum seekers, acting on behalf of the Home Office for National Asylum Support Services (NASS). Asylum seekers constitute 0.2 per cent of the region's population, compared with 0.9 per cent for England as a whole (Census 2001) and the city of Newcastle has the fifth highest number of asylum seekers in NASS accommodation (Home Office 2005). It is estimated that there are 5,170 asylum seekers in the North East accounting for over 100 different nationalities.

The connections between the local and the global is seen here where events happening in one part of the world have a direct impact upon people and place in another as war, terrorism, 'ethnic cleansing' and genocide underlie many forced population movements. The changing portrait of the North East has been further complicated since 2004 with the rise of the Accession states to the European Union. So-called 'A-8' migration from the initial eight former Communist countries has seen new population movements from Baltic states such as Lithuania, nations such as Poland and newly formed states such as Slovakia and Slovenia which were formerly part of Czechoslovakia and Yugoslavia respectively. This is giving way to an increasingly mobile and 'hyper-diverse' population that extends well beyond the older post-colonial BME communities in the North East. We could aptly describe the new terrain as an 'ethnoscape' (Appadurai 1990) comprising an undulating landscape of ever mobile subjects that includes tourists, refugees, asylum-seekers, exiles, guest-workers, temporary and permanent immigrants.

School-based Ethnography

The material on which this chapter is based draws upon ethnographic research undertaken with children and young people in two schools in North East England.[1] Emblevale Middle School is a mixed, multi-ethnic institution of 405 students (aged 9–12 years), around a quarter of which are from black and minority ethnic backgrounds. The majority of these are South Asian (including those from Indian, Pakistani and Bangladeshi backgrounds) with a small fraction of African-Caribbean and East Asian students. Snowhill Comprehensive is a large state school with a predominantly white intake (aged 11–18 years). At the time of the research Snowhill had 1,936 students of whom 1,869 were white British citizens. Observations, interactions and semi-structured interviews with groups and individuals were conducted in each of these sites. The aim was to understand how children and young people feel about whiteness, racism and anti-racism. For the purposes of this chapter it is worth noting that an approximate number of girls and boys were observed and interviewed from 9–10 years, 11–12 years and 16–17 years. The older participants were self-selecting, while form teachers were instructed to provide a diverse and wide-ranging sample derived from the two cohorts of younger students. Interviews were conducted by way of focus group sessions with 3–5 students in each discussion group. The method was an attempt to engender serious engagement on race without recourse to making individuals feel overly uncomfortable. Rather than construct individuals as 'racist' or 'anti-racist' the research adopted a discursive approach to reveal how young subjects have a remarkable propensity to move between racist and anti-racist discourses according to context and situation. This perspective captures the mobility of race relations in young lives and draws attention to contradictory practices. The fluid and malleable nature of these relations also offers potential for anti-racist policy and practice to be effective.

Young People's Perceptions of Racism and Ethnicity

In the mainly white region of the North East whiteness exists as a taken-for-granted social norm. When discussions of race, ethnicity and multiculture take place, the assumption is that it is something to do with visible minority ethnic groups. The young people I spoke with had little conception of themselves as having an ethnicity or being implicated in race-making practices. Instead whiteness was seen as a homogenous category deemed ordinary and unremarkable to students and teachers alike. This meant that few white youth had any investment in the schools' occasional attempts to host multicultural events or celebrations. Moreover, a number of white working-class students I spoke with did not perceive whiteness to be a privilege but rather a cultural disadvantage in classroom contexts.

> Anoop: *Are there any advantages to being white in this school?*
> Nicola: Well, no.
> Michelle: Cos coloured people can call us [names].
> James: It's not fair really cos they can call us like 'milk bottles' and that, but us can't call them.
> Sam: The thing is in this school, is like if you're racist you get expelled or something, but they [black students] can call us names and the teachers don't take any notice of it.
> James: They take no notice. [Group discussion: 11–12 years]

In this case the school's sensitivity to racist harassment appears to bolster white injustice among respondents, and create a feeling that such forms of 'moral' anti-racism are 'not fair' (see also MacDonald et al. 1989; Hewitt 1996). That teachers are said to ignore name-calling from black students, yet expel white students for using racist taunts, affirms a sense of white defensiveness (see Gillborn 1996). This led some white youth to make charges of 'reverse racism' in name-calling disputes with black peers in which they presented themselves as 'victims' of racism.

> Anoop: *So what do the name-callers say?*
> Michelle: Things like 'milk-bottle'.
> James: And 'whitey'.
> Michelle: And 'milky way' and things. [11–12 years]

Such discussions reveal the unspoken grievances that some white youth may harbour and their acute sensitivity to any perceived forms of unfairness. Alongside the opinion that anti-racism was 'unfair' to the needs of white youth, ran an overwhelming feeling that black students had an identifiable culture that they could draw on which was denied to the Anglo-ethnic majority. A positive assertion of this culture by minority ethnic youth would tend to be sceptically interpreted by white students as a deliberate act of exclusion.

Sam: What I don't like is all the Pakistani people all talk in their language and you dunno what they're talking about. Used to be this lad in our class, Shaheed, he would talk to his mate Abdul, half in English, half in another language.

Nicola: If they wanna talk about you they can talk in another language.

Michelle: If we wanna talk about them, they know what we're saying. [11–12 years]

Revealingly white students were keen to make a careful distinction between racism as a discourse of power available to them through regimes of representation (in language, speech, metaphors and imagery); and racism as a 'chosen' subject position that was explicitly ideological and practised in daily, vehement exchanges. Whereas the former stance offered a latent potential for racist enactment, triggered only at certain moments, the latter position was more readily condemned as explicitly racist and anti-egalitarian. It is this 'unevenness' of racism in young people's lives that became increasingly apparent. The grainy line separating what white students said to their black peers in certain situations, and how they felt towards them more generally, became a source of tension when episodes of racism surfaced in classroom contexts. Most specifically in fraught and heated personal exchanges between students, racist name-calling offered an inviting mode of redress.

Sam: We canna say anythin' cos they [black students] can get us annoyed and it's hard not calling them a racist name or somethin'. I never bin racist cos I don't think it's right but some people jus' think it's hard to not call them a racist name if an argument starts.

The student responses listed here question why white racial epithets such as 'whitey', 'milk-bottle' or 'milky way' are not construed as forms of racist name-calling. As other researchers have implied the meanings carried in white, derogatory terms rarely carry the same weight as anti-black racist terminology (Back 1990; Troyna and Hatcher 1992a). Troyna and Hatcher (1992a, 1992b) argue that *racial* insults such as 'white duck' or 'ice-cream' must be carefully distinguished from *racist* terms such as 'Paki', which are saturated with ideological power. In Troyna and Hatcher's (1992a, 1992b) structural definition, it is precisely because black and white students occupy different positions of dominance and subordinance in race relations, that white epithets are considered 'racial' name-calling forms and black epithets are viewed as 'racist' name-calling terms. Here, there is no equivalence between black and white name-calling as ultimately, 'Racist attacks (by whites on blacks) are part of a coherent ideology of oppression which is not true when blacks attack whites, or indeed, when there is conflict between members of different ethnic minority groups' (1992b, 495).

While some students may have engaged in a 'white backlash' against moral forms of anti-racism, others disclosed a more complex understanding of power. Ema, a 16-year-old, white, working-class student explained how if someone used a term like 'black bastard', 'I'd say something and get 'em done', indicating she would report the remark to a teacher. However she continues:

Ema: If someone says, 'She's just called me "white trash"', I'd say, 'And what's wrong with that?' I'd probably think, 'Well maybe it would hurt them, but to me it wouldn't be anything to say "white"'. I'd be proud of it.

Ema makes a qualitative distinction between using a black or white racial epithet before an insult. She indicates that white has a neutral or even positive signification that cannot be easily overturned ('I'd be proud of it'). As Troyna and Hatcher (1992b) would have it, the prefix 'white' does not draw on an historical, 'coherent ideology of racism' (slavery, imperialism, apartheid, discrimination, xenophobia, nationalism) in the ways that a term 'black bastard' might.

Although awareness to the qualitatively different racialised experiences of black and white youth remains pertinent, contemporary definitions of racism have been further extended. While Troyna and Hatcher's definition foregrounds the 'asymmetrical power relations' (1992b, 495) between blacks and whites and is a welcome improvement on liberal, power-evasive models of racism, there remain potential short-comings with this anti-oppressive framework. To begin with, there is an immediate reification of race as an insurmountable point of embodied difference that too readily equates whiteness with power and blackness with oppression. Here, whites are endowed with the privilege of being the central architects of history, and the key agents of social change (Bonnett and Nayak 2003). The multiple positions that blacks and whites may come to occupy – and how these subjective locations are nuanced by class, gender, sexuality and generation – are subsequently condensed into a racial dichotomy of powerful/ powerless. Subjectivities, however, cannot be regarded as fixed, coherent and unambiguous. Furthermore, the tendency to construe racism across a black/ white binary may in turn occlude other examples of racist hostility such as anti-Semitism; the 'ethnic cleansing' that has taken place in parts of Eastern Europe; and the hostility meted out to Romany travellers, asylum-seekers, the Irish, Poles or white skinned Muslims. Indeed, an engagement with whiteness beyond racial polarities may allude to a complex understanding of racism that may invoke aspects of nationhood or religion as further points of discrimination. Placing the issue of racism and ethnicity in a wider context of global change, unequal development and the legacies of 'new' and 'old' diasporic movements suggests that black/white binaries of race, however convenient or strategically useful they may appear in the short-term, are strictly limited in light of new global migrations.

Young People's Perceptions of Anti-racism

Anti-racism has an interesting if somewhat ambiguous place in the history of the North East. From the 1790s heated disputes had arisen concerning the ethics of using sugar in tea on account of human exploitation abroad. In 1823 the Newcastle upon Tyne and Wear Anti-Slavery society formed to combat the injustices that continued after the traffic of slaves was made illegal. A decade later 6,288 women from Newcastle and Gateshead protested for the abolition of slavery in Africa and the Caribbean. After attacks upon Arab seamen in South

Shields in 1919, seamen from the Yemen and the North East stood alongside one another in resistance to the imposition of a rota system for employment that would effectively divide the labour force along 'colour' lines (Lawless 1995). The British Union of Fascists headed by Sir Oswald Mosely and his intimidating 'Blackshirts' may have drawn large crowds in central Newcastle but their activities were restricted by local communities, Tyneside Socialists and a patchwork of anti-Fascist groups. In 1933 large public meetings in Newcastle and Sunderland were observed condemning Nazi anti-Semitism in Germany. The traditional hospitality associated with the North East was witnessed in 1937 when Britain admitted some 4,000 Basque children but offered no state support. A number of these child refugees, displaced from the Basque region as a consequence of the Spanish War, which was later eclipsed by World War II, were dispersed along the Northeast coastline. Here, families in Tynemouth, Gateshead and North Shields aided by the miners' lodge, church groups and other community groups provided the essential financial relief, education and housing refused by the state. Protest marches would later be instigated by the Tyneside Campaign Against Racial Discrimination against Enoch Powell's inflammatory 'River's of Blood' speech in 1968. Today organisations such as the Tyne and Wear Anti-Fascist Association continue to monitor, expose and protest against far-Right candidates in the area. These brief examples offer alternative portraits to the construction of the region as inherently racist by also acknowledging a longstanding tradition of anti-racism. It further suggests that seemingly white preserves such as the North East also contain fascinating if much neglected histories of race, migration and ethnicity (Nayak 2003).

Despite such activism anti-racism is regularly subject to a local and national 'backlash'. In research with anti-racist practitioners in Devon and Tyneside Bonnett's (1993a, 1993b) respondents opted for what he terms a 'softly softly' approach to anti-racism that was highly conscious of offending white sensibilities. This suggests that the types of anti-racist practice which may be successful in a large multi-ethnic metropolis may have little meaning in ostensibly white locales. The fear of 'lashing back' was particularly acute in the late-eighties and early-nineties where a rabid Right-wing press instigated a series of 'moral panics ... orchestrated around "looney-left councils" supposedly banning black dustbin liners, insisting on renaming black coffee "coffee without milk", and banning "Ba-ba black sheep" [a children's nursery rhyme] from the classroom – scares which turned out to rest on complete fabrications' (Media Research Group 1987 cited in Rattansi 1993, 13). These fabrications, however fictional they may be, continue to preside in the geographical imagination and are articulated in the accounts of new generations.

Nicola: I've got this book from when I was little, it's called *Little Black Sambo*. It's got Black Mumbo and Black [giggles] Jumbo.
Sam   Oh, I had that, I used to 'ave that.
Michelle: It's been banned. You're not allowed to say, 'Baa Baa Blacksheep'.
Nicola: And you're not allowed to ask for a black coffee.
Michelle: Aye.

Anoop: Who says?
Michelle: On the news.
Sam: So we go round singin', 'Baa-Baa Multicoloured Sheep'!' [11–12 years]

While the majority of media examples above are taken out of context, exaggerated or simply invented they continue to exert a strong emotional pull on the psyche. They form part of a collective, imagined geography that creates a vision of anti-racism as 'gone mad'. In this rendering anti-racism is conveyed as abstract, arbitrary and entirely ineffectual. It serves to demonise such 'innocent' objects and rituals as nursery rhymes, children's stories and hot drinks. What is also apparent is that anti-racism appears an invasive, regulatory device that censors behaviour and can even result in literature being 'banned'. This notion of a 'politically correct' state censorship that suppresses 'free will' was a recurring theme in children's accounts.

Nicola: There's these dolls you're not allowed to 'ave.
Sam: Gollywogs.
Nicola: Aye. And on the news now [it says] every child has gotta have a black doll.
Anoop: *Hold on, are you saying that every child by law has got to have a black doll?*
Nicola: Yeah, so they grow to accept black people.
James: And black Action Man [male doll].
Michelle: Aye, black Action Man. [11–12 years]

The language utilised in children's descriptions of anti-racist practice suggest they interpret it as a largely proscriptive, negative set of values. They refer, incorrectly, to items that have been 'banned', artefacts you have 'gotta have' and things you are 'not allowed' to say or do. In essence, anti-racism is transformed into a broader 'discourse of derision'. There is little indication that these initiatives have benefits for all and certainly not whites. The feeling among school students is that white ethnicity has to be regulated and British anti-racism is a disciplinary tool for enacting this. At best, anti-racism is random and nitpicking – focusing upon dolls, gollywogs, coffee, nursery rhymes or dustbin liners – at worst it is downright 'unfair' and even prejudice against whites!

James: They don't go on about 'Baa Baa Whitesheep'.
Nicola: That's even more racist.
James: Them [black people] could be banned for buyin' white milk.
Michelle: [laughing] Well ya can't buy black milk. [11–12 years]

Because of the striking manner in which children in Emblevale Middle School viewed these pedagogies I was keen to know if older students in Snowhill Secondary School shared these beliefs. In both institutions anti-racism was consistently perceived as an 'unfair', anti-white practice. Some white youth in Snowhill even regarded themselves as the primary victims of racism.

Anoop:	Do you think that blacks can be as racist as whites?
Lucy:	I think it works both ways.
Chris:	More, I would say.
Anoop:	More racist?
Chris:	Er, they're like bitter against the way they've been treated like slavery and that. They feel like somehow they've been hard done by.
Lucy:	It works both ways. [16–17 years]

The perception that blacks are 'bitter' against whites and feel 'hard done by' draws upon a familiar trope that gives way to the belief that they have a 'chip on their shoulder'. As with younger Emblevale students there is a belief that racism 'works both ways', it is something that blacks and whites commit alike. Chris, whom we heard from above, went on to describe the 'reverse racism' he perceived in multi-ethnic cities such as Leicester. 'There's a lot of racism', he recounted, 'but it's like different, it's from the blacks against the whites, you know what I mean?'

In my discussions with children and young people, anti-racism and multiculturalism were regarded as invasive, regulatory devices. Rather than eradicating inequality they are seen as discriminatory technologies. Instead of being pedagogies of liberation they are regarded as modes of oppression. Ultimately, the fact that white youth felt alienated from the overarching philosophy of anti-racism meant that it had little chance to succeed or make any lasting inroads upon their public consciousness. This was not because young people held entrenched racist opinions but rather because the message had not been made meaningful to them and the history and geography of living in a mainly-white area. Furthermore, some white working-class youth are likely to view anti-racism as an irrelevance or even an assault upon their identities. Interestingly the word anti-racism is seldom found in current government policy where more opaque terms such as 'community cohesion' have come to the fore. In order to prise open some new geographical imaginings – and reveal the relevance of anti-racism – I shall attempt to develop some alternative cultural pedagogies of place.

Cultural Pedagogies of Place

To challenge widespread perceptions of anti-racism as an anti-white practice I want to invoke one of the core themes in human geography: people and place. By exploring these issues possibilities exist to develop cultural pedagogies that make anti-racism meaningful to young people. While a number of teachers in the schools where I did the research tended to treat multiculturalism as the celebration of Otherness,[2] I found it more productive to place whiteness, local history and social class at the heart of anti-racist projects. As Raphael Samuel has noted, 'Local history also has the strength of being popular ... People are continually asking themselves questions about where they live, and how their elders fared' (1982, 136–7). Embracing the popular in this way may entail a clearer understanding of the cultural specificities of white, English ethnicities, and engender a perspective that is more sensitive to marginal working-class experiences.

Exposing the diverse histories of white students could be a productive exchange where the immediate heritage of respondents is discussed. The recognition that 'English' identities had changed over time allowed these students to feel less threatened by the prospect of black British settlement. Projects directed by young people which draw on familial life-history accounts may be of particular interest to teachers and pedagogues, who may wish to share their own racially inscribed biographies. Encouraging students in mainly white preserves to sensitively trace their ethnic and social-class lineage remains a fruitful way of deconstructing whiteness, as Catherine Nash (2003) has shown through cultural geographies of relatedness related to genealogy. However, we should remain mindful that firstly, 'no strategy is likely to be completely successful, and second, that an effective strategy in one context, may fail in another context or at another time' (Gillborn 1995, 89). With these provisos in place, the approach deployed is sensitive to the local culture of the community and subject to my particular relationship with students. I found that imploding white ethnicities offered a way of contextualising anti-racism, and helped to develop an interest amongst students in race relations they felt they could have a personal stake in. Such a pedagogy is imperative for as Stuart Hall has noted, 'We are all, in that sense, *ethnically* located and our ethnic identities are crucial to our subjective sense of who we are' (1993, 258). The value of cultural pedagogies of place became known when new inflections upon an assumed, coherent English ethnicity began to unfold. While most students identified as part of the Anglo-ethnic majority at the beginning of the research the monolithic conflation of whiteness with Englishness became increasingly difficult to sustain against this rich tapestry of interweaving personal and familial histories and geographies. In discussions of belonging and identity students reflected upon their own local heritage and connected this up to a more 'global sense of place' (Massey 1991).

Danielle: Mine parents were born in Germany, cos me nanna used to travel o'er
            abroad.
Brett: I used to have an Italian granddad.
Alan: Me next name's O'Maley an' that isn't English.
[...]
Nicola: I tell yer I'm English, but I'm part German. My granddad came over as a
            Prisoner of War, he was working over at Belsey Park and my grandma was
            teaching.
[...]
James: Some white people have got black people in their family. Like say my aunty
            married a black person and had babies.
Michelle: I've got one in my family.
[...]
Sam: I'm a quarter Irish, a quarter Scottish, a quarter English and a quarter
            Italian. [11–12 years]

The responses generated indicate that the implosion of white ethnicities offers alternative geographical trajectories that many students hold a fascination with. Tracing their familial past is a means of personalising geography, making it relevant to their life experiences to date. In the course of this process it was not unusual for students to also refer to generational elements of racism within their family lineage. Many students mentioned parents or grandparents with pronounced racist opinions, allowing for further points of critique and discussion between young people. Although many white people may lay claim to the identity of white-Englishness, the narratives illustrate how these ethnicities are discursively constituted in the present situation. Shattering the impermeable veneer that binds whiteness to Anglo-identity enables students to view themselves as 'post-race' subjects with a genuine stake in anti-racist and multicultural issues (Nayak 2006). The deconstruction of white identity became, then, a means of splicing Englishness, whiteness and ethnicity.

Contemporary debates on cultural hybridity have tended to focus upon the dynamic and rhizomatic exchanges between black and white cultures. However, there is a danger in reifying blackness and whiteness as 'pure objects' from which other actions proceed so that it is only in the complex melange of semiotic interplay that we see produced new hybrid forms. Instead, it is possible to consider these race signs as 'always and already' hybrid. For example, although many young people self-presented as 'white English' closer scrutiny often revealed a mixed and textured account of ethnic belonging. Many students reflected upon the fragmented, 'hyphenated' identities of being say, Anglo-Irish, Scotch-Irish or Anglo-Italian. Englishness could also be deconstructed, for example when Alan (10 years) pointed to the hybrid history of English identity remarking, 'It's all a mixed breed [*sic*] in England cos we've had the Vikings and the Saxons come across ... France, Denmark, them places'. In turn I could further share knowledge of the locality by discussing the longstanding migrant settlement in the region and the resistance to Fascism during the interim war period and beyond. Conversations concerning local identities – 'Geordies', 'Makums', 'Charvers', 'Smoggies' (see Nayak 2003 for further discussion) and so on – along with debates on the labouring heritage of the area appear capable of captivating young people and providing new insights on place and identity.

With older students it is even possible to engage in a critical dialogue with whiteness. Ema and Jolene (both 16 years) each identified as working-class young women. Ema's father was in the army and during the fieldwork period she too decided to sign-up in preference to completing her Sixth Form education. Jolene's father, meanwhile, had an equally masculine occupation that reflected the depleting infrastructure of the local post-industrial economy, working as a bailiff. Both Ema and Jolene engaged in an appraisal of their own racial identities, which at times disrupted the association of Britishness/Englishness with whiteness.

Jolene: There's different colours of white.
*An*oop: What d'you mean?
Jolene: Like Chinese. Do you know what I mean, what colour are they? There isn't
           a colour – we're not proper white.

Ema:	Different shades really.
Jolene:	There's Chinese; there's other people; there's us; naturally dark skins who are white.
Ema:	People say like, 'I'd hate to be black' and everything but when they go on the sun-bed and get tanned, they love to be tanned.
Jolene:	Yeah! People go on the sun-bed just to get browned.

The critical deconstruction of whiteness undertaken by Ema and Jolene fractures the homogenous conception of white identity with the knowledge that 'whites' are comprised of 'different shades really'. Instead of seeing white as colourless (Dyer 1997), the young women introduced a wide spectrum of colour symbols which at its most extreme included bronzed, sun-tanned figures who still manage to 'claim' the elusive emblem of whiteness. As Bonnett asserts, 'To subvert "blackness" without subverting "whiteness" reproduces and reinforces the "racial" myths, and "racial" dominance, associated with the latter' (1996, 99). The question of what 'colour' Chinese people are, further disrupts the fixed polarisation of race as a discourse shared solely between black and white citizens. Indeed, Dikotter (1992) has shown how Chinese people at particular moments identified as 'white' before interactions with Western powers constituted them, precariously, as 'yellow'. Similarly, in their respective research, Ali (2003), and Song and Parker (1995) have shown how mixed race identities can be configured in oscillating ways that may give rise to 'post-race' identifications, passing or misrecognition.

In these readings whiteness is not an accurate mode of racial classification but an arbitrary sign that varies over time and place (Nayak 2007). The social construction of whiteness is also made apparent where students recognise that strictly speaking they are not 'proper white' (whatever that might be). Using alternative cultural pedagogies of place, based on local histories and geographies, has the potential to transform anti-racism from a purely symbolic to a material practice. Locating anti-racism within the geographies of young people can elicit the undoing of race categories. Deconstructing the identities of ethnic majorities with as much purpose and vigour as that of minority groups remains a vital component of anti-racist practice in the changing local-global context of mainly-white areas.

Notes

1   The research was conducted as part of my doctorate research on whiteness and white identities undertaken in the years between 1995–1999. A full methodological description of the work can be found in Nayak (1999, 2003). Pseudonyms have been used to protect the identities of people and places.
2   For example when the school held special multicultural events this included Indian cookery demonstrations, henna dying and the playing of South Asian musical instruments. Although well-meaning in their intention, these colourful demonstrations tended to construe ethnic minorities as inhabiting a culture that is set apart, different, exotic and self-contained. This fixed presentation of 'culture' reifies difference and

suggests that multiculturalism exists outside the mainstream and is essentially about non-white Others.

References

Ali, S. (2003), *Mixed-Race, Post-Race: Gender, new ethnicities and cultural practices* (Oxford: Berg).

Amin, A. (2002), 'Ethnicity and the Multicultural City: Living with diversity', *Environment and Planning A* 34, 959–80.

Appadurai, A. (1990), 'Disjuncture and Difference in the Global Economy', *Theory, Culture and Society* 7, 295–310.

Back, L. (1990), *Racist Name-calling and Developing Anti-racist Initiatives in Youth Work* (Coventry: Centre for Race and Ethnic Relations, University of Warwick).

Back, L. (1996), *New Ethnicities and Urban Culture: Racism and multiculture in young people's lives* (London: UCL Press).

Bonnett, A. (1993a), *Radicalism, Anti-Racism and Representation* (London: Routledge).

Bonnett, A. (1993b), 'The Formation of Public Professional Radical Consciousness: The example of anti-racism', *Sociology* 27:2, 281–97.

Bonnett, A. and Nayak, A. (2003), 'The Territory of Race', in K. Anderson, M. Domosh, S. Pile and N. Thrift (eds) *Handbook of Cultural Geography* (London/California/Delhi: Sage).

Burgess, J.A. (1985), 'News from Nowhere: The press, the riots and the myth of the inner city', in J.A. Burgess and J.R. Gold (eds) *Geography, the Media and Popular Culture* (Kent: Croom Helm).

Derbyshire, H. (1994), *Not in Norfolk: Tackling the invisibility of racism* (Norwich: Norwich and Norfolk Race Equality Council).

Dikotter, F. (1992*)*, *The Discourse of Racism in Modern China* (London: Hurst).

Dyer, R. (1997), *White* (London: Routledge).

Dwyer, C. (1993), 'Constructions of Muslim Identity and the Contesting of Power: The debate over Muslim schools in the United Kingdom', in P. Jackson and J. Penrose (eds) *Constructions of Race, Place and Nation* (London: UCL Press).

Dwyer, C. (1999), 'Contradictions of Community: Questions of identity for young British Muslim women' *Gender, Place and Culture* 6:1, 5–26.

Gaine, C. (1987), *No Problem Here* (Stoke on Trent: Trentham Books).

Gaine, C. (1995) *Still No Problem Here* (Stoke on Trent: Trentham Books).

Gillbourn, D. (1995), *Racism and Antiracism in Real Schools* (Buckingham: Open University Press).

Gillborn, D. (1996), 'Student Roles and Perspectives in Antiracist Education: A crisis of white ethnicity?', *British Educational Research Journal* 22:2.

Gilroy, P. (1987), *There Ain't No Black in the Union Jack* (London: Hutchinson).

Hall, S. (1993), 'New Ethnicities', in J. Donald and A. Rattansi (eds) *'Race', Culture and Difference* (London: Sage/Open University Press).

Hewitt, R. (1996), *Routes of Racism: The social basis of racist action* (Stoke-on-Trent: Trentham).

Home Office (2005), *Asylum Statistics: 4th Quarter 2004 United Kingdom.*

Hopkins, P. (2007), 'Young People, Masculinities, Religion and Race: New social geographies', *Progress in Human Geography* 31:2, 163–77.

Jackson, P. (1988), 'Street Life: The politics of carnival', *Environment and Planning D: Society and Space* 6, 213–27.

Jay, E. (1992), *Keep Them in Birmingham: Challenging racism in the South West* (London: Commission for Race Equality).

Keith, M. (1991), 'Policing a Perplexed Society? No-Go Areas and the Mystification of Police-Black Conflict', in E. Cashmore and E. McClaughlin (eds) *Out of Order? Policing Black People* (London: Routledge).

Keith, M. (1993), *Race, Riots and Policing: Lore and disorder in a multi-racist society* (London: UCL Press).

Keith, M. (2005), *After the Cosmopolitan: Multicultural cities and the future of racism* (London: Routledge).

Lawless, R.I. (1995), *From Ta 'izz to Tyneside: An Arab community in the North East of England during the early twentieth century* (Exeter: University of Exeter Press).

MacDonald, I., Bhavnani, R. Khan, L. and John, G. (1989) *Murder in the Playground: The Burnage Report* (London: Longsight).

Massey, D. (1991), 'A Global Sense of Place', *Marxism Today*, June 1991, 24–9.

McGuinness, M. (2000), 'Geography Matters? Whiteness and Contemporary Geography', *Area* 32:2, 225–30.

Nash, C. (2003), '"They're Family!": cultural geographies of relatedness in popular genealogy', in S. Ahmed, C. Casteneda, A.M. Fortier and M. Sheller (wds) *Uprootings/ Regroundings: Questions of home and migration* (Oxford: Berg).

Nayak, A. (1999), 'White English Ethnicities: Racism, anti-racism and student perspectives', *Race, Ethnicity and Education* 2:2, 177–202.

Nayak, A. (2003), *Race, Place and Globalization: Youth cultures in a changing world* (Oxford: Berg).

Nayak, A. (2006), 'After Race: Ethnography, race and post-race theory', *Ethnic and Racial Studies* 29:3, 411–30.

Nayak, A. (2007), 'Critical Whiteness Studies', *Sociology Compass* 1:2, 737–55.

Nizhar, (1995), *No Problem Here: Race issues in Shropshire* (Telford: Race Equality Forum for Telford and Shropshire).

Peach, C. (Ed.) (1975), *Urban Social Segregation* (London: Longman).

Phillips, D. (2006), 'Parallel Lives? Challenging Discourses of British Muslim Self-segregation', *Environment and Planning D: Society and Space*, 24, 25–40.

Rattansi, A. (1993), 'Changing the Subject? Racism, Culture and Education', in J. Donald and A. Rattansi (eds) *'Race', Culture and Difference* (London: Sage/Open University Press).

Robinson, F. (1988), *Post-industrial Tyneside an Economic and Social Survey of Tyneside in the 1980s* (City of Newcastle upon Tyne, Newcastle).

Samuel, R. (1982), 'Local History and Oral History', in R.G. Burgess (ed.) *Field Research: A sourcebook and field manual* (London: George Allen and Unwin).

Smith, S. (1987), 'Residential Segregation: A geography of English racism', in P. Jackson (ed.) *Race and Racism in Social Geography* (London: Allen and Unwin).

Song, M. and Parker, D. (1995), 'Cultural identity: Disclosing commonality and difference in in-depth interviewing', *Sociology*, 29:2, 241–56.

Troyna, B. and Hatcher, R. (1992a), *Racism in Children's Lives: A study of mainly white Primary schools* (London: Routledge/National Children's Bureau).

Troyna, B. and Hatcher, R. (1992b), 'It's Only Words: Understanding "racial" and racist incidents', *New Community* 18:3 493–6.

Watt, P. (1998), 'Going Out of Town: Youth, race and place in the South East of England, *Environment and Planning D: Society and Space*16, 687–703.

Chapter 20

Investigations into Diasporic 'Cosmopolitanism': Beyond Mythologies of the 'Non-native'

Divya P. Tolia-Kelly

Introduction

In this chapter my aim is to investigate the arguments for a cosmopolitan turn in the social sciences when researching and writing on race and racialised peoples. When researching the UK's South Asian diaspora, the embracing of a definition of 'cosmopolitan' for this community is critical in challenging categorisations of race and definitions of 'native' and 'non-native'. All of these categorisations rely heavily on cultural definitions within the social and natural sciences that define taxonomies in both the realms of human and non-human species. The philosopher Lorraine Code (2006) has argued for 'ecological thinking' in respect of our ethical and social approaches to research with the 'other'. In particular it is important to acknowledge the historical and ethnocentric nature of the epistemologies we use in social sciences research and research dissemination. Code demonstrates the need for responsible taxonomies. In this chapter I investigate the possibilities for responsible taxonomies within the discourses of 'native and non-native' species as well as developing those which are truly against 'race' (Gilroy 2004). I examine these possibilities by engaging with Ulrich Beck and Natan Sznaider's call for a 'cosmopolitan turn' (2006) in the social sciences and considering this notion in two pieces of research; one situated in the arboretum landscape in Burnley and the other in a South Asian home. Through these sites I will argue that 'cosmopolitanism' as a discourse and definition is an appropriate term for defining the political positioning of diasporic populations living in the UK. This chapter is an investigation into challenging the usual categories within which racialised populations in Britain are framed and thus metaphorically locked into limited possibilities spatially, socially, culturally and academically.

Virtue in Practice: Redefining *Cosmopolitanism*

For many social scientists cosmopolitanism is 'an intellectual and aesthetic stance of openness towards divergent cultural experiences' (Hannerz 1990, 239). It is intended as an orientation towards acknowledging in a positive way, the nature of exchanges and experiences across cultures, borders and societies. However, the lens of 'cosmopolitanism' is a term that often has a limited field of encounter and yet has been treated optimistically by key authors (Beck 2000; Cheah 1998). It is frequently used to narrate the phenomenon of intensely mobile individuals or groups, whose practices and narratives operate in a cultural mode in the public sphere, which are at the sites of global cities and metropolitan capitals (Binnie and Skeggs 2004). The cosmopolitans we are drawn to in social science research, are globally migrant figures traversing in apparently evenly globalised communication, transport and cultural networks. Often these figures are non-marked (sexually, racially or through physical ability), masculine individuals enjoying the freedom to traverse, shape, be affected and affective in their new geographical and social spheres of contact. Cultural theorist and feminist, Mica Nava argues that some formulations of a 'cosmopolitan identity remain Eurocentric, and privileged and that:

> Much of the recent debate about cosmopolitanism has been concerned with questions of cosmopolitan democracy and global governance or with travel and migration. The historical trajectory of the cosmopolitan imagination and vernacular expressions in everyday local life and culture has, on the whole, been neglected. (Mica Nava 2002, 81)

Nava's argument is made with the aim of challenging distinctive western imaginaries in this realm of understanding cosmopolitanism beyond these narrow conceptualisations and to engage with narratives and practices around 'difference'. 'Transnationalism' (see for example Jackson et al. 2004) has been a new way of considering mobility, race and networks in a culturally fluid and globalising world. It is a process of envisioning connections, networks, material and cultural exchanges *between* nations and peoples. However, this conceptualisation of transnationalism has retained a notion of 'difference' that is situated within a western lens of often ossified characterisations of 'national' or 'community' practices (see Modood 1990; 1994). These ossified characterisations of 'other' cultures risk reasserting 'difference' that resonates with a neo-orientalist lens of categorising *difference* upon global cultures and practices. Thinking 'difference' through an account of cosmopolitanism that embraces transnational peoples as exemplifying of cosmopolitan communities and as having 'cosmopolitan virtue' (Turner 2002), challenges the stereotypes of diasporic communities as victims of geographical ghettoisation, and thus, immobile, and politically, socially and culturally homogeneous citizens of post-Imperial political rule. Instead a framework of those communities as 'cosmopolitan' enables us to transcend notions of singularity, immobility and a stereotyping of these communities as culturally ossified racialised groups; it also challenges euro- and ethno-centric interpretations of the term.

Both of the diasporic groups that I have engaged with in my research are members of the South Asian Diaspora; the British Muslim community in Burnley and the South Asian community in London. Both communities are figured both in academic accounts and within public media as seemingly homogeneous and immobile within the narrative of 'Asian' or 'Muslim' especially post the 9/11 attacks (see Amin 2002). What a critical engagement with cosmopolitanism can offer is a conceptualisation which does not deny or override 'the complex structures of local heritage and what are perceived as the more progressive values of nationhood' (Nava 2002, 87). The two communities engaged with in this chapter are examples of diasporic communities in the UK that have complex relationships with the nation as post-colonial citizens of the British Empire and as citizens of many other states and locales. Their historical situation through heritage, memory and social and political identities is embraced through my re-defining them as cosmopolitan citizens. This redefinition is based on the challenges made by Beck (2002), Turner (2002) and Clarke (2002) respectively on the complacency and exclusionary nature and use of the term 'cosmopolitan' in social science research. Beck himself has a greater political project in mind; an effective cosmopolitan world politics, based in a transnational framework. However, for a successful truly internationalist approach to culture and politics we do need to address the inherently parochial orientation of contemporary cultural theory on race and identity; hampered by a *methodological nationalism* (Beck and Sznaider 2006, 3). In Keith's (2005) view the means of de-centring this epistemological constraint and to extend our ways of knowing, seeing and experiencing an alternative ontological eye, is through situating the *cosmopolitan* and using a critical lens of urbanism to produce an 'analytical and theoretical object' which is free to transcend 'national boundaries' that are assumed to be central to citizenship and practices of belonging. This is a critical approach in thinking about diasporic citizenship in the UK.

A cosmopolitanism in research praxis has also been highlighted as a necessary step in doing truly 'international' research. What I have tried to do in this research is to take seriously Beck's concerns about an inherent *methodological nationalism* (2006) within research. For example research on 'community', 'ethnicity', 'national governance', 'the state' and 'cultural politics' often is embedded within an ontological and epistemological commitment to a narrow *national* sensibility or politics. For him *methodological nationalism* it is not a methodology based on an explicit nationalism but a socio-ontological given in social science research; one that 'equates societies with nation-states'. As he argues:

> A national view on society and politics, law, justice, memory and history governs the sociological imagination. (Beck 2006, 5)

This national perspective, embedded in the structures of thought, theory and method requires a turn to a 'cosmopolitan social science' which

> entails the systematic breaking up of the process through which the national perspective of politics and society, as well as the methodological nationalism of political science, sociology, history and law, confirm and strengthen each other in their definitions of reality. (Beck 2006, 5)

A lack of a reorientation towards an *internationalist* sensibility risks a parochial and defunct social science research contribution. A cosmopolitan research practice as well as cultural lens is thus required.

In my research I use this lens to figure the social groups that I am researching but also to challenge notions of racialised taxonomies of 'native' and 'non-native' which are deeply embedded in cultural geographies of Britain and the structures of feeling around national identity. Cultural theory remains one of the critical disciplines used to challenging prejudicial scientific definitions of race groups – Gilroy names this the re-emergence of a *biologism* (2004), based on a colonial affirmation of citizenship, nation and civilised cultural categories that *belong* or not to a British iteration cultural *nationalism*. These values are deeply embedded in the histories of anthropology, geography, biology and the natural sciences. My first case study will show how a cosmopolitanism has been embraced to challenge the mythologies of race and notions of the 'native' and 'non-native'.

Mythologies of the 'Non-native'

In the UK the national culture of England and Englishness have dominated the traditionally singular national story of Britain; resulting in a culturally and morally exclusive cultural discourse. This is a cultural phenomenon itself embedded within the mythologies of 'nation' (Mitchell 2001, 277). The 'national' landscape is central to constructions of Englishness, which in turn becomes implicated in certain moral geographies of rurality that serve to exclude (Agyeman and Spooner 1997; Matless 1998; Neal 2002) and marginalise 'mobile' citizens such as those not defined as 'rooted' and 'bounded' within the national discourse(see Kinsman 1995; Tolia-Kelly 2006). Excluded are the landscape values of the translocal, mobile, British citizen. Racialised British populations, including Afro-Caribbean, British Asian, Chinese, Eastern European people, are commonly understood as residing in urban spaces, rather than as part of the Romantic landscape space of national parks (Darby 2000).

These ethnicised communities are of course here as a result of traversings across several continents, nations and represent a diverse community with heterogeneous relations with nations and the British Empire. They are representative of dialogic communities with active relationships with nations and cultures across the globe. What is overlooked in contemporary references to particular identities is 'the dynamic interconnection between identification and distantiation – between desire and repudiation – which are complexly at play in the production of the cosmopolitan imagination' (Nava 2002, 88). What is often focussed on is the psychosocial experience of not-belonging, exclusion and marginality. What is not often engaged with is the positive enfranchisement with other nations, landscapes, international networks of 'modernity'. The dynamism of a diasporic community, their connection with the global articulations of modernity and their role in shaping the cultures of modernity need to be embraced in new conceptualisations of race and translocality (Appadurai 1996). Considering diasporic communities in this way acknowledges that socially localised communities become *internationalised*. These communities also have varied roles in shaping and affecting the geographical landscape in which translocality situates them (see Conradson and McKay 2007).

Spaces of Cosmopolitanism A: Lilly's Home

For diasporic cultures Turner argues that 'these global communities require ironic membership if the modern world is to escape from the vicious cycle of ethnic conflict and retribution – 'characterised by cool loyalties and thin patterns of solidarity' (Turner 2002, 58). Are 'cool loyalties' such as a lesser commitment to a 'homeland', 'religious constituency' or 'nation state' possible for Diaspora positionings? A sense of security is essential and a sense of hope for cosmopolitan, political, cultural and natural laws. My first example is one of Lilly's home.[1] This is positioned as a site of cosmopolitan living which through the textures and fabrics present in the home, operates as a site which holds evidence of many other citizenships and many other practices of 'being at home' which have occurred as part of being a member of a cosmopolitan diaspora. For Lilly, a sense of security is enabled through the material cultures that she has in her home. Lilly was born in India in the 1950s and has been a resident in over a dozen different countries including Canada, America, Germany, Thailand, Malaysia and Iraq. For Lilly, her home is a collection of artefacts that link her to a culture of living in all of these countries, not simply as souvenirs of these places but a collage of things that she has taken to each place to make her feel 'at home', and enfranchised. These are materials that resonate with security and confidence and have kept her strong in alienating circumstances as an 'other' body out of place abroad and in the UK.

In previous research I have focussed on material cultures of South Asians including Hindu mandirs (Tolia-Kelly 2004). I have argued that individually, these are dynamic sites whose content, size and aesthetics shift over time. The shrines hold religious relics such as water from the Ganges, and vibhuti – the sacred dust from incense burning from pilgrim sites in India or from other temples, chunis – small pieces of embroidered cloth used on statues (murtis) from temples. Murtis themselves have been purchased sometimes at sites of pilgrimage, the Asian cloth shops, grocers, or even sweet shops. Alongside these are chromolithographs (Pinney 1995) – vividly coloured, paper images of icons. Here, I will argue that the presence of these shrines, contribute to heritage-practices and a sense of 'cosmopolitan identity'. The images in the shrines are iconographic significations locked into a connection with place, therefore they act as a religious sign for Hinduism and a sign for a territorial sense of being, belonging and citizenry.

In Lilly's home there are several material objects that operate as an iconic symbol of her parents' home and her homes in India, Canada, Thailand and Iraq. The visual grammars and aesthetics, with which each object signifies a particular space-time shifts and is dynamically linked to her present sense of identity; through biological history, heritage or cultural nationalism. Through having these characteristics, any individual object or image moves between encompassing religious, spatial, and historical iconography (Pinney 1997, 111). For Lilly these signs are part of a collective, visual vocabulary for her sense of belonging both to the British Asian community, her social status there, and her sense of being an international player with business and family connections all over the globe.

On arriving in Lilly's home I asked her to point out three objects that were important to her sense of 'home' in Britain. The first object that Lilly talks about is an artificial mini palm plant. I was really surprised to be taken to a plastic plant. But for Lilly this was the heart of the family home. A long time ago, in the 1980s after moving to Canada, Lilly's removal firm lost the mini 'Palm France', she recalls the distress of losing it and trying to find a replacement.

> I couldn't find it in Montreal. I looked for it in Toronto, I couldn't find and finally in Ottawa I found. And there were two, so I took two.
>
> It's a Polynesian favourite and this is a tree. I used to have loads in my father's farm and our house also. And wherever I went if I could I planted those trees 'fresh'. In every garden I made a similar corner, otherwise I would have this artificial one. It's a palm fern. You know a mini-palm? (Personal interview, Lilly, 27 March 2000)

The value of Lilly's possessions is linked to her diasporic sense of disconnection and loss of leaving India and are part of celebrating new homes and identities. They form a body of cultural heritage for Lilly and her family. The objects that Lilly encounters in her home each day are the remnants, the physical debris of both biographical and social heritage. They are solid, tangible points of connection to the past.

Figure 20.1 Lilly's shrine

Source: Author's own collection.

My interest in the shrine is the way that it is a site of collecting, a site where memories are collated and stored. The shrine is a valuable site, whose form fits the experience of migration. Shrines are dynamic; they allow growth of a collection of pieces that are sacred and blessed. These are not limited to sacred religious objects and icons. They are considered to be at the heart of family relations. Sometimes they are literally sites where the family genealogy can be traced. The shrine becomes collaged, continually superimposed with objects reflecting intimate moments and sacred life moments which are preserved and treasured. This process of collage is an accrual of the sacred, and that which is emotionally valued by the family. Embroidered onto these textures of the object is a set of relationships between biographical and national and/or cultural identifications. Over time the shrine accumulates layers of meaning, it is believed to emanate protective vibrations; it is a spiritual place first before it is a purely religious site. Its significance grows with time along with its representativeness of family biography; the objects form a significant collage of embedded events, moments, and aesthetic imprints.

Within the home Lilly has other *sacred* objects, sacred not for religious reasons, but to her sense of belonging to the world. There is a golden face of Buddha. For Lilly, Buddha recreates India, her adolescence and connects her to all her other homes.

> ... Buddha, I bought it in 1983, in Bangkok. I was a philosophy student, my main subject was Buddha doctrine. I really believed in him. And it was just so serene. We always take him wherever we travel. I bought him because I grew up with this type of furniture. This is like my childhood home, carved furniture ... we had our house like this when we were working in the Middle East. And Buddha again comes from my parents, at the moment they have it outside in the garden. It's a sitting Buddha, it's this big bronze. You know you like see all these things and I think ... I think like you're trying to connect, to home basically. (Personal interview with Lilly, 27 March 2000)

Through these material cultures Lilly expresses both her connections with a cosmopolitan *Englishness*, expressed through diasporic cultures, embedded within the cultures of everyday living. What is being formed in her home is a citizenry that accumulates, both bounded to a cultural nationalism and a mobile, empowered cosmopolitan identity. These new formations are linked dialectically to both past modes of engagement with cultures of being and belonging abroad and now at *home*. This is not a new mode for the culture of post-colonial citizens, which is, and has been, situated throughout the colonial territories and rememorialised through experiences in the physical landscapes of England. Mobile bodies, imaginaries and cultures of landscape have historically shifted and shaped cultures of cosmopolitan living. The aesthetics, histories and cultures of cosmopolitanism of the British Asian Diaspora stretch out to all the spaces of the colonies and beyond. Their cultural and citizenry practices are locked into a continuing process of transformation. Thus, diasporic*ness* has been historically mobile, empowering and retained currency through a cultural politics of being 'mobile' and 'modern'. Cool loyalties for Lilly herself, and these communities are a necessary part of living, being and negotiating the contemporary geopolitical scene, thus making them truly cosmopolitan beings in a form of *dynamic stasis* in the UK.

Spaces of Cosmopolitanism B: Placing the *Non-Native* in the Burnley Millenium Arboretum

The Black Environment Network and Non-native Cultures of UK Natures

The Black Environment Network is an organisation that is committed to raising issues of race equality in environmental realms and has a programme of 'Ethnic Environmental Participation'. The Network's interests extend to critiquing cultural and natural categorisations which are disingenuous to British landscape history and which may exemplify questionable practices that constitute 'ecological racism'. In their 'Trees of Time and Place Campaign' they warn against the casual use of the terms 'native' and 'alien' as they resonate with racist rhetoric that situates the racialised populations of Britain equally discordant and repulsive to a *national* landscape:

> In an era in which heart-rending pieces of news inform us of racist murders and ethnic cleansing, putting mortal fear into the hearts of every ethnic person, the distorted associations with these terms are extremely undesirable (Wong 2007)

This is an example of an environmental organisation, embracing a 'cosmopolitanism' philosophy and practice towards human-nature relations in the British landscape scene.

As an advocator of natural and cultural cosmopolitanism, cultural theorist, Bryan Turner (2002) argues that to achieve this sense of cosmopolitanism in the modern world, requires a 'scepticism and distance from one's own tradition' forming the basis of 'an obligation of care and stewardship for other cultures'. Turner's definition if applied to a national landscape culture would be one where in Britain we would attend to evolving a cosmopolitan attitude to cultural landscape, nationality and essentially embrace those citizens and natures classified as 'non-native'. It would be one that considered 'mythologies' of landscape and nature within the realms of cultural classifications and different senses of time (see Hirsch and O'Hanlon 2006; Garner 2004). The millennium arboretum at Burnley is a site where that care and stewardship has been embraced on the small scale, by arboriculturalist, Phil Dewhurst. Phil Dewhurst was appointed for three years by the local authority to oversee this government funded 'millennium' project as the Burnley town aboriculturalist in 1999. The now planted arboretum at Burnley is a landscape of *organic cosmopolitanism* (see Tolia-Kelly, in press) where notions of circulation of landscape values and natures are embedded in the ideological foundations for the design and practices of creating the arboretum. This arboretum situates notions of *planting* cosmopolitanism within a multi-racial district of Britain which is not normally considered within the realms of 'studying cosmopolitan landscapes' (Söderström 2006, 553).

The arboretum has become the site of urban landscape where a cosmopolitan vision of landscape, ecology and culture is growing within the landscape of Burnley. It is a material artefact which in its dynamic form, is an attempt to challenge bounded notions of citizenship, through enrolling memory and the transnational ecological values of contemporary Burnley folk. I conclude with a re-evaluation of cosmopolitanism in the form of the arboretum and its aim to garner civic pride amongst the Burnley community.

The British Muslim community, from Burnley in Lancashire,[2] has been doubly displaced as a result of British colonial rule in India and what is now Pakistan. During the 1947 partition of India, this group was forcibly moved from Gujarat state in North West India to the Himalayan foothills of Pakistan (also now named 'Gujarat'). These Burnley Muslims have been 'British' from birth, holding UK passports as a result of residency in British India. Their sense of British identity has also been shaped by movement to northern Britain since the 1960s. Particular families came to the northern textile towns around Lancashire, including Manchester, Leeds, Halifax and places such as Burnley. During the 1970s many of these textile mills closed down, leaving a residue of unemployment, poor housing and limited infrastructure. More recently, in 2001, these communities experienced racial tension and increased socio-economic marginalisation (see Amin 2002). Dewhurst describes his experience of racial tensions in Burnley and his belief systems around human nature itself and a framework of thinking that incorporates a philosophy embedded in ecological time-lines. The arboretum is devised and designed within these 'structures of feeling' that operate beyond narrow definitions of *English* landscape. Through materially connecting environmental memories and citizenships through the tree planting process, the arboretum has aimed to bridge societies within Burnley as part of the programme of *racial and community cohesion*.

In Burnley itself there are moves to repair tensions between 'white' and 'black' communities and to challenge the basis of racist violence which is embedded in territorial language of belonging and non-belonging. The results of these tensions were experienced by the Burnley community in the riots of 2001. There have been several government initiatives to address these rifts. There is currently an emphasis on community cohesion which is in itself problematic, in that it assumes fixed minority ethnic identities. Amin (2004) has argued that prosaic sites of cultural exchange and transformation are critical sites of challenge to the limited narrations of the 'White legacy' of national belonging in Britain. Alongside prosaic sites there is a need for 'plural and contested senses of place' (Amin 2004, 959) which contribute to the politics of ethnicity and identity. The Burnley millennium Arboretum initiative is a site where these come together. Since the riots in Burnley, Oldham and Bradford the government has set up various initiatives to increase racial cohesion in the form of 'cultural' rejuvenation (see Amin 2002, 2004) rather than economic investments in housing stock, infrastructures, public transport, and entrepreneurship. In Danes House ward, a key site of tensions during the riots and at the heart of Asian community, there is a significant level of poor quality social housing, and low levels of social and community facilities. It is Phil Dewhurst's vision that community cohesion can be addressed and greatly enhanced through

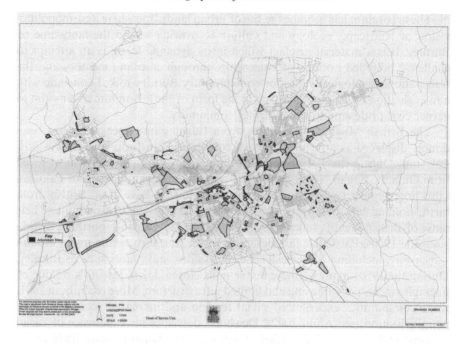

Figure 20.2 Map of Burnley Millenium Arboretum

Source: Burnley Borough Council.

the community planting of trees, and the physical landscape that is ultimately left as a legacy for the folk of Burnley in future centuries.

Phil Dewhurst's vision of a future morphology of Burnley has been inspired by the majesty of the trees and their ability to contribute to civic pride and to the social economy. These are both buoyed through their aesthetic presence and their cultural capital. Because trees grow in different timescales, (beyond the re-election schedules of local or national government or indeed individual lifelines), over time the arboretum will provide a texture of landscape, which in scale and size will occlude the degraded urban scene that is Burnley today. Their huge presence within the landscape 'will force you to be respectful' (Interview with Phil Dewhurst 2000), and intrinsically, through aesthetics and presence, instil civic pride. Identity, community and the politics of belonging are thus addressed through the material force of the new morphology, but also through the new set of genetic properties of species planted.

In the millennium arboretum, the commitment to 'native' and 'non-native' is reflected in Phil Dewhurst's ideology and in the aesthetic and cultural relationships that the Burnley community have with its sites, features and textures. The planting of non-natives is a radical idea that is controversial and generally rejected within ecological thinking and policy formation. The Department of the Environment, Transport and the Regions and organisations such as *English Nature* have well established policies on Biological Translocations:

The invasions (of non-native species) can have serious economic and ecological consequences. The ecological cost may be the irretrievable loss of native species and ecosystems, including loss of characteristic local distinctiveness. (English Nature 2005; see also Maskell et al. 2006)

However, in Burnley borough council debates, Dewhurst won the right to plant an arboretum that included 3,000 non-native trees. He argued that the arboretum should reflect the multicultural landscape within which it was to be planted. By involving both sets of communities in the planting process, he sought to build a common heritage, a heritage that is shared, but which has distinctive elements on both sides of the race divide. Dewhurst's ideological imperative reflects the fact that he contests the 'fixity' of definitions that are embodied in the philosophy of ecology of dominant agencies. For example the *Joint Nature Conservation Committee* advice on non-native species, is not dissimilar to the rhetoric used by far right campaigners in Burnley or Oldham in respect of the British Muslim community:

Non-native species may displace native organisms by preying on them or out-competing them for resources such as for food, space or both. In some cases this has led to the elimination of indigenous species from certain areas. Occasionally non-native species can reproduce with native species and produce hybrids, which will alter the genetic pool (a process called genetic pollution), which is an irreversible change (http://www.jncc.gov.uk/page-1532)

These types of discourses are deeply embedded in the ecological policy making literatures which are defining landscape design and planting practices in public spaces such as Burnley Parks.

When I asked how he defended his position to local authority and others resistant to planting non-native species, Phil explained that in his view any ecosystem has a natural equilibrium that it will attain, by ensuring an ecological mix, and through competition a natural balance will win out. For him resistance to non-natives in ecological debates is a response to the problems of a narrow sense of 'time'. The key question is 'at what point in time do we start with our taxonomy of native species?' Also in all ecosystems, there are thousands of species that have become 'naturalised, these are adopted species, those species seen to have now been integrated and thus now welcome to stay. This is based on a 'cultural' calculation, and not a scientific 'truth'.

We are a genus of one species with many subdivisions, there are divisions of that species that have predisposition that could prove fatal to the other subdivisions of that species– in the big picture – the exposure of that species to all there is will give it there is to be exposed to, ultimately gives it greater strength … As an individual you stand on your own two feet and are confronted with the world. And the claim that you shouldn't be there is just wrong. You might not like the fact that there are gay people or those that have a different skin colour – these are elements of the human condition but being exposed to these can only help you – that would enable us to retain integrity for the species – I would argue that this is the same for any other organism. (Interview with Phil Dewhurst, 2 July 2004).

This statement suggests that by embracing an 'organic cosmopolitanism' new civic materialities are possible for the people of Burnley. These in turn will lead to new cultural landscapes and new civic political landscapes. Ecological bridge building occurs at many levels, in this episode, there is a legacy of multi-species heritage for the 'white' and 'black' communities of Burnley; both the present and future generations.

Conclusions

Lilly's 'cool loyalties' towards each of the nations that she has lived in does not mean that she has no singular *cultural nationalism* at any one moment, simply that her lived experience of negotiating cosmopolitan citizenship all over the world opens up the possibilities of a *cosmopolitics* situated in the domestic scene in the diasporic sites and lives of British Asians in London. The Burnley arboretum also, has provided some evidence of the possibilities for creating new civic materialities that address the problematics of the 'native' and 'non-native' debate in both human and natural realms. The millennium arboretum recognises that cosmopolitan and transnational identifications are necessary for inclusivity, and politically enfranchising cultures of citizenship. Our task as cosmopolitan researchers and writers is thus to continually challenge and iterate cosmopolitan epistemologies and continue to espouse plurality rather than fixity in our cultural and biological definitions of racialised British Asian communities in Britain.

Cosmopolitanism has often seemed to 'claim a universality by virtue of its detachment from affiliations that constrain nation-bound lives ... the term should be extended to transnational experiences that are particular, rather than universal' (Robbins 1998). In this chapter I have given two examples of the particular experiences of cosmopolitanism that is lived within the UK away from the usual sites and situations iterated and privileged in geographical accounts of the cosmopolitan landscape. Here, the racialised 'non-native' emerges as the 21st century mobile, occidental and world citizen with a dialogic connection across transnational cultural, economic and political borders. In this research articulations of diasporic cosmopolitanism have been effectively made through various forms within both the domestic spaces of home and the public landscape of the arboretum.

Notes

1   Lilly's home interview was conducted in 2000 as part of a research process entitled 'Iconographies of diaspora'. This research which was based on twenty-two interviews with women in their homes and twelve focus groups held at women's centres in North West London, to investigate the nature of relationships that Asian women had with

citizenship in Britain, landscape and nature, as diasporic British citizens. In this research the material and visual cultures were considered artefacts of social history and evidence of memory-history narratives in the domestic space.

2   The research with the Burnley Muslim group was conducted as part of a broader research project in collaboration with artist Graham Lowe. We held twelve workshops with around 18–24 people in the summer of 2004, recruited through the Pakistani Welfare Association in Nelson with the assistance of the Community Cohesion Unit in Burnley.

References

Amin, A. (2002), 'Ethnicity and the Multicultural City: Living with diversity', *Environment and Planning A*, 34, 959–80.

Amin, A. (2004), 'Multi-ethnicity and the Idea of Europe', *Theory, Culture and Society* 21, 1–24.

Agyeman, J. and Spooner, R. (1997), 'Ethnicity and the Rural Environment', in P. Cloke and J. Little (eds) *Contested Countryside Cultures: Otherness, marginalization and rurality* (London: Routledge).

Appadurai, A. (1996), *Modernity at Large: Cultural dimensions of globalisation* (Minneapolis: University of Minnesota Press).

Beck, U. (2000), 'The Cosmopolitan Perspective: Sociology of the second age of modernity', *british Journal of Sociology* 51, 71–105.

Beck, U. (2002), 'The Cosmopolitan Society and its Enemies', *Theory, Culture and Society* 19:1, 17–44.

Beck , U. (2006), *The Cosmopolitan Vision* (Cambridge: Polity Press).

Beck, U. and Sznaider, N. (2006), 'Rooted Cosmopolitanism', in U. Beck, N. Sznaider and R. Winter (eds) *Global America: The cultural consequences of globalization* (Liverpool: Liverpool University Press).

Binnie, J. and Skeggs, B. (2004), 'Cosmopolitan Knowledge and the Production and Consumption of Sexualised Space: Manchester's gay village', *Sociological Review* 52:1, 39–61.

Cheah, P. (1998), 'Introduction Part II: The Cosmopolitical Today', in P. Cheah and B. Robbins (eds) *Cosmopolitics: Thinking and feeling beyond the nation* (Minneapolis, MN: University of Minnesota Press).

Code, L. (2006), *Ecological Thinking* (Oxford: Oxford University Press).

Conradson, D, and McKay, D. (2007), 'Translocal Subjectivities: Mobility, connection, emotion', *Mobilities* 2:2. 167–74.

Clarke, N. (2002), 'The Demon-Seed: Bio invasion as the unsettling of environmental cosmopolitanism', *Theory, Culture and Society* 19:1/2, 101–25.

Darby, W.J. (2000), *Landscape and Identity: Geographies of nation and class in England* (Oxford: Berg).

English Nature, (2005), *Position Statement on Invasive Non-native Species*, 9 November 2005, <http://www.englishnature.org.uk.text_version/news/statements> (accessed May 2007).

Garner, A. (2004), 'Living History: Trees and Metaphors of Identity in an English Forest' *Journal of Material Culture* 9:87, 87–100.

Gilroy, P. (2000), *Against Race* (Cambridge, MA: The Belknapp Press of Harvard University Press).

Hannerz, U. (1990), 'Cosmopolitans and Locals in World Culture', *Theory, Culture and Society* 7:2/3 237–51.

Hirsch E. and O'Hanlon, M. (2006), 'Landscape, Myth and Time', *Journal of Material Culture* 11:1/2, 151–65.

Joint Nature Conservation Committee, <http://www.jncc.gov.uk/page-1532> (accessed May 2007).

Jackson, P., Dwyer, C. and Crang, P. (2004), *Transnational Spaces* (London: Routledge).

Kinsman, P. (1995), 'Landscape, Race and National Identity: The photography of Ingrid Pollard', *Area* 27:4, 300–10.

Keith, M. (2005), *After the Cosmopolitan* (London: Routledge).

Malik, S. (1992), 'Colours of the Countryside – a Whiter Shade of Pale', *Ecos* 13:4, 33–40.

Matless, D. (1998), *Landscape and Englishness* (London: Reaktion).

Maskell, L.C., Firbnank,, L.G., Thompson, K., Bullock, J.M. and Smart, S.M. (2006), 'Interactions between Non-native Plant Species and the Floristic Composition of Common Habitats', *Journal of Ecology* 94,1052–60.

Mitchell, D. (2001), 'The Lure of the Local: Landscape studies at the end of a troubled century', *Progress in Human Geography* 25:2, 269–81.

Modood, T. (1990), 'Muslims, Race and Equality in Britain: Some post-Rushdie reflections', *Third Text* 11, 127–34.

Modood, T. (1994), 'Political Blackness and British Asians', *Sociology* 28, 859–76.

Nava, M. (2002), 'Cosmopolitan Modernity: Everyday imaginaries and the register of difference', *Theory, Culture and Society* 19:1–2, 81–99.

Neal, S. (2002), 'Rural Landscape, Representations and Racism: Examining multicultural citizenship and policy-making in the English countryside', *Ethnic and Racial Studies* 25:3, 442–61.

Pinney, C. (1995), 'Moral Topophilia: The significations of landscape in Indian oleographs', in E. Hirsch and M. O'Hanlon (eds) *The Anthropology of Landscape: Perspectives on place and space* (Oxford: Clarendon Press).

Pinney, C. (1997), *Camera Indica: The social life of Indian photographs* (London: Reaktion).

Robbins, B. (1998), 'Actually Existing Cosmopolitanism', in P. Cheh and B. Robbins (eds) *Cosmopolitics: Thinking and Feeling Beyond the Nation* (Minneapolis: Minnesota Press).

Söderström, O. (2006), 'Studying Cosmopolitan Landscapes', *Progress in Human Geography* 30:5, 553–8.

Tolia-Kelly, D.P. (2004), 'Processes of identification: Precipitates of re-memory in the South Asian Home', *Transactions of the Institute of British Geographers* 29, 314–29.

Tolia-Kelly D.P. (2006), 'Mobility/Stability: British Asian cultures of landscape and englishness', *Environment and Planning A* 38:2, 341–58.

Tolia-Kelly D.P. (in press), 'Organic Cosmopolitanism: Challenging cultures of the non-native at the Burnley Millennium Arboretum', *Garden History*.

Turner, B. (2002), 'Cosmopolitan Virtue, Globalization and Patriotism Theory', *Culture and Society* 4:19, 45–63.

Wong, J. (2007), 'The Native and Alien Issue', *Ethnic Environmental Participation* 2,<http://www.ben-network.org.uk/resources/publs.aspx> (accessed 25 May 2007).

Chapter 21

Afterword:
New Geographies of Race and Racism

Peter Jackson

The editors asked me to write an 'afterword' to this exciting collection of essays on the new geographies of race and racism as someone who was involved in an earlier generation of geographical research on this topic. When I edited *Race and Racism: Essays in social geography* (Jackson 1987), there was comparatively little work on the geography of racism. Twenty years ago, the field was dominated by empirical studies of residential segregation (sometimes referred to rather disparagingly as 'spatial sociology') and there was relatively little interaction between social geography and other disciplines such as cultural studies. Social construction theory was still quite novel, though it became the norm within a few short years (Jackson and Penrose 1993), and it was considered politically rather daring to use our academic work to challenge the historically entrenched geographies of race and racism. If categories like race were socially constructed, we argued, then we could no longer simply map and measure pre-given 'racial' categories. Our job, instead, must be to trace the specific geographies and histories that had given rise to these racialised categories and to examine their social consequences in terms of what David Sibley (1985) insightfully referred to as 'geographies of exclusion'. Acknowledging its political salience, the language of 'race' had been recuperated in the 1980s (albeit in scare quotes to highlight its contentiousness), having been previously submerged in the politer discourse of ethnicity. Where patterns of ethnic segregation had previously been the focus, addressed through the rather clinical lens of dissimilarity indices and other measures of residential difference, new methods were being introduced to explore the more embodied and politicised aspects of our racialised identities. These included ethnographic fieldwork, studies of visual representation, vexed questions of positionality and a welter of other approaches inspired by a range of post-structuralist theory designed to address the discursive construction of 'race'.

Looking at the essays in the current volume shows just how much has changed since the 1980s in terms of academic approach and understanding as well as the external forces shaping academic research. Most obviously, perhaps, there is far less emphasis on mapping and measuring residential segregation than in the 1980s, though, as we shall see, the theme remains politically important and has called forth a range of new studies in the context of recent debates about the extent to which different communities in Britain are leading 'parallel lives'.

In terms of theoretical approach, the inspiration of Chicago School sociology has given way to a range of other sources from Bourdieu to Benjamin and from Simmel to Deleuze. There is greater historical depth to geographical research than in the 1980s and more interest in the social construction of 'whiteness' among the majority British-born population and among more recent arrivals from the European Union's accession states. There is a very welcome interest in the geographical imaginations that underpin contemporary constructions of racialised difference, explored here in both textual and visual terms. And, perhaps most challengingly, there is a critique of social constructionism itself which has formed the theoretical underpinning for so much recent geographical research on race, a cosy consensus that is now rightly being questioned.

In making these observations, I have commented on all of the chapters to varying degree. But I have departed somewhat from the editors' preferred tripartite structure of 'Racing Histories and Geographies'; 'Race, Place and Politics'; and 'Race, Space and 'Everyday' Geographies' to try and identify some significant cross-cutting themes.

Changing Contexts

Since the 1980s, the context of geographical research on race and racism has changed in some very dramatic ways. Most obvious has been the impact of the 'global war on terror' and the events of 11 September 2001 in the US and the 7 July 2005 London bombings. These events have thoroughly re-shaped our geographical imaginations in both popular and academic terms. To the language of 'race' has been added a lexicon of religious extremism, with an urgent and sometimes sensationalist interest in Islamic fundamentalism. A religious question was added to the 2001 Census and new research programmes have been funded with a contentious focus on the 'radicalisation' of British Muslims.[1] Numerous organisations have reported a significant rise in 'Islamophobia' while there has been a rapid growth in research on Muslim identities, particularly focusing on youthful masculinities. While, in the 1980s, it was young Black men who were thought to pose the most pressing threat to law and order (Cashmore and Troyna 1982), it seems that radicalised Muslim youth are now equally if not more a focus of public disquiet.

There was relatively little discussion of 'globalisation' in the 1980s, at least in these terms. Today, the term is prevalent in public policy and media debate as well as in academia, reflected in the current volume's interest in new sources of immigration, related particularly to the expanded membership of the European Union. But couching the discussion in terms of the contemporary forces of 'globalisation' can have the unintended effect of down-playing longer historical trends and continuities with the past. So, for example, in *Bonnett*'s chapter, taking a longer historical perspective demonstrates how earlier processes of globalisation led to a 'reinvention' of many of our key concepts, focusing here on the crisis of 'whiteness' in the late nineteenth century and the emergence of a political discourse about 'the West'. So, too, as *Bressey* demonstrates in her chapter, a longer history of political action to counter racism (focusing in this case on the work of the League of Coloured Peoples in the 1930s and 40s) can be deployed to counter those who claim that contemporary forms of anti-racism can be dismissed as 'mere' political correctness. Present-day debates about the cultural privileges of whiteness, prompted by new sources of migration to Britain from Eastern Europe, are also usefully set in context by *McDowell*'s work on the relative invisibility of earlier generations of migrants, focusing here on European Volunteer Workers from Latvia. The politics of visibility is also discussed by *Nagel and Staeheli* in the context of British Arab activists in London, Liverpool, Birmingham and Sheffield where, they argue, cultural markers of difference (like veils and beards) are now as important as skin colour and facial features were in previous generations. Nagel and Staeheli also argue that 'Arabness looks different in the working-class neighbourhoods of Liverpool and Sheffield than it does in [the cosmopolitan neighbourhoods of] Kensington and Knightsbridge' (p. 92) – a point about place-difference to which we return below. Besides their historical depth, these chapters are also united by their commitment to a relational approach to the construction of racialised difference – where what happens 'here and now' is profoundly affected by what happens at other times and in other places. Relational geographies were not on the agenda in the 1980s, at least defined as such, and these chapters all show the political and ethical benefits of adopting such an approach.

Segregation, Segregation, Segregation

While studies of residential segregation are much less prevalent now than in the 1980s (see, for example, Jackson and Smith 1981; Peach et al. 1981), it remains an important and politically-charged issue. It is therefore encouraging to note that geographers have responded with skill and sophistication to recent policy debates including the suggestion by the Chair of the Commission for Racial Equality Trevor Phillips in 2005 that Britain was in danger of 'sleepwalking' towards North American-style segregation. Following major disturbances in several northern cities in 2001, the Cantle report had drawn attention to the depth of polarisation in British cities, suggesting that different communities might now be leading 'parallel lives'. Further prompted by the London bombings in 2005, the government responded by establishing a Commission on Integration and Cohesion and by extending its policy of 'managed migration'. In these circumstances, disputes about the measurement of trends in ethnic segregation in British cities became a major issue of public concern (cf. Simpson 2004). The debate continues in the present volume, with *Poulsen and Johnston* using a typological approach and a finer geographical scale of analysis than most analyses to question the current received wisdom about decreasing levels of segregation in British cities. They challenge these findings taking issue with conventional indices of segregation which, they claim, average out important geographical differences within cities.

The intensity of these debates serves to confirm that segregation is a highly-charged political discourse in contemporary Britain – a point that is underlined by *Phillips*'s work on Oldham, Rochdale and Bradford where she finds little evidence of a desire for exclusion and self-segregation among the cities' diverse Muslim communities. The political implications of ethnic segregation are also explored in different contexts by *Glynn*'s analysis of the waning support for New Labour by Bengalis in London's East End and by *Keith*'s account of the rising tide of support for the British National Party among white working-class communities in Barking and Dagenham.

Place Matters

Most of the chapters in this book contribute to the argument that racism takes different shape and form in different times and places – that 'geography matters' in Massey and Allen's (1985) succinct formulation. The argument is built up in a particularly compelling way in several chapters on the distinctiveness of Scottish racism. The conventional wisdom is that, compared to the pervasive and deep-seated nature of racialised prejudice and discrimination south of the border, racism is notable by its relative absence in Scotland. This process of racialised denial has complex roots as *Hopkins* shows in his discussion of young Muslim men and asylum-seekers in Edinburgh and Glasgow where their distinctive experience is explained in terms of demographic differences (the diversity and structure of the Black and minority population in Scotland), Scotland's distinctive pattern of politics of governance (its unique legislative framework) and the distinctive formation of Scottish national identities. This argument is taken up by *Penrose and Howard* with reference to the ambivalent cultural politics of the *One Scotland, Many Cultures* policy. They expose the faulty logic of the 'demographic deficit' argument for Scotland's distinctive racialisation (fewer visible ethnic minorities than England) and go on to show that apparently well-intentioned anti-racist campaigns, based on the 'Russian doll' analogy and the symbolism of a multi-ethnic 'tug of war' where all pull together to raise the national flag, are capable of more complex and contested interpretations through their connection to problematic notions of a unified national identity.

Through a close analysis of the 2004 Citizenship Referendum, *Crowley, Gilmartin* and *Kitchin* demonstrate that a specific form of racialised politics is beginning to emerge in Ireland as the country has shifted from being a nation of net emigration to one of net immigration. The rapid rise of in-migration from Poland, Lithuania and Latvia, in particular, gave rise to a racialised moral panic which, following a campaign in support of 'Common Sense Citizenship', led to an abandonment of the long-established right to citizenship by birth. Racial politics are even more complicated in Northern Ireland where, until recently, the sectarian divide between Protestants and Catholics has tended to over-shadow all other forms of racialised difference. *Geoghegan*'s chapter analyses the struggle to articulate an effective Race Relations policy in these difficult circumstances, ultimately transformed into a Good Relations policy, designed to encompass both Community Relations (across the sectarian divide) and 'race relations' more generally. Ironically, he argues, some of the bureaucractic manoeuvres that were considered necessary to implement these policies (including communal registration and active reallocation of those who declined to declare a religious identity) led to a reinscription of the very identitarian politics they were designed to eradicate.

New Theorisations of Identity and Difference

The next set of essays signal some genuinely new directions in geographical studies of race and racism. These include studies of the intersection between race and sexuality in Bassi's analysis of Birmingham's gay Asian club scene and Lim's exploration of South Asian masculinities which involves a close analysis of a single momentary encounter in a London pub. The emphasis on gender and sexuality and an interest in momentary, fleeting encounters are both new directions that have been taken by geographers, following the discipline's 'cultural turn' in the 1980s. Both chapters are based on theoretically-informed ethnographic work.

In *Bassi's* case, the study is rooted in a Gramscian analysis of the radical potential involved in collective re-appropriations of the traditional rituals of Punjabi wedding festivals where gay men and lesbians actively re-work the sounds and symbols associated with bhangra and Bollywood to achieve a temporary and fragile suspension of racist, patriarchal and homophobic conventions. The emancipatory potential of these events is constantly in danger of being undermined by the 'burdensome stereotyping' of these young South Asian consumers. Moreover, as Bassi's analysis reveals, the burdens do not fall equally on all participants, impacting most severely on the 'butch lesbians' and 'rent boys' whose conspicuous appearance threatens to undermine the precarious acceptability and hard-won respectability of less visible members of the South Asian gay and lesbian community.

While Bassi focuses on the moments of radical possibility in club nights like Ultimate Karma and Saathi, *Lim* focuses on even more transient encounters which, after Deleuze, he describes in terms of the 'event'. This turn to the fleeting experiences of everyday life seems, at first, to run counter to the book's earlier emphasis on charting longer historical trajectories of race and racism. Like Bassi, however, Lim insists on historicising individual events, noting how the passing encounter he observed in a suburban London pub had been shaped by social memories of the London bombings and wider collective histories that have transformed popular understandings of South Asian men from 'passive, soft and weak' to stereotypically aggressive, hard and threatening forms of militant masculinity. The chapter explores the gap between discourse and practice through an analysis of a specific embodied encounter where the glance of an eye or the turn of a body are suggestive of much more enduring geographies of affect.

These ideas are further explored in *Swanton's* account of 'everyday multiculture' in the Bradford suburb of Keighley. Here, too, fleeting encounters and embodied geographies (where 'pupils dilate' and 'adrenaline surges') are freighted with more deep-seated cultural meaning, rooted in the imaginative geographies of the 'global war on terror'. Swanton's analysis is situated as a critique of social construction theory and its 'deadening effect' on forms of understanding that seek to go beyond an empirical focus on narrative constructions and textual representations. The chapter follows those, like Saldanha (2006), who have sought to re-ontologise race, tracing its material constitution beyond the 'merely' representational and discursively constructed. These are provocative arguments, in the best sense of the word, and deserve to be discussed at greater length.

More-than-Representational Geographies of Race?

The dangers of re-ontologising such an explosive term as 'race' should not be under-represented and risk being hijacked by those who would argue for a return to the concept's supposedly 'real', biological roots. That is not, of course, what is intended here. While I am attracted to Swanton's agenda of taking seriously the materialities through which race is constituted in moments of intercultural encounter, I am dubious about the lack of ethnographic depth that is suggested by reducing such encounters to lists of co-present things ('souped-up cars, veils, street names, graffiti, skin, calls for prayer, Lollywood posters, pubs, kebab skewers, salwar kemeez, newspaper headlines and so much more') – a strategy which is repeated throughout the chapter and seems to act as a substitute for more systematic social analysis. I do not see why such experiments in 'non-representational' analysis must be couched as an alternative to, or in opposition to, constructionist accounts. Not all constructionists are 'abolitionist' in their approach to race, denying its material consequences or ignoring the embodied nature of our social identities. Indeed, many constructionists have sought to provide a detailed empirical, embodied and materialist account of specific processes of racialisation. This might greater attention to the momentary and transient geographies of affect (as described by Swanton and others) as well as longer-term, historicised accounts of the development of particular social formations. While constructionist accounts have undoubtedly tended to privilege the discursive and the social over the material and the embodied, this is, to my mind, not intrinsic to the approach. Striving towards a more holistic account of the emergence of race – how it surfaces immanently in particular momentary encounters and how racialised meanings 'stick' to some bodies more than others – together with more historicised accounts of the *longue durée* of racialised discourse is, in my view, a key intellectual challenge for future geographies of race and racism. What Swanton calls the 'temporary fixings' (p. 244) and 'affective intensities' (p. 246) of race are clearly worthy of examination. But so, too, are the underlying social constructions that give shape and meaning to these particular events. In this, I find myself aligned with those, like Lorimer (2005), who advocate a 'more than representational' rather than a 'non-representational' approach to understanding social life.

The possibilities for such an analysis are outlined in *Clayton*'s exploration of the everyday geographies of marginality and encounter in Leicester which draws rather less on non-representational theory and more on Doreen Massey's (2005) analysis of the 'throwntogetherness' of cities and urban life. While Leicester is often regarded as a model of successful 'race relations', Clayton argues that white working-class youth struggle to find a space to call their own in the city's racially-coded neighbourhoods. This is a theme that *Nayak* takes up in the context of his ethnographic work in the mainly white areas of North-East England. Here, Nayak argues, 'whiteness' is a taken-for-granted social norm among young people even in more ethnically mixed inner city schools. For anti-racist initiatives to be effective in such areas, Nayak insists, ways must be found to make the message meaningful to their diverse audiences. Nayak has himself experimented with new cultural 'pedagogies of place', inviting white students to trace their own genealogies. In many cases, this process revealed surprisingly heterogeneous biographies among these supposedly homogeneous 'white' students, giving them more of a personal stake in current debates about British multiculturalism.

Probably the most personal account of racialised identity in this volume is *Pollard*'s touching visual essay, centred on images of her father's hands from family photo albums and other 'domestic' memorabilia. Pollard's father was a printer who migrated to London and sent letters home to his wife in Guyana, writing in his well-schooled copperplate script. Pollard skilfully interweaves extracts from these letters with family photos, revealing her father's richly ambivalent experience of post-war London which both celebrates 'how happy London can be' in the semi-public spaces of the metropolis (in cinemas and on buses, for example) but also how cruel the city can be ('lost in a sea of white faces', 'a stranger in a strange land'). Pollard recalls walking to school for the first time with her father's hand on her shoulder, a reassuring physical presence in an alien world. The intimate observation of hands and handwriting serves as a powerful lens through which to observe the wider geographies of post-war immigration and settlement. While the essay is strikingly visual in form, focusing on her father's distinctively long thin hands, it chimes with several other themes in the book concerning the embodied and performative nature of racialised identities, the ambiguities of city life for racialised minorities and the ambivalences of belonging in what is often perceived as a very hostile and unfamiliar world. These ideas are extended in *Tolia-Kelly*'s analysis of the nature of memory-objects among post-colonial migrant communities in London and Burnley. By focusing on a range of material objects – from ornamental trees to domestic shrines – she is able to challenge the stereotypical emphasis on non-belonging, marginality and exclusion among these communities to develop alternative and creative readings of diasporic identity, cosmopolitanism and 'cool belonging' where individual and collective loyalties are shared across more than one nation or place of attachment.

Taken together, then, these essays demonstrate how much has changed within the last ten or twenty years of geographical research on race and racism in Britain. While some core concerns remain, in terms of understanding the social and political significance of residential segregation, for example, there are many new areas of interest and new approaches, emphasising everyday experience and embodied identities, encompassing majority groups and definitions of 'whiteness' as well as the racialisation of diverse minority 'ethnic' communities. The context of research has shifted significantly, too, with new patterns and sources of migration, new political arrangements in Britain's devolved national administrations and the impact of wider events such as the 'global war on terror'. New theoretical approaches are challenging the previous consensus around the social construction of race, demanding new methodological strategies to capture the geographies of affect that are at the heart of this engagement with the more-than-representational. Like racism itself, geographical studies of race are constantly changing shape in response to external forces and internal challenges. These are exciting times for geographers of race and racism as the essays in this volume amply confirm.

Note

1   I am referring, in particular, to the controversy that surrounded the 'New Security Challenges' research programme, co-funded by the Economic and Social Research Council (ESRC), the Arts and Humanities Research Council (AHRC) and the Foreign and Commonwealth Office (FCO). This programme replaced an earlier FCO research initiative on 'Combating terrorism by countering radicalisation' which was withdrawn following widespread criticism from British academics. The UK and Commonwealth Association of Social Anthropologists passed a resolution condemning the proposed initiatives, arguing that such research, designed to inform UK counter-terrorism policy overseas, was 'prejudicial to the position of all researchers working abroad' including those with no connection to these programmes (see *Anthropology Today* February 2007).

References

Cashmore, E. and Troyna, B. (1982), *Black Youth in Crisis* (London: Allen and Unwin).
Jackson, P. ed. (1987), *Race and Racism: Essays in social geography* (London: Allen and Unwin).
Jackson, P. and Penrose, J. (eds) (1993), *Constructions of Race, Place and Nation* (London: UCL Press).
Jackson, P. and Smith, S.J. (eds) (1981), *Social Interaction and Ethnic Segregation* (London: Academic Press).
Lorimer, H. (2005), 'Cultural Geography: The busyness of being "more-than-representational"', *Progress in Human Geography* 29, 87–94.
Massey, D. (2005), *For Space* (London: Sage).
Massey, D. and Allen, J. (eds) (1985), *Geography Matters!* (Cambridge: Cambridge University Press).
Peach, C., Robinson, V. and Smith, S. (eds) (1981), *Ethnic Segregation in Cities* (London: Croom Helm).
Saldanha, A. (2006), 'Re-ontologising Race: The machinic geography of phenotype', *Environment and Planning D: Society and Space* 24, 9–24.
Sibley, D. (1985), *Geographies of Exclusion* (London: Routledge).
Simpson, L. (2004), 'Statistics of Racial Segregation: Measures, evidence and policy', *Urban Studies* 41, 661–81.

Index

Prof. Richard Baker, Chairman and Information phase operation.
R. Legal activities of skill legal and function Technical Journal.
Verlag GmbH, Raschingerstraße 34, 80333 München, Germany.